Medicinal and Aromatic Plants of the World

Volume 9

Medicinal and Aromatic Plants (MAPs) have been utilized in various forms since the earliest days of mankind. They have maintained their traditional basic curative role even in our modern societies. Apart from their traditional culinary and food industry uses, MAPs are intensively consumed as food supplements (food additives) and in animal husbandry, where feed additives are used to replace synthetic chemicals and production-increasing hormones. Importantly medicinal plants and their chemical ingredients can serve as starting and/or model materials for pharmaceutical research and medicine production. Current areas of utilization constitute powerful drivers for the exploitation of these natural resources. Today's demands, coupled with the already rather limited availability and potential exhaustion of these natural resources, make it necessary to take stock both of them and enrich our knowledge regarding research and development, production, trade and utilization, and especially from the viewpoint of sustainability. The series Medicinal and Aromatic Plants of the World is aimed to look carefully at our present knowledge of this vast interdisciplinary domain, on a global scale. In the era of global climatic change, the series is expected to make an important contribution to the better knowledge and understanding of MAPs.

Budapest, Prof. Dr. Ákos Máthé.

Ákos Máthé • Irfan Ali Khan
Editors

Medicinal and Aromatic Plants of India Vol. 2

Springer

Editors
Ákos Máthé
Department of Water and Environmental Sciences, Albert Kázmér Faculty
Széchenyi István University
Mosonmagyaróvár, Hungary

Irfan Ali Khan
Nawab Shah Alam Khan Centre for Post Graduate Studies and Research
Affiliated to Osmania University
Hyderabad, Telangana, India

ISSN 2352-6831 ISSN 2352-684X (electronic)
Medicinal and Aromatic Plants of the World
ISBN 978-3-031-44913-0 ISBN 978-3-031-44914-7 (eBook)
https://doi.org/10.1007/978-3-031-44914-7

This Springer imprint is published by the registered company Springer Nature Switzerland AG
The registered company address is: Gewerbestrasse 11, 6330 Cham, Switzerland

Paper in this product is recyclable.

Preface

This book is the ninth volume of the series Medicinal and Aromatic Plants of the World (MAPW), with a focus on the medicinal and aromatic plants (MAPs) of India.

India is one of the few countries to be spread over 15 agro-climatic zones with a rich wealth of biodiversity. Amongst the documented 17,000–18,000 flowering plant species, more than 7000 species are medicinal and aromatic plants, which maintain a bridge between traditions and holistic well-being. The AYUSH System of Medicine and its components Ayurveda, Unani, Siddha and Homoeopathy, established centuries ago, flourish even today.

Allopathic drugs were introduced during the British regime and resulted in a gradual setback for the Indian traditional medicine. Modern technological advances have further contributed to the expansion and progress of allopathic medicines, which has curbed the growth of traditional medicine. With an estimated 70% of the India population being rural, traditional medicine still constitutes an integral component of primary healthcare.

This scenario underlines the importance of pioneering initiatives by the Government of India to promote research with the ultimate aim of exploring the scientific basis of these ancient Indian sciences. Initiatives to validate drugs from medicinal and aromatic plants will surely flourish their use in the millennia to come.

Several chapters of the present volume bear witness to above said. Besides dealing with biodiversity issues, the chapter on Holy Basil (*Ocimum sanctum* L.) can be regarded an eloquent example of medicinal and aromatic plants that bridge traditional and present-day/prospective uses.

The present volume offers up-to-date and comprehensive information on selected pharmacologically important plant families (Apiaceae, Fabaceae, Lamiaceae), highlighting their traditional uses, in India.

Present-day knowledge on the use of medicinal and aromatic plants in the treatment of widespread diseases, like vitiligo and wound healing, are likely to be of interest for many readers.

The chapter on *Artemisia annua* offers a comprehensive review on the pharmacological properties of this species that has acquired fame in the fight against malaria. As it happens, recent research has explored, artemisinin, the main component of this species that seems to be active in killing human breast cancer cells.

It is anticipated that the current volume as well as the ones to follow will deal with endemic or less studied MAP species characteristic of India.

Editors sincerely hope that the 15 chapters compiled in this volume will shed light on the extreme wealth and conservation of India's medicinal plant diversity, as well as the rich traditions and present achievements in medicinal and aromatic plant research and innovative utilization that will surely contribute to the well-being of mankind.

Mosonmagyaróvár, Hungary Ákos Máthé

Hyderabad, Telangana, India Irfan Ali Khan
November 2022

Contents

About the Editors

Ákos Máthé Habil. Dr., Prof. Dr. MÁTHÉ, Ákos has 40 years' background of teaching and research in plant ecophysiology and agricultural botany. He is Professor Emeritus at the Albert Kázmér Faculty of Széchenyi István University, Mosonmagyaróvár, Hungary

He was two times Fulbright Scholar (1986 and 1995) and visiting professor at University of Veterinary Medicine, Vienna (1995–1997).

Teaching/Research/Consulting and Publication activity of Professor Máthé focus on ecophysiology, plant domestication/introduction, production of MAPs, new crops, new uses of plants, including feed-additives.

Serving as president of ICMAP (2014–2019) and Chairman of ISHS Section MAPs (2006–2014) he was founding secretary of Hungarian Medicinal Plant Association.

Professor Máth collaborated in international projects: FEED SEG, CEEPUS, ERASMUS+, HERBAID, GOOD HERBS, Herbs4Youth, EOHUB, etc.

He has also authored some 100 publications and was former editor of *Herba Hungarica*, *Acta Agronomica Hungarica a* convener/speaker at international scientific congresses, conferences. Network co-ordinator of ESCORENA MAP (http://www.agrowebcee.net/map/).

Irfan Ali Khan obtained his MSc from Aligarh Muslim University and PhD in Botany from Osmania University, Hyderabad, specializing in 'Genetics and Plant Breeding'. Professor Khan is the Former Director of Nawab Shah Alam Khan Centre for Post Graduate Studies and Research (Affiliated to Osmania University), Anwarul Uloom College Campus, Mallepally, Hyderabad. Presently he is the Managing Director of Ukaaz Publications, Hyderabad. He has published 163 research papers in the reputed national and international journals and is now on the panel of 'Experts on Mungbean' for all countries of the South-East Asia and the Middle East. Professor Khan has been the editor of *Frontiers in Plant Science*, has edited 74 reference books and has co-authored 3 textbooks with his wife, Professor Atiya Khanum. He is a Fellow of the Indian Society of Genetics (F.I.S.G.). Besides this, he is the Editor-in-Chief of *Annals of Phytomedicine-An International Journal*. Professor Khan is the senior author of the famous textbook *Fundamentals of Biostatistics*, by Khan and Khanum, which has been released by world renowned agricultural scientist Dr. M.S. Swaminathan on February 13, 1994 in Hyderabad. This textbook has been included as a textbook and also as reference book in more than 400 universities and research institutes in India and abroad. Besides this, he has given a formula of LSD (Least Significant Difference) with suitable examples which is more or less a substitute for Student's 't' test to compare to two treatments.

Chapter 1
Medicinal and Aromatic Plants in India: A Bridge from Traditions to Modern Wellbeing

Ákos Máthé and Irfan Ali Khan

Abstract Traditional medicine, particularly herbal medicine, is still an important source for health, especially in the rural and remote areas of India. Ayurveda, Siddha and Unani are highly esteemed and applied in several ways. Traditional medicinal plants frequently provide the source for new drugs and drug development. Evidence based incorporation of Indian traditional medicine in clinical practice can be regarded as a bridge arching from Traditions to modern (holistic) wellbeing with the aim of offering safe, efficient quality healthcare to all. The use of herbs, as natural products, is mostly regarded comparatively safe, eco-friendly. Their further advantage is that they are frequently locally available. Ayurvedic system of medicine can be regarded as the most widely used system in traditional Indian medicine. Ancient Charaka Samhita describes all aspects of Ayurvedic medicine. The development of Ayurvedic drugs is gaining importance, especially in cases where synthetic medications fail to bring the desired results. In the modern societies, the increasing uses of Ayurvedic drugs seem to reveal new aspects of safety, efficacy and quality.

Keywords Aromatic plants · Ayurveda · Clinical practices · India · Policy · Medicinal plants · Pharmacovigilance · Traditional system of medicine

1.1 Introduction

The frequently cited assessments of the World Health Organization (WHO) state that about 70 percent of the world's population relies on plants for their primary health care (Akerele and Heywood 1991). This fact should be related to the some

Á. Máthé (✉)
Department of Water and Environmental Sciences, Albert Kázmér Faculty, Széchenyi István University, Mosonmagyaróvár, Hungary

I. A. Khan
Nawab Shah Alam Khan Centre for Post Graduate Studies and Research, Affiliated to Osmania University, Hyderabad, Telangana, India

Á. Máthé, I. A. Khan (eds.), *Medicinal and Aromatic Plants of India Vol. 2*,
Medicinal and Aromatic Plants of the World 9,
https://doi.org/10.1007/978-3-031-44914-7_1

35,000–70,000 plant (occasionally implying medicinal fungi) species that are recorded as medicinal plants.

The use of herbs and herbal extracts can be traced back to earliest times of myths, traditions and writings that witness the pain-easing or disease-treating properties of plants.

Parallel with human evolution, the evolution of plant-based medicine systems took place, also in India. The Materia Medica of India contains information on ethnic folklore practices and traditional aspects of therapeutically important natural products. Indian traditional medicine has its roots in various systems, including Ayurveda, Siddha and Unani (ASU). These are primarily, based on the observation and use of plants within a given local area.

In India, healing with the traditional medicinal systems has not only survived but todate it enjoys much support and attention at a state level: even an independent Ministry of AYUSH (Ayurveda, Yoga, Unani, Siddha and Homeopathy) has been created. It has the vision to revive the deep knowledge of ancient Indian systems of medicine and to ensure the optimal development and propagation of the Ayush systems of healthcare.

Regarding the rich diversity of Medicinal and Aromatic Plants forming the basis of Ayush systems of healthcare, it can be stated that in India, Medicinal and Aromatic Plants have not only constituted, but will constitute a bridge from Traditions to holistic wellbeing.

1.2 Historical Background of Indian Systems of Medicine

Indian special literature sources on the traditional uses of Medicinal and Aromatic Plants agree that Ayurveda can be considered as one of the most esteemed traditional systems of medicine that has survived and flourished from ages till date (Jaiswal and Williams 2017). It is based on the vast knowledge of nature-based medicine, i.e.: on the knowledge of relationship of human body constitution and function versus both nature and the elements of the universe. These are considered acting in coordination and affecting the living beings. Rooted in nature, this system of medicine is expected to continue to flourish also in the centuries still to come.

It is common knowledge that Ayurveda has an age-old history. Its origin can be traced back to the second Century BC., to the epoch of renowned schools of "Hindu Philosophical teachings named "Vaisheshika", as well as the School of Logic named as "Nyaya". It is also said to be related to "Samkhya", the manifestation framework, in the period when the schools of both "Nyaya" and "Vaisheshika" flourished.

Ayurveda is considered to be of divine origin, from the Hindu God, Brahma who is called as the creator of the universe (Heyn 1990). Tradition has it that the Creator of the universe passed on this holistic knowledge of healing to the sages to serve the well-being of mankind. The knowledge of traditional medicines was endowed to the disciples and finally to the common man by various writings and oral narrations. As regards the herbs, the information about their healing properties was composed in the form of poems, called "Shlokas". Sages used Shlokas to describe the use of medicinal plants.

The Hindu system of healing is believed to be based on the following four – on their own – eminent compilations of knowledge (Vedas): Yajur Veda, Rig Veda, Sam Veda, and Atharva Veda. Rig Veda, the perhaps most well-known of them, describes 67 plants and 1028 Shlokas. As a comparison, by the Atharva Veda and Yajur Veda 293 and 81 medicinally useful plants are described, respectively. The practice of Ayurveda is based upon the knowledge gained from these Vedas.

The writings in Rig Veda and Atharva Veda are attributed to "Atreya", a renowned scholar. He is believed to have acquired this knowledge from Lord Indra, who in turn initially received it from Lord Brahma. Another early writer in medicine, Agnivesha, a scholar of Atreya, compiled the knowledge from the Vedas: the work called "Charaka Samhita". was edited by Charaka and some other scholars. Charaka Samhita describes all aspects of Ayurvedic medicine; whereas Sushruta Samhita describes the "Science of Surgery". These ancient texts have been translated into different languages (like Tibetan, Greek, Chinese, Arabic and Persian). It is used by practitioners of traditional medicine, even today. Charaka Samhita is the most respected among all known related compilations / records (Nighantu Granthas, Madhava Nidana and Bhava Prakasha, etc.).

In the course of time, knowledge of systems from different ethnic origins has brought about the interchange, integration of the different systems. To date, the medicinal practice(s) which are either thought to be of Indian origin or may have come to India from the outside world and got absorbed into Indian culture, is recognized as traditional medicine of India (Adhikari and Paul 2018).

Figure 1.1 illustrates the main health conditions or disorders for which drugs have been developed of plant origin.

1.2.1 Attempts to Comply with Healthcare Programs

Quasi to comply with the emerging interest of the world, the Government of India has initiated several projects to explore the possibility of evaluating the therapeutic potential of traditional systems and to incorporate them in the national healthcare programs. In this process, the Ministry of Health and Family Welfare, Government of India, has excelled with initiatives for the development and preservation of these important aspects of cultural heritage.

Presently, it is the Department of AYUSH that regulates education and research in these systems. The National Medicinal Plant Board [NMPB] deals with conservation and research issues in botanicals.

As a central body in the field of Yoga and Naturopathy, the NMPB is actively engaged in promotion, propagation, research, education, training and publication work.

Fig. 1.1 Drugs developed for various disorders from plant sources. (Source: Adhikari and Paul 2018)

1.3 Medicinal Plants in Ayurveda (the Science of Life)

Ayurveda is the most widely used system in traditional Indian medicine. With a preference on holistic medicine, it deals with the body, mind, and spirit. It strives to establish harmonious coexistence with nature through which physical, mental, and emotional health can be reached (Shi et al. 2021).

The name Ayurveda can be translated as the "The Science of Life" and is composed of the Sanskrit words: "ayur" (life) and "veda" (science or knowledge).

Being not merely a system of medicine, Ayurveda involves a holistic convention of harmonious living. Records say it that its beginning can be traced back to ancient information in Rigveda and Atharva veda. The source of Ayurveda has not survived to the present. However, its ideas and methodologies have been idealized in between 2500 and 500 BCE, in India (Adhikari and Paul 2018).

Ayurveda deals with happy and unhappy life. Telling the appropriate from the inappropriate in relation to the life, it measures the life expectancy and the quality of life. It constitutes a holistic arrangement of medical services that regards the human body as a network of seven fundamental tissues ("Rasa," "Rakta," "Mansa," "Meda," "Asthi," "Majja," and "Shukra") and the waste-results of the body (e.g.: excretion, urine, and sweat). The latter are derived by the five fundamental components fire, water, air, ether, and earth, as well as three dynamic energies or functional philosophies "vata, pitta, and kapha" (Tridosha). Disturbance (either unevenness or unsettling influence) to these fundamental standards of the body is likely to causes disease.

Medicinal plants are valued by Ayurveda for their biological activities, since they serve as remedies for numerous diseases. Procurement/collection is crucial regarding the **quality** of medicinal plants. According to certain sources, ca. 40% of approved pharmaceutical products in use today have their origin from natural substances. Ayurveda recognizes several factors to be considered before plant collection. These include the followings: place of origin, collected plant part, method and time of collection, which are equally significant aspects for obtaining best qualities of medicine. Further important factors (e.g. Guna, Desha, Kala, Pakva-apakvaavastha, nav-purana avastha, Prayojyanga, Karma and Disha) should also be considered, since they also affect the quality of drugs in the course of collecting the plants materials (Mukherjee et al. 2017).

1.3.1 Quality of Ayurvedic Preparations

According to scientific research, collection conditions affect the drug potency through several factors. These include the ecological conditions of the habitat (temperature, rainfall, duration of daylight, altitude), methods of cultivation, time and method of collection, processing and storage. They have a significant impact on the secondary metabolites (Tavhare and Nishteswar 2014). Best quality (Ayurvedic) drugs can be obtained by applying appropriate collection techniques.

The National Medicinal Plants Board (NMPB), Ministry of Ayush, Government of India has launched a "Voluntary Certification Scheme for Medicinal Plants Produce (VCSMPP)", in 2017, to boost the Good Agricultural Practices (GAPs) and Good Field Collection Practices (GFCPs) in the production of medicinal plants. The VCSMPP has expanded not only the availability of the certified quality medicinal plants' raw material within the country but it has also improved their export potential increasing India's share in the global export of herbs ("National Medicinal Plants Board (NMPB)| Ayush Next" n.d.).

Most of the Ayurvedic drugs and phytomedicines are so called 'poly-herbal formulations': i.e., the whole extracts of various herbs that are generally believed to exert their therapeutic activity in a synergistic way. In contrast to the present trends to isolate single constituents from herbs, in poly-formulations, the therapeutic efficacy is a result of the synergy (combined effect) of constituents, even if being present only at low concentrations in the herbs (Williamson 2001; Mukherjee et al. 2011). Ayurvedic texts mention thousands of single or poly-herbal formulations.

Despite their long term therapeutic use, farther pharmaco-epidemiological evidence is needed to support their safety and efficacy (Mukherjee et al. 2017).

1.3.2 Safety of Ayurvedic Preparations

Modern scientific methods and technologies can be used to further explore and validate Ayurvedic drugs mentioned in the ancient texts of traditional Indian system of medicine. By evaluating identity, purity, stability, physical and biological properties etc., their therapeutic efficacy is enhanced to high quality.

The development of Ayurvedic drugs is gaining momentum, especially in cases where synthetic medications fail to bring the desired results. In the modern societies, the increasing uses of Ayurvedic drugs seem to reveal new aspects of safety, efficacy and quality.

1.3.3 Efficacy of Ayurvedic Preparations

Most Ayurvedic drugs and phytomedicines are available in the market as 'polyherbal formulation'. Generally, this indicates the whole extracts prepared from herbs. These formulations are thought to exert their therapeutic activity in a synergistic way, as the a result of the combined effect of constituents present at low concentrations in the herbs (Williamson 2001 and Mukherjee et al. 2011).

Several successful attempts have been reported about the design multi-ingredient formulations (e.g.: the joint support supplement "Artrex" to treat rheumatoid arthritis and osteoarthritis (OA)). "Artrex" has been validated by in a randomized, double-blind, placebo-controlled clinical trial. This and similar formulations have proved to be effective when compared to relevant synthetic drugs: as such, they have been patented both in India and other countries (Mukherjee et al. 2017).

1.4 Present Status of Ayurveda and Other Indigenous Systems of Medicine in India

Ayurvedic Medicine, as the basis of indigenous Indian medicinal systems, renders alive the unique theories and traditions of Ayurveda. It makes it possible that they are still accessible to the complementary health practitioner of today. Modern medicine, however, needs more evidence-based studies in order to comply with the requirements on Quality, Safety and Efficacy of both crude drugs and preparations thereof.

The World Health Organization (WHO) and the Government of India signed an agreement to establish the WHO Global Centre for Traditional Medicine, in 2022. The Center aims to globally harness the potential of traditional medicine through modern science and technology, to improve the health of people and the planet. Its reach encompasses both ancient practices (e.g. acupuncture, ayurvedic medicine and herbal mixtures) and modern medicines (who.int/initiatives/who-global-centre-for-traditional-medicine/).

Around 18 major Indian states have independent Directorates to supervise ISM related issues. In 6 states the ISM is administrated under the Health Directorate of the State, whereas in six smaller states and Union Territories an "Officer in-charge" exerts the same tasks. There are around 6.11 lakh practioners of ISM & H, in India. The number of Hospitals and dispensaries in this sector is more than 26,000 (http://www.indianmedicine.nac.in). There are approximately 8500 manufacturers of Ayurvedic drugs with around 1 billion US dollars gross turnover in all the ISM & H systems, in India. Drug manufacturing in this sector is regulated by the Drugs and Cosmetic Act (1940) and Rules (1945) (Jain 2001).

1.5 Modernization and Integration of Herbal Medicine in Clinical Practice, in India

In India, the national policy on traditional and alternative medicine was introduced in 1940, in the form of Drug and Cosmetic Act 1940 and Drug and Cosmetic Rule, which has been updated in several states. Since that time, due to the combined efforts of the public and government sector, the promotion of herbal medicine has yielded considerable recognition, in both India and internationally.

A detailed description of the various modernization stages have been summarized by Sen and Chakraborty (2017) and others.

As a most recent step, the Central Council of Indian Medicine (CCIM) was established, in 1970. This central governmental body was involved in the framing and implementing different regulations including the curricula and syllabi in the Indian System of Medicine (ISM): i.e.: Ayurveda, Siddha and Unani. Following the incorporation of Sowa Rigpa system of medicine into Central Council of Indian Medicine (CCIM), in 2012, the Department of Indian Medicine and Homeopathy (ISM & H) was formed with the objective to develop the ISM. Subsequently, in 2003, this Department was renamed as Department of Ayurveda, Yoga and Naturopathy, Unani, Siddha and Homoeopathy (AYUSH). In 2014, the separate ministry on AYUSH was established.

1.5.1 Research and Development

Regarding Ayurvedic medicine, several research initiatives and the elaboration of supportive databases have been reported in various domains, e.g.: clinical research including the clinical trial of ayurvedic preparations and drug research with medico-botanical surveys, cultivation of medicinal plants, pharmacognostic studies, phytochemical studies, drug standardization, pharmacological and toxicological studies, etc. (Sharma et al. 2002; Billore et al. 2004; Satyavati 2005; Mishra 2004; Gupta and Tandon 2004); Wealth of India series (2000); Dahanukar et al. (2000). Recently, Ayurvedic medicine, its usage and global penetration etc. were reviewed by Ravishankar and Shukla (2007).

1.5.2 Education

Recognizing the importance of education for both present and future generations, the Central Council of Indian Medicine (CCIM) was established by the Central Council of Indian Medicine Act of 1970. It is mainly responsible for managing the teaching and practice of ISM. In 2016, there were already more than 125 graduate schools in India (76 Ayurveda, 8 Unani, 3 Siddha, and 40 homeopathy schools). The top institutions of Indian medical education are the following: National Ayurveda Institute of India (Jaipur), National Naturopathy Research Institute (Poona), Ayurveda Graduate Institute of Teaching and Research (Jamnagar), National Unani Medical Research Institute (Bangalore), and National Siddha Medical Research Institute (Chennai) (Shi et al. 2021).

Remarkably, the contents of curricula have been greatly improved in traditional pharmacovigilance education (e.g.: Ayurveda). Its teachers, doctors, and medical assistants play an important role in the supervision and research of traditional pharmaco-vigilance.

The Indian Government promotes the education of Indian System of Medicines and Ayurveda globally by setting up AYUSH Academic Chairs with foreign Universities/institutes. The ministry of Within the framework of an AYUSH fellowship scheme annually 104 scholarships are offered to foreign nationals from 99 countries to pursue undergraduate, postgraduate and Ph. D courses in AYUSH systems, in premier Institutes, in India. In addition to promoting the presence of Indian experts at international meetings, the Government of India provides also incentives to AYUSH drug manufacturers, entrepreneurs, AYUSH institutions to take part in international exhibitions, conferences, workshops, trade fairs, etc. with the aim of generating public awareness about the AYUSH systems of medicine and to register AYUSH products with the regulatory authorities of the foreign countries. To date 50+ products (Unani and Ayurveda) have been registered in 08 countries (including Kenya, USA, Russia, Latvia, Canada, Oman, Tajikistan and Sri Lanka) and 33 "AYUSH Information Cells" have been set up in 31 countries.

It has also been reported that the Ministry of AYUSH has signed an MoU with Ministry of Railways for the establishment of AYUSH wings at 5 Railway Zonal Hospitals and with Ministry of Defence to integrate Ayurveda under the health establishments of Ministry of Defence/Directorate General of Armed Forces Medical Services (https://pib.gov.in/PressReleasePage.aspx?PRID=1656379).

1.5.3 Intellectual Property Right Protection

It is frequently cited that India has 8% of the world's plant resources. It is also postulated that *ca.* 44+ percentage of this could be developed into drugs. As a significant number of traditional drug patents has been plagiarized, the Indian government has initiated a series of practices to protect and rationally develop traditional medical knowledge. These include the new revision of the patent law, the collection and filing of traditional medicine knowledge, the establishment of a registration and invention patent system, and the establishment of a special investment foundation. These measures are expected to provide a more comprehensive protection for Indian traditional medicine patents.

1.5.4 Pharmacovigilance System

With the increased recognition of traditional medicine in India, the rate of reported adverse reactions to Ayurveda drugs has also dramatically increased, therefore the safety of traditional medicine drugs has also received increasing attention.

The idea of an Ayurvedic pharmacovigilance system, different from the national pharmacovigilance system, was first raised in 2006, by the clinical pharmacologists, Urmilla Thatte and Vaidya Supriya Bhalerao, in Mumbai. This was followed by the Protocol for National Pharmacovigilance Program for Ayurveda, Siddha, and Unani Drugs (PNPP for ASU). This was issued by the Traditional Medicine Administration of the Ministry of Health and Family Welfare of India to supervise the safety of the three types of traditional medicines, in 2008.

In 2009, the National Pharmacovigilance Program was renamed into the Pharmaco-vigilance Program of India (PvPI). To date, it is jointly implemented by the Central Drugs Standard Control Organization and the Indian Pharmacopoeia Commission to supervise all types of drugs in India (Shi et al. 2021).

1.5.5 Policy Issues

Ayurveda focuses on the management, education, regulation, development, and growth of Indian pharmaceutical systems (ISM) in both India and abroad. The main objectives of state policies (issued in 2002) declare that Ayurveda should be used for the health benefit and health care of local people (priority should be given to people who have modest access to medication and medical facilities). By providing affordable Ayurvedic services and medicines, the systems should ensure that the supply and authentic products of APIs meet the Pharmacopoeia standard requirements (Shi et al. 2021).

In India, there are several autonomous institutions that function in various forms including research committees, professional committees, pharmacopoeia laboratories, national research institutes, academies of sciences, and hospitals. Participating institutions in Ayurvedic research activities include the Central Drug Research Institute (CDRI), Central Institute of Medicinal and Aromatic Plants (CIMAP), National Botanical Research Institutes (NBRI). There are also R & D centers attached to Ayurvedic drug manufacturing firms (Kurup 2004) (Ravishankar and Shukla 2007).

1.6 Conclusions

Despite significant advances in modern science, technology and allopathic medicine, providing quality healthcare to all populations has remained to be a problem. Parallel with efforts to authenticate herbal products, traditional medicine, in particular herbal medicine, continues to be a major healthcare provider in India, especially in rural and remote areas. Indian traditional medicinal systems like Ayurveda, Siddha and Unani have a very rich history of their effectiveness; more and more already known uses are being verified by modern research, including clinical trials. Traditional medicinal plants also frequently provide the source for new drugs and drug development.

In a pioneering way, several steps have been taken in India to promote traditional medicine, the use of herbal medicine and to integrate them into clinical practice. Evidence based incorporation of Indian traditional medicine in clinical practice forms a veritable bridge from Traditions to modern (holistic) wellbeing that will provide safe, efficient quality healthcare to all.

We cannot escape from nature because we, as human beings, are part of nature. It is especially in India that the blind dependence on synthetics seems to be over and people are returning to the natural systems of medicine in the hope of safety and security. As herbs are natural products, their use is mostly regarded comparatively safe and eco-friendly. Their further advantage is that they are frequently locally available. Consequently, there is a need to promote them to save the human lives like it was worded in the Chiang Mai Declaration of 1988. "Save Plants to Save Lives". In these ways, medicinal plants will be able to serve as a bridge from traditions to the holistic wellbeing of mankind.

References

Adhikari PP, Paul SB (2018) History of Indian traditional medicine: a medical inheritance. Asian J Pharm Clin Res 11(1):421. https://doi.org/10.22159/ajpcr.2018.v11i1.21893

Akerele O, Heywood VH (1991) The conservation of medicinal plants: proceedings of an international consultation, 21–27 March 1988 Held at Chiang Mai, Thailand. Cambridge University Press. Retrieved from https://books.google.hu/books?hl=hu&lr=&id=mZZOAAAAIAAJ&oi=fnd&pg=PR11&dq=guidelines+on+the+conservation+of+medicinal+plants&ots=ouyFrvnIxh&sig=yxjNl3qyGK75bAgxlcK1hkWl9EE&redir_esc=y#v=onepage&q=guidelines

Billore KV, Yelne MB, Dennis TJ, Chaudhari BG (2004) Database on medicinal plants used in Ayurveda, vol 6. Central Council for Research in Ayurveda and Siddha, New Delhi, pp 110–132

Gupta AK, Tandon N (eds) (2004) Reviews on Indian medicinal plants, vol I–III. Indian Council of Medical Research, New Delhi

Heyn B (1990) Ayurveda: the Indian art of natural medicine and life extension. Inner Traditions/ Bear & Co, Rochester

Jain NK (2001) A textbook of forensic pharmacy. Vallabh Prakashan, Delhi

Jaiswal YS, Williams LL (2017) A glimpse of Ayurveda – the forgotten history and principles of Indian traditional medicine. J Tradit Complement Med 7(1):50–53. https://doi.org/10.1016/j.jtcme.2016.02.002

Kurup PNV (2004) In: Roy CR, Muchatar RU (eds) Ayurveda in traditional medicine in Asia, vol 2002. WHO- Regional Office for South East Asia, New Delhi, pp 3–16

Mishra LC (ed) (2004) Scientific basis for Ayurvedic therapies. CRC Press, New York

Mukherjee PK, Ponnusankar S, Pandit S, Hazam PK, Ahmmed M, Mukherjee K (2011) Botanicals as medicinal food and their effects on drug metabolizing enzymes. Food Chem Toxicol 49(12): 3142–3153. https://doi.org/10.1016/j.fct.2011.09.015

Mukherjee PK, Harwansh RK, Bahadur S, Banerjee S, Kar A, Chanda J et al (2017) Development of Ayurveda – tradition to trend. J Ethnopharmacol 197:10–24. https://doi.org/10.1016/j.jep.2016.09.024

National Medicinal Plants Board (NMPB)| Ayush Next. (n.d.). Retrieved October 16, 2022, from https://ayushnext.ayush.gov.in/detail/news/national-medicinal-plants-board-nmpb

Ravishankar B, Shukla VJ (2007) Indian systems of medicine: a brief profile. African Journal of Traditional and Complement Alternate Medicine 4(3):319–337. https://doi.org/10.4314/ajtcam.v4i3.31226

Satyavati GV (2005) History of pharmacology of medicinal plants in India in topics in the history of pharmacology. In: Patil PN, Gulati OD, Balaraman R, Goyal RK (eds) Ahemdabad. B.S.Shah Prakashaz, India

Sen S, Chakraborty R (2017) Revival, modernization and integration of Indian traditional herbal medicine in clinical practice: importance, challenges and future. J Tradit Complement Med 7(2): 234–244. https://doi.org/10.1016/j.jtcme.2016.05.006

Sharma PC, Yelne MB, Dennis TJ (2002) Data base on medicinal plants used in Ayurveda, vol 4. Central Council for Research in Ayurveda and Siddha, New Delhi, pp 341–357

Shi Y, Zhang C, Li X (2021) Traditional medicine in India. J Tradit Chin Med Sci 8:S51–S55. https://doi.org/10.1016/j.jtcms.2020.06.007

Tavhare SD, Nishteswar K (2014) Collection practices of medicinal plants-vedic, ayurvedic and modern perspectives. Int J Pharm Biol Arch 5(5):54–61. Retrieved from www.ijpba.info

Williamson E (2001) Synergy and other interactions in phytomedicines. Phytomedicine 8(5): 401–409. https://doi.org/10.1078/0944-7113-00060

Chapter 2
Holy Basil (*Ocimum sanctum* L.): An Important Indian Medicinal and Aromatic Plant: Its Properties, Utilization and Genetic Improvement

Amarapalli Geetha, Saidaiah Pidigam, Pandravada Someswar Rao, Vishnukiran Thuraga, Suhail Khan Mohammad, and J. Suresh

Abstract Holy Basil or Tulsi (*Ocimum sanctum* L.) is one of the most sacred plants and is under cultivation, in India, for more than 3000 years. It's medicinal and as well as therapeutic properties are well described in the Indian Material Medica. *O. sanctum* has multiple pharmacological properties. With huge biological potential, Tulsi is a scientifically, traditionally, and clinically proven efficient medicinal herb for therapeutic applications. Application of breeding protocols have resulted several high yielding varieties, rich in essential oils and developed both in Sri Tulsi and Krishna Tulsi. In India, two institutes are engaged actively in Tulsi research: ICAR-NBPGR (in germplasm collection and conservation) and CSIR-CIMAP (in evaluation and development of varieties). Two hybrid Tulsi varieties have been released for commercial cultivation. Research has been conducted on chemotypes

A. Geetha (✉)
Department of Crop Physiology, College of Agriculture, Professor Jayashankar Telangana State Agricultural University, Hyderabad, Telangana, India

S. Pidigam
Department of Genetics and Plant Breeding, College of Horticulture, Sri Konda Laxman Telangana State Horticultural University, Hyderabad, Telangana, India

P. S. Rao
Division of Plant Genetic Resources and Economic Botany, ICAR-National Bureau of Plant Genetic Resources Regional Station, Hyderabad, Telangana, India

V. Thuraga
Post Doctoral Fellow, Department of Plant Breeding, Swedish University of Agricultural Sciences, Uppsala, Sweden

S. K. Mohammad
Department of Biotechnology, University School of Biotechnology, Guru Gobind Singh Indraprastha University, Dwaraka, New Delhi, India

J. Suresh
Department of Genetics and Plant Breeding, College of Agriculture, Professor Jayashankar Telangana State Agricultural University, Hyderabad, Telangana, India

Á. Máthé, I. A. Khan (eds.), *Medicinal and Aromatic Plants of India Vol. 2*,
Medicinal and Aromatic Plants of the World 9,
https://doi.org/10.1007/978-3-031-44914-7_2

for enhancing essential oil production. *In vitro* tissue-culture techniques are also applied for chemotype development. The investigations on whole nuclear and chloroplast genomes will be very useful in evolutionary studies to identify genes underlying the secondary metabolite-synthesis. The genomic resources and molecular tools investigated will be also useful in medical research.

Keywords Holy basil · *Ocimum tenuiflorum* L. · *Ocimum sanctum* L. · Utilization · Genetic improvement · Chemotypes · Breeding · Therapeutic uses · Genome

2.1 Introduction

Tulsi is native to India; in particular, based on phyto-geography and molecular data on *O. tenuiflorum*, North-Central India is concluded as the geographical origin (Bast et al. 2014; Mishra et al. 2014). It is a cultivated plant and an escape, spread over the Indian sub-continent ranging up to 1800 m (the Himalaya and the Andaman and Nicobar Islands) in the South (Kirtikar and Basu 1984). This plant is described as "The Queen of Herbs," "The Incomparable One" and "The Mother Medicine of Nature" in the Ayurvedic text of Charaka Samhita. The variants of holy basil with coloured flowers having purple or green calyces and purple or white corolla are present. India and adjoining regions are having two main cultivated types of Tulsi (Kothari et al. 2005). *O. sanctum* green type is regarded as Sri Tulsi *or* Ram tulsi (Fig. 2.1a, b). It is the most common type in cultivation, whereas the second one is Krishna Tulsi or Shyama Tulsi by virtue of bearing purple or dark leaves. Krishna tulsi is common for commerce due to its higher drug potency. Vana Tulsi (*Ocimum gratissimum*) is also claimed as one more variant of holy basil (Siva et al. 2016). For proper and correct identification of the types, expert guidance and officially approved herbariums can be referred. However, in contrast, the reports of Carovic et al. 2006 revealed that the presence of diverse cultivars and chemotypes within species do not differ morphologically. Variation in leaf pigmentation, size and shape of various basil species are due to their cultivation for centuries together (Nazim et al. 2009). Indistinguishable morphotypes augmented from various ecologies possessing different chemical constituents are reported by Ali and Ali 2012. Seedling morphological investigations revealed the relatedness among the taxonomy of the genus *Ocimum* (Singh et al. 2012).

2.2 Taxonomic Characteristics

By virtue of vivid medicinal properties, *Ocimum* spp., belongs to the important essential oil producing and medicinal plants of India. Genus Ocimum belongs to Lamiaceae, whose species are widely distributed among tropical and warm temperate parts of the world. Around 160–185 species are reported in the genus, spread over the warm regions of the world (Charles and Simon 1990). The species *Ocimum*

Fig. 2.1 (**a** and **b**)Variants of *Ocimum sanctum,* Sri Tulsi and Krishna Tulsi. (Source: Malav et al. 2015)

sanctum is known by the name Basil, a royal plant as derived from the greek word "Basilica". Inter-specific hybridization coupled with polyploidy among the species of this genus led to confusion in taxonomic classification (Tucker 1986). Molecular based phylogenetic investigations have given clue that the tribe Ocimeae is evolved in tropical Asia and then introduced elsewhere of cultivated regions (Paton et al. 2004). Studies on the tribe Ocimeae and sub-tribe Ociminae from South-east Asian countries were reviewed by Suddee et al. 2005). The genus has two major groups *viz., Ocimum basilicum* (French basil, Sweet basil, Common basil) and *Ocimum sanctum* (Sacred basil, Holy basil) based on morphological, geographical reference, cytology and chemical constituents. The first group, basilicum group, which has basic chromosome number of 12 with four species namely, *O. canum* Sims. (2n = 24, 26), *O. basilicum* L. (2n = 48), *O. americanum* L. (2n = 72) and

O. kilimandscharicum Guerke (2n = 76). All the group members are cultivated as herb or shrub possessing essential oil constituents such as linalool, camphor, methyl chavicol, methyl cinnamate, eugenol and citral etc.

Sanctum group has basic chromosome number of eight with six species *i.e.*, *O. sanctum* L. (2n = 32), *O. gratissimum* L. (2n = 40), *O. viridae* Wild (2n = 40), *O. suave* Wild (2n = 64), *O. carnosum* (2n = 48) and *O. micranthum* (2n = 48). All the group members are cultivated as shrubs especially for eugenol, elimicin and methyl isoeugenol. *Ocimum sanctum* Linn (2n = 32) is locally known as Tulsi (Hindi, Gujarati, Sanskrit). At full bloom stage, whole herb is employed for extraction of aromatic oil. It survives as a biennial or triennial shrub. The leaves of this plant on steam distillation result in volatile oil of bright yellow in colour with pleasant odour. Eugenol (71%), eugenol methyl ether (20%), carvacrol (3%) and minute fractions of nerol, selinene, camphor, caryophyllene, ß-pinene, linalool, cineole etc. are the essential oil components present in Tulsi. Eugenol is the active chemical principle in leaves that attribute to therapeutic potential as blood glucose level reducer in type-2 diabetics and painkiller, as well (Prakash and Gupta 2005). The essential oil find diverse uses in perfumery and cosmetic industries including indigenous medicine systems. Tulsi is also grown as a pot herb and its leaves are consumed as condiment in salads and in various other foods (Mondello et al. 2002; Prakash and Gupta 2005). Most commonly, the holy basil is utilized in the preparation of herbal tea preservatives (Kirtikar and Basu 1984) and healing remedies (Anbarasu and Vijayalakshmi 2007).

2.3 Morphological Description

The Holy basil, *O. sanctum* is an erect, herbaceous and multi-branched plant that attains a height of 30–75 cm. The growth form of Tulsi varies from herbs to trees with large variations in leaf shape, size, hairs, glands and other morphological peculiarities. The flowers are purplish or crimson in colour, which are produced in racemes. The fruits are slightly compressed, sub-globose or broadly ellipsoid, smooth, pale brown or reddish. *O. sanctum* is distinguished from the other species by spreading pedicels and presence of internally glabrous calyces.

2.4 Ecological Requirements

O. sanctum is native to the Indian subcontinent and cultivated throughout Southeast Asian tropics. It has widest distribution ascending upto 1800 m in the Himalayas and in Andaman and Nicobar Islands, as well. In India, the Tulsi is still a dominant household plant. It is one of the most sacred plants and has been grown in India for more than 3000 years. Except its susceptibility to root rot disease that too occasionally under water stagnated conditions, which remained as a hardy species with almost zero threatening pests and diseases. The plant can grow well in a wide range of habitats and in a wide variety of soils and climates. In a shorter period of time, due to larger consumption of fresh leaves, the area under cultivation has spread across the country, from the South Indian temples. States, such as Uttar Pradesh, Maharashtra, Bihar, Madhya Pradesh, Telangana, Andhra Pradesh, Jammu and Assam West Bengal, etc. have sizeable area of *Ocimum* spp. in India, which is cultivated over an area of 25,000 ha with annual production of about 250–300 tonnes of oil.

2.5 Major Chemical Constituents and Bioactive Compounds

The medicinal properties of *Ocimum sanctum* L. are well described (Bhargava and Singh 1981; Pandey and Anita 1990; Rajeshwari 1992; Mnadal et al. 1993; Sen 1993; Sarkar et al. 1994; Ray 1995). Tulsi is in use for treating convulsions, arthritis, fever, bronchitis *etc*. Phytochemicals, chemical compounds and pharmacological properties and of *Ocimum sanctum* are reviewed (Table 2.1 and Fig. 2.2) by Siva et al. 2016. Various *Ocimum sanctum* plant parts, such as root, leaves, flowers, stem, seeds etc. posess therapeutic potentials. *Ocimum sanctum* L. is taken traditionally as herbal tea, dried leaf power, fresh leaf *etc*.

Exhaustively documented therapeutic potentials of *Ocimum sanctum* L. are as anti-asthmatic and anti-kaphic drugs. In Indian Materia Medica, various uses of extracts of *Ocimum sanctum* leaves in aqueous, methanolic and hydro-alcoholic forms in bronchitis, rheumantism and pyrexia are sufficiently described.

New Fatty Acid Derivatives

The leaves of *Ocimum sanctum* L. contains vanillic acid, oleiyl glucoside, β-sitosterol glucoside, diglucosyl oleate and *Ocimum* napthanoic acid, in addition of which sanctumoic acid and benzoyl gluco-oleate, the two fatty acid derivatives were isolated in methanolic extract in the Os grown in Delhi region. The new reported derivatives are anticipated as responsible for drug medicinal properties (Ali and Ali 2012). The derivatives are identified as dotriacont-20-en-14-ol-1-oic acid and 4′-benzoylglucopyranosyl octadec-9-enoate (Fig. 2.3) (Ali and Ali 2012).

2.6 Traditional Use Part(s) Used and Common Knowledge

Over the years, *Ocimum* species have been traditionally exploited to treat various ailments in Indian Ayurveda and traditional African, Chinese and European medicine. Tulsi extracts have strong pharmacological actions and applications in human ailment treatments (Table 2.2 and Fig. 2.4).

The extracts are used in the treatment of fever, bronchitis, arthritis, convulsions *etc*.; as expectorant, analgesic, anti-cancer, anti-asthmatic, anti-emetic, diaphoretic,

Table 2.1 Phytochemicals, chemical compounds and pharmacological properties of *Ocimum sanctum*

Part of the plant	Phytochemicals	Chemical compounds	Pharmacological properties
Leaf	Flavonoids, alkanoids, saponins, tannins, phenols, anthocynins, terpenoids, steroils	Eugenol, eugenal, urosolic acid, carvacol, linalool, caryophyllene, limatrol, caryophyllene, methyl carvicol, anthocyans	Anti stress, antichronic, anti hypolipidemic, anti-oxidant, anthelmintic, anti malarial activity (against plasmodium vivex), anti-fungal (against ring worm and also skin diseases), anti fertility activity, anti cancer (carsinigenic), antiviral activity
Stem	Phenols, saponins, flavo-noids, triterpenoids, tannins	Rosmarinic acid, apigenin, cirsimaritin, isothymusin, isothymonin.	Genitourinary system disorders.
Seeds	Fatty acids, sitosterol.	Sugars (xylose and polysaccharides).	Reduced blood and uri-nary uric acid level in albino rabbits.
Whole plant	Flavonoids, alkanoids, saponins, tannins, phenols, anthocynins, flavonoids, triterpenoids, tannins		Control diabetes mellitus, anti dot for dog bite, scor-pion bite and insects bite
Flower			Antispasmodic agent (as smooth muscle relaxant).
Root			Decoction of root acts as a diaphoretic in malarial fever, anti larvicidal (against to mosquitoes), antifungal (*aspergillus Niger*)

Source: Siva et al. (2016)

anti-diabetic, anti-fertility, hepatoprotective, hypotensive, hypolipidimic and antistress agents in traditional medicinal practices; some plant extracts in Ayurvedic remedies for stomach disorders, inflammation, common colds, heart disease head-aches, poisoning and malaria (Sen et al. 1992; Singh et al. 1996; Mauli et al. 1997; Mediratta et al. 2002); even as therapeutic agents as anti-inflammatory, anti-oxidant and immune-modulatory and anti-stress properties (Devi 2000; Sethi et al. 2003). Tulsi is also acclaimed as vitalizer and enhances physical endurance (Siva et al. 2016).

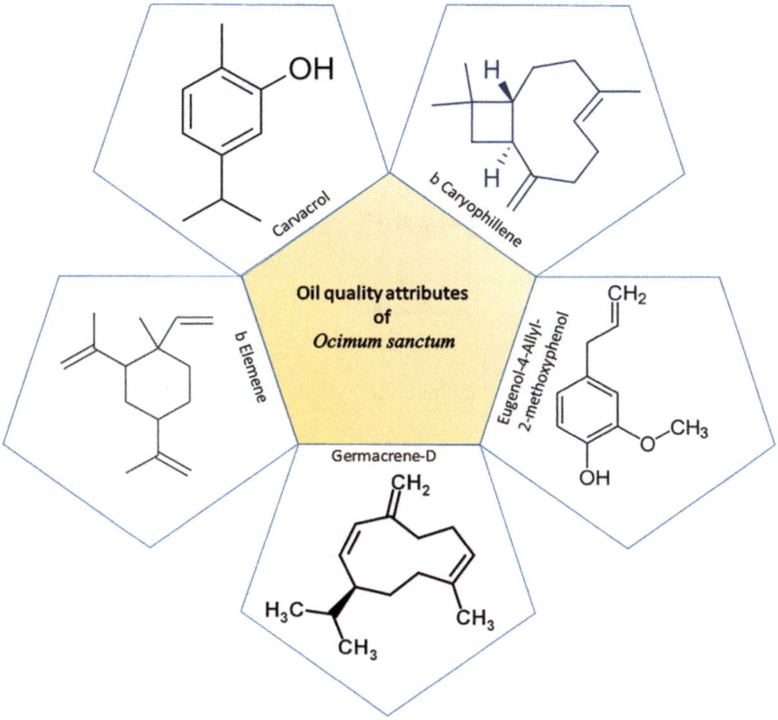

Fig. 2.2 Oil quality attributes of *Ocimum sanctum* varieties (Eugenol, ß-elemene; ß-Caryophyllene; Germacrene-D)

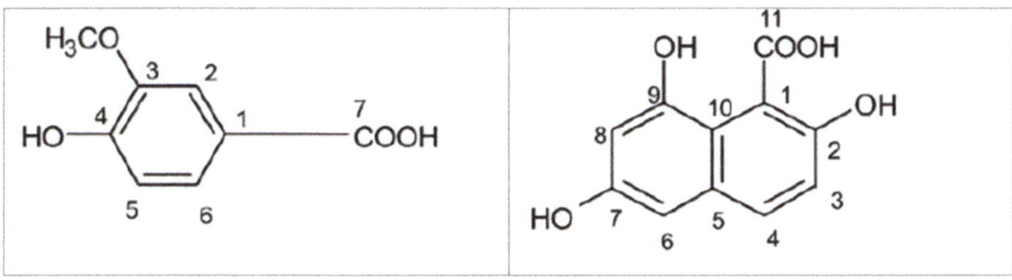

Fig. 2.3 The chemical structures of vanillic acid and ocimumnapthanoic acids of *Ocimum*

2.7 Modern Medicine Based on Its Traditional Medicine Uses

There are several domains wher modern medicine can make use of the traditional uses of MAPs. Table 2.2 is meant to serve as an example showing various applications of different organs, as well as drugs prepared of Ocimum sanctum.

Table 2.2 Utilization of *Ocimum sanctum* in various applications

Property reported	Plant organ used	References
Antimicrobial, antidiabetic, antiviral, anticancer, antimalarial, antidiarrhoeal, radiation protective, analgesic, anti-inflammatory, anti-hyperlipidemic activities, *etc.*	Various parts of the plant	Ali et al. (2021), Santos et al. (2021)
Multiple pharmacological properties, use of essential oil in aromatherapy	Essential oil	Dada et al. (2020)
Insecticidal, trypanocidal, fungicidal, nematicidal, larvicidal properties	Essential oil	Chowdhary et al. (2018), Bhavya et al. (2021)
Phytoremediation potence (removal of dyes, crude oil, pesticide residue and heavy metals)	Entire plant	Ramírez-Sandoval et al. (2011), Choden et al. (2021), Dada et al. (2020), Lakshmanraj et al. (2009)
Treating gastric and hepatic disorders	Aqueous decoction of Tulsi leaves	Rajeshwari (1992), Pandey and Anita (1990)
Effect on viral hepatitis	Herbal preparations of tulsi	Rajeshwari (1992)
Treating glaucoma, cataract, chronic conjunctivitis and other painful eye diseases using eye drop preparations in Ayurvedic	The leaf juice	Rajeshwari (1992)
To treat chronic fever, dysentery, hemorrhage and dyspepsia	Juice of fresh leaves	Rajeshwari (1992), Pandey and Anita (1990)
Remedy for cold	Decoction of Tulsi	Rajeshwari (1992), Pandey and Anita (1990)
Check vomiting and anthelmintic	Tulsi leaves	Sen (1993)
Prophylactic against malaria	Fresh Tulsi leaves with black pepper	Pandey and Anita (1990)
Antimalarial activity against *plasmodium vivex* and *Pl. falciparum*	Ayurvedic preparation with tulsi, garlic, pipelli and turmeric	Rajeshwari (1992)
Diaphoretic in malarial fever	Decoction of the root of Tulsi plant	Pandey and Anita (1990)
Antimalarial effect	Tulsi extracts and essential oil	Rajeshwari (1992)
Antifungal activity against *aspergillus Niger*	Entire plant	Rajeshwari (1992)
Effect on viral encephalitis	Aqueous extract of Tulsi	Rajeshwari (1992)

(continued)

Table 2.2 (continued)

Property reported	Plant organ used	References
Lowering the blood sugar level	Aqueous decoction of whole plant	Ray (1995)
Treatment of ringworm and other skin diseases	Tulsi paste	Sen (1993), Rajeshwari (1992) Pandey and Anita (1990)
Antidote for dog bite, scorpion bite and insect bite in traditional medicine	Plant	Sen (1993), Rajeshwari (1992), Pandey and Anita (1990)
Antispasmodic agent	Fresh leaves and flower tops	Sen (1993), Rajeshwari (1992)
Genitourinary system treatment	Seeds	Rajeshwari (1992)
Anti-stress, analgesic, anti-hyperlipidemic, antioxidant potentials in experimental animals	Leaves	Rajeshwari (1992), Bhargava and Singh (1981), Ray (1995), Mnadal et al. (1993), Sarkar et al. (1994)
Reduction of blood and urinary uric acid level in albino rabbits	Leaves and seeds	Sarkar et al. (1990)
Gastric ulceration and secretion in albino rats	Plant	Sen (1993), Mnadal et al. (1993), Sarkar et al. (1990)
Abortifacient effect in women	Leaves	Chopra et al. (1993), Kirtikar and Basu (1965), Batta and Santhakumari (1971)

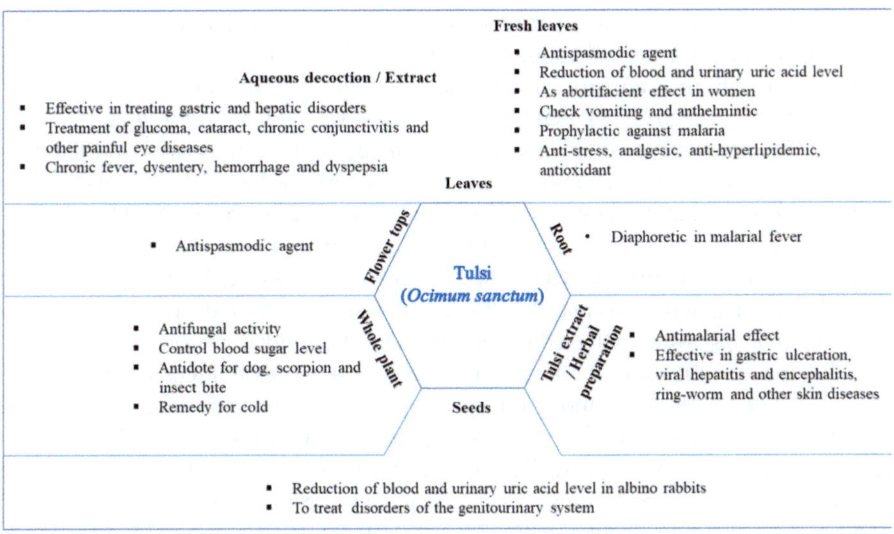

Fig. 2.4 The pharmacological actions of Tulsi

2.8 Cultivation Practices

Sacred basil thrives well on a wide range of soils. Rich loam, poor laterite, saline and alkaline to moderately acidic soils are also well suited for its cultivation. It flourishes well under fairly high rainfall and humid conditions. Long days and high temperatures have been found favourable for plant growth and oil production. It can grow up to an altitude of 900 m. The plant is moderately tolerant to drought and frost. *Tulsi* is propagated through seeds. Tulsi is found to be infested with few insect pests and diseases. Leaf rollers and Tulsi lace wing are *the major pests.* The plant is susceptible to *Oidium spp.*, *Rhizoctonia solani* and *Rhizoctonia bataticola.* The crop is to be harvested at full bloom stage to obtain maximum and best quality oil. The first harvest comes at 90–95 days of planting and harvested at every 65–75 days interval. About 5 tons of fresh herbage per hectare can be obtained by two to three harvests in a year. The oil yield varies with type, season, and place of origin. The whole herb contains 0.1–0.23% essential oil and an oil yield of 10–23 kg can be obtained per hectare. The variations, in respect of essential oil, are reported when grown in different parts of India with a range of 40–71%. Seasonal variations associated with amount of eugenol are also recorded.

2.9 Germplasm of *Ocimum sanctum*

ICAR-NBPGR is playing a key role in plant exploration, evaluation, and conservation of *Ocimum sanctum* collections from India. The minimal descriptors (Fig. 2.5) used for germplasm evaluation are also prepared by NBPGR only.

During the year 2015, NBPGR has collected several *O. tenuiflorum* germplasm, while in the same year; two *Ocimum tenuiflorum* germplasm lines are imported from Uzbekistan. As a part of virus-indexing, a mandatory protocol of bulk material on import by directorate of plant protection, quarantine, and storage (DPPQS), Faridabad, Haryana, tested the seed sample of *Ocimum* from Germany in isolation and under controlled conditions, where the seedlings were checked for Pepino mosaic virus, Broad bean wilt virus and Alfalfa mosaic virus by EM and ELISA techniques. Five *Ocimum* spp. were collected under the externally funded project by scientists of NBPGR for identification of elite species to develop agrotechniques for cultivation. By preliminary evaluation among four germplasm lines, the spike length (cm) range of 8.0–9.1 cm with an average of 8.42 cm; plant height (cm) range of 44.0–77.0 cm and average of 54.31 cm, while fresh herbage yield per plant (g) range of 54.06–85.06 g with the average of 72.26 g and average plant canopy (cm^2) of 1352.21 with the range between 967.33 and 2292.24 cm was reported. Among the several species of *Ocimum* evaluated, *Ocimum tenuiflorum* germplasm named as IC599368 registered the highest herbage yield of 661.0 g, while the highest essential oil yield per plant (g) of 1.7 g was extracted from two elite germplasm lines *viz.*,

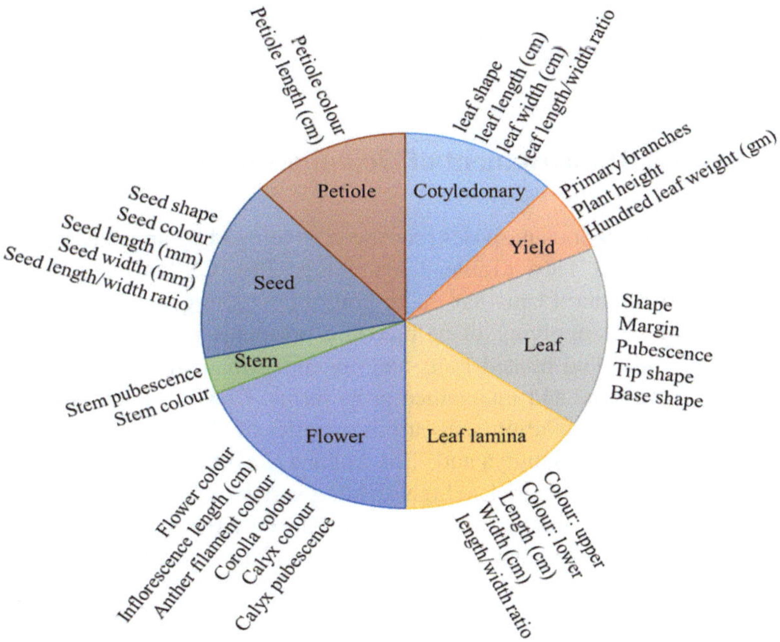

Fig. 2.5 The minimal descriptors for germplasm evaluation of *Ocimum sanctum* in India

IC589192 and IC599368. Germplasms of Tulsi are maintained under field gene banks at NBPGR, New Delhi (Anonymous 2016).

As a new record, *Ocimum kilimandscharicum* Guerke (Lamiaceae), maintained and used as local plant-based pesticide by tribes and new to the Indian flora, has recorded in natural habitats in its wild habitat in Odisha (IC599299, IC599345) and its wild occurrence was found to be a new record for peninsular India. The seed is conserved in National gene bank, New Delhi including preservation as herbarium materials (Anonymous 2016).

The two major morphotypes, Rama tulsi and Shyama tulsi are highly distinct in their oil quality especially for eugenol and methyl eugenol contents (Philip and Damodaran 1985; Pushpangadan and Bradu 1995; Kothari et al. 2005; Archana et al. 2013). Malav et al. (2015) characterized 49 germplasm accessions of cultivated *Ocimum tenuiflorum* L. augmented from 4 Indian phytogeographical regions with high degree of variation exhibiting rich diversity within the populations among the morphotypes. The study also revealed that irrespective of the distinctiveness of morphology of Rama and Shyama types, both were seen scattered among clusters from the phytogeographical regions. Within Rama type, the genotypes from NE region were very distinct plant types (Malav et al. 2015).

The clustering pattern of the genotypes showed a high degree of genetic diversity among different accessions of *O. tenuiflorum* (Ahmad and Khaliq 2002). In the phenotypic relationship, among the basil genotypes, UPGMA cluster analysis was used to analyze the similarity matrix to draw genetic relationships and phylogeny among genotypes (Shazia et al. 2011).

2.10 Genetic Improvement of *Ocimum sanctum*

Several varieties of various purposes have been developed by application of various breeding protocols in Tulsi (Table 2.3 and Fig. 2.6). CSIR-Central Institute of Medicinal and Aromatic Plants has major contributions in genetic improvement of *Ocimum sanctum* by applying plant breeding protocols of significance, majorly targeting the yield, yield related traits and eugenol enhancing properties from the genetic stocks available and maintained at its farms. CIMAP, Lucknow, India has released for farmer's cultivation, the various varieties of *Ocimum sanctum* including CIM series such as CIM Saumya and CIM Angana (Lal et al. 2003, 2008). Lal et al. (2003) developed eugenol enriched high yielding variety, CIM Ayu using traditional breeding method, while another variety which is the dark purple pigmented and high yielding variety, CIM Angana belongs to Shyam tulsi is also developed by Lal et al. (2008).

In India, though, rich morphotype variability and decades of history of holy basil cultivation are there, the efforts to characterize the morphological diversity are meager. Several researchers studied the germplasm characterization followed by selection (Singh et al. 2004, 2014a, b; Srivastava et al. 2018), variability and correlation (Patel et al. 2015a, b, Singh and Sehgal (1999)), yield and biomass stability (Lal 2014; Patel et al. 2015a, b, 2016) and eugenol content as well (Lal

Table 2.3 Characteristics of varieties released from CSIR-CIMAP in sacred basil

Variety	Breeding method	Characters	Yield	Quality	Year of release
CIM-Ayu	Mass selection	Tall, green leaves, high eugenol content in essential oil	Herb yield: 200 q/ha; dry leaves yield: 15.85 q/ha. Oil yield: 110 kg/ha	Oil content: 0.72%; oil quality: Eugenol 83%, ß-elemene 7.47%	2003
CIM-Kanchan	Selection	Medium tall, leaves light green and slightly undulated, with light	Oil quality: Methyl eugenol 70%, ß-elemene 7.6%, ß-caryophyllene 15.7%		2004
CIM-Angana (Krishna Tulsi)	Half-sib selection	Tall, leaves light to dark purple in colour. Suitable for tea	Herb yield: 181 q/ha; dry leaf yield: 14.33 q/ha; oil yield: 91.71 kg/ha	Oil quality: Eugenol 40.42%, ß -elemene 14.11%; ß -Caryophyllene 9.07%; Germacrene-D 16.65%.	2007

Fig. 2.6 Genetic improvement protocols used in *Ocimum sanctum* for varietal development

2014). Study carried out on morphological variability in *Ocimum* spp. showed significant influence of genotype and environment (Agarwal et al. 2013). In the advancement of molecular systematic abundance era also, the morphological characterization is basic for phylogenetic studies (Lee 2006; Bruce et al. 2007).

2.10.1 The Genome of Ocimum sanctum

More than 80% of the patents filed or granted on *Ocimum spp*. are focused on chemical extracts or use of plant parts for essential oil attributes only. So, there is a need to unravel the metabolic potential of this sacred herb. To meet this target, sequencing of the whole nuclear genome and chloroplast genomes was undertaken (Rastogi et al. 2015). This is the first medicinal plant whose genome is sequenced and completely analyzed.

Rastogi et al. 2015 revealed highlights of the genome of *Ocimum sanctum* using assembled *de novo* with genome size of approximately ~600 Mb. The data generated out of the 4 libraries of which, two libraries of long and short reads were of Illumina

HiSeq2000, one was of 454 GS FLX and one was the matepair library of SOLiD 5500XL. This data revealed the total of 107,785 contigs, which resulted into 22,776 scaffolds, 9059 super-scaffolds and 136 proteins homologous to five important plant genomes were also identified in addition to SSR markers. Perhaps, it is the smallest nuclear genome in Lamiaceae and smallest chloroplast genome in the Order, Lamiales. Phylogenetically, *S. miltiorrhiza* is most similar to *O. sanctum. O. sanctum* genome (386 Mb) is compact with relatively fewer repeat sequences with small plant genome and 1.5 times to that of the model plant, *Arabidopsis thaliana* (~135 Mb), whereas, of same size as that of paddy (*Oryza sativa)* (~420 Mb). In comparison with other gene model prediction plants such as tomato (*Solanum lycopersicum)* (~900 Mb), tobacco (*Nicotiana tabacum)* (~4567 Mb), Tulsi (*O. sanctum)* genome (~386 Mb) is smaller. The research also assembled chloroplast genome, showing maximum similarity to that of *S. miltiorrhiza*, an important medicinal plant of traditional Chinese medicine. Both are rich in phenylpropanoid derivatives and have therapeutic activities. The investigation also reported large number of copalyl diphosphate synthases (CPS) homologs in *basil* genome scoping for discovery of newer diterpenes of bioactivity of potence, which is yet to implicated (Rastogi et al. 2015). In the same year, another research team, Upadhyay et al. (2015) also revealed draft genome of *O. tenuiflorum* L using paired end and mate-pair sequence libraries and the assembly, which resulted in a draft genome of 374 Mb size. Genomic information reported in this investigation useful for evolutionary studies, to identify genes underlying the secondary metabolite-synthesis (Rastogi et al. 2015) and also genes in respect of specific pathways. The molecular tools and genomic resources investigated will aid in basil's utility in medical community and molecular breeding (Rastogi et al. 2015).

2.10.2 Heterosis Breeding for Ocimum *Improvement*

Heterosis yield enhancement has been successfully demonstrated in many crops. There is a greater scope for its exploitation in *Ocimum*. But, there are a very few reports available in this direction. Singh et al. 2015 studied heterotic crosses for oil yield and other quantitative traits in basil. The hybrid EC388896 × IC369247 exhibited positive and significant heterobeltiosis for oil yield, fresh and dry herb yield. Another cross, EC387893 × IC326711 registered significant and positive heterosis for oil content. Inter and intra-specific hybridization among basil species has resulted in morphotypic and chemotypic variation (Carovic et al. 2006; Nazim et al. 2009; Tilwari et al. 2013). Till recent, no hybrid in either Sri Tulsi or Krishna Tulsi is released for cultivation. However, in holy basil, for betterment of heterosis breeding for oil yield improvement, the better combining ability lines are required. But in Vana tulsi, *Ocimum gratissimum* L. hybrids were released. Regional Research Laboratory, Jammu has released two hybrids namely. 'Clocimum' and 'Clocimum-3c' targeting high eugenol content utilizing recurrent selection method of plant breeding. "Clocimum" variety produces 60–65% eugenol, while "Clocimum-3c" contains 90–95% eugenol, both these varieties were extended to private companies and farmers for commercial cultivation in the trade name of RRL-og-1as eugenol rich varieties.

2.10.3 Chemotypes of Holy Basil

Based on the chemical compounds in the form of relative concentration of the essential oil, many chemotypes are identified in *Ocimum* species (Kumar et al. 2019; Varga et al. 2017). Grayer et al. (1996) has described the chemotypes based on presence of compounds upto 20% of the total essential oil, while, Varga et al. (2017) reported upto 10%. Among the published articles related to *Ocimum* species chemical diversity, in *O. tenuiflorum* is 10%. *O. tenuiflorum* chemistry is well studied due its distribution over worldwide. On the regional basis, India has 75 various chemotypes from eight *Ocimum* species, of which, a total of 24 *O. tenuiflorum* chemotypes are registered. The major essential oil chemotypes from *O. tenuiflorum* are characterized by a large amount of eugenol (23–77%) and methyl eugenol (36–93%), β-caryophyllene, methyl eugenol (Raina and Misra 2018; Rana and Blazquez 2015; Piras et al. 2018), new chemotypes β-elemene (Kitchlu et al. 2013), β-bisabolene (Carovic-Stanko et al. 2011), α-cubebene, α-humulene (Saran et al. 2017) *etc.*

2.10.4 Strategies of Chemotype Improvement in O. tenuiflorum

Various classical and advanced genetic improvement methods like tissue culture, genetic engineering and genetically engineered plants are utilized to improve plant herb yield, oil yield and oil chemical composition in *Ocimum*-species. The main tissue culture techniques applied in *Ocimum sanctum* improvement are summarized in Table 2.4.

Using metabolic engineering, key genes of metabolite biosynthesis were characterized in *Ocimum* species. Anand et al. (2016) have characterized eugenol synthase (*egs*) involved in the phenylpropanoid biosynthesis from several *Ocimum* species (Lange and Ahkami 2013). Over expression of OkHMGR in Tulsi enhanced terpenoid accumulation leading to essential oil content. Silencing of *OS4CL* through RNAi in Sacred basil reduced the eugenol level (Bansal et al. 2018). Around 40 TF families (MYB, WRKY, bZIP, bHLH, HB, NAC, etc.) *via*. Transcriptome sequencing in basil (Rastogi et al. 2014) was done. Several enzymes of phenylpropanoid pathway from three *Ocimum*-species including sacred basil were discovered (Rastogi et al. 2020). The anatomical structures are crucial for essential oil synthesis and storage in *Ocimum*-species, which can be useful to improve chemotype contents.

Table 2.4 Tissue culture techniques applied in *Ocimum sanctum* improvement

Tissue culture technique	References
In vitro propagation for regeneration of *Ocimum* species	Dode et al. (2003), Manan et al. (2016), Rady and Nazif (2005)
Chemotype development through over production of metabolites	Namdeo (2007)
Influence of Tulsi callus culture in betulinic acid and in vitro derived leaves	Pandey et al. (2015)
Influence of light on phenylpropanoid biosynthesis	Nadeem et al. (2019), Nazir et al. (2020)
Differentiated plantlets or organ culture for metabolite production with stable essential oil yield	Karuppusamy (2009)
Study of *in vitro* grown leaves and somatic embryos for eugenol quantitiy compared to field-grown Tulsi leaves	Bhuvaneshwari et al. (2016)
Superiority of cell culture for metabolites production	Nitzsche et al. (2004)
Terpenoid content in cell culture in the elicitor presence in comparison to field-grown Tulsi plants.	Mathew and Sankar (2014)
Accumulation of total phenylpropanoids with elicitors in suspension cell cultures in Tulsi	Vyas and Mukhopadhyay (2018)
The hairy roots from the plant's aerial part for metabolite production mediated by *Agrobacterium rhizogenes*	Srivastava and Srivastava (2007), Murthy et al. (2008)
Hairy root cultures of *O. tenuiflorum* with ursolic acid and eugenol using elicitors	Sharan et al. (2019)
Biotransformation to accumulate metabolites utilizing cell or organ culture	Giri et al. (2001)
Somaclonal variations improve the essential oil profile of *Ocimum*-species	Krishna et al. (2016)

Maurya et al. (2019) reported the dependency of essential oil content with density and size of peltate and capitate glandular trichomes basil species. Transcriptomic analysis of *O. basilicum* and *O. tenuiflorum* was carried out to identify genes involved in the glandular trichome development concerning the essential oil biosynthesis. Most of the transcripts belonged to the TF families, such as bHLH, C2H2, R2R3MYB, and R3MYB, which regulate trichome development were identified. The large size and high density of trichomes lead to oil accumulation. Enhancing the anatomy and density of glandular trichomes in *Ocimum* species for essential oil enhancement using biotechnology applications for new chemotypes are the need of the hour. *In vitro* tissue-culture techniques are also applied for chemotype development in *Ocimum*-species.

2.10.5 Potential Causes of Chemodiversity in Ocimum *Species*

The existence of different chemotypes in *Ocimum* species could also be attributed to cross-pollination leading to intra and inter-specific hybridization, resulting in higher variation in the chemical profiles (Gurav et al. 2020; Varga et al. 2017). Other factors, such as plant habit, can also influence specialized metabolism. For instance, the popular sanctum group of *Ocimum* species prominently harbors phenylpropanoid biosynthesis because of perennial woody habit. The possible explanation for this is phenylpropanoid biosynthetic pathway generates monolignol alcohols required for lignin biosynthesis. Also, natural evolutionary events, polyploidy, and selective breeding can play a significant role in chemical diversification as observed in different *Ocimum* species (Iijima et al. 2004). The abiotic stresses have profound effect on the oil yielding quality. The main principle such as eugenol and methyl eugenol were decreased in their levels in tulsi under water scarce and abundant environments and salt stress (Rastogi et al. 2019). There are certain adverse health effects of some metabolites such as, methyl eugenol, and chavicol and camphor of essential oil of *Ocimum* species which are genotoxic or carcinogenic at some concentrations revealed by in vitro studies in model organisms (Zuccarini 2009). The aqueous extracts or organic extracts of *Ocimum* species of a tissue or whole plant were less toxic. The terpenoids registered cytotoxic effects on plasma membrane, ROS, mitochondrial function, and/or lipid peroxidation (Agus 2021). In the liver, the hepatotoxicity exhibiting metabolites were induced at metabolism in the liver (Zárybnický et al. 2018). Hence, the plant-based natural molecules are advised to be used in formulations but not in their pure form.

2.11 Conclusions

Ocimum sanctum (Os) is an indigenous Ayurvedic medicinal plant of Indian origin with a cultivation history of more than 3000 years. This herb is extensively utilized in pharmaceutical preparations, food and perfumery industries. Tulsi is known for its strong therapeutic activities based on its phytochemical and medicinal properties. Eugenol is the main active ingredient with therapeutic potential, while another chemical, methyl eugenol is used as a flavouring agent. The leaves, stems, seeds and roots are attributed with antispasmodic, stomachic and carminative attributes, in native medicine. Tulsi is already acclaimed as an adaptogen and astringent with applications to cure inflammation, stomach disorder, common colds, headaches, malaria heart diseases and various forms of poisoning. ICAR-NBPGR (germplasm collection and conservation), CSIR-CIMAP (evaluation and development of varieties) and Regional Research Laboratory, Jammu (commercial varieties for cultivation) are the Indian institutes actively involved in Tulsi research. Considering the metabolic and therapeutic potential of this wonder herb, CSIR-Central Institute of

Medicinal & Aromatic Plants, Lucknow, has completed the whole genome sequencing of *Ocimum sanctum*. This will help to understand and unravel the secrets of this 'mother of all herbs'.

References

Agarwal C, Sharma NL, Gaurav SS (2013) An analysis of basil (*Ocimum* sp.) to study the morphological variability. Indian J Fundam Appl Life Sci 3(3):521–525

Agus HH (2021) Terpene toxicity and oxidative stress. In: Patel VB, Preedy VR (eds) Toxicology: oxidative stress and dietary antioxidants. Academic, Cambridge

Ahmad SD, Khaliq I (2002) Morpho – molecular variability and heritability in *Ocimum sanctum* genotypes from northern Himalayan region of Pakistan. J Biol Sci 5(10):1084–1087

Ali A, Ali M (2012) New fatty acid derivatives from *Ocimum sanctum* L. leaves. Indian Drugs 49(11):13–18

Ali HM, Nguta JM, Mapenay IO, Musila FM, Omambia VM, Matara DN (2021) Ethnopharmacological uses, biological activities, chemistry and toxicological aspects of *Ocimum americanum* var. americanum (Lamiaceae). J Phytopharmacol 10:56–60

Anand A, Jayaramaiah RH, Beedkar SD, Singh PA, Joshi RS, Mulani FA, Dholakia BB, Punekar SA, Gade WN, Thulasiram HV, Giri AP (2016) Comparative functional characterization of eugenol synthase from four different *Ocimum* species: implications on eugenol accumulation. Biochim Biophys Acta Proteins Proteom 1864:1539–1547

Anbarasu K, Vijayalakshmi G (2007) Improved shelf life of protein-rich tofu using *Ocimum sanctum* (tulsi) extracts to benefit Indian rural population. J Food Sci 72:M300–M305

Anonymous (2016) Annual Report of the ICAR-National Bureau of Plant Genetic Resources 2015–16, NBPGR, Pusa Campus, New Delhi, India, 195 + x p

Archana PR, Ashok K, Dutta M (2013) Chemical characterization of aroma compounds in essential oil isolated from "holy basil" (*Ocimum tenuiflorum* L.) grown in India. Genet Resour Crop Evol 60:1727–1735

Bansal S, Narnoliya LK, Mishra B, Chandra M, Yadav RK, Sangwan NS (2018) HMG-CoA reductase from camphor Tulsi (*Ocimum kilimandscharicum*) regulated MVA dependent biosynthesis of diverse terpenoids in homologous and heterologous plant systems. Sci Rep 8:1–15

Bast F, Pooja R, Devendra M (2014) Chloroplast DNA phylogeography of holy basil (*Ocimum tenuiflorum*) in Indian subcontinent. The Scientific World Journal, Hindawi Publishing Corporation:847482

Batta SK, Santhakumari G (1971) The antifertility effect of *Ocimum sanctum* and *Hibiscus rosa-sinensis*. Indian J Med Res 59:777–781

Bhargava KP, Singh N (1981) Anti-stress activity of *Ocimum sanctum* Linn. Indian J Med Res 73: 443–451

Bhavya ML, Obulaxmi S, Devi SS (2021) Efficacy of *Ocimum tenuiflorum* essential oil as grain protectant against coleopteran beetle, infesting stored pulses. J Food Sci Technol 58:1611–1616

Bhuvaneshwari K, Gokulanathan A, Jayanthi M, Govindasamy V, Milella L, Lee S, Yang DC, Girija S (2016) Can *Ocimum basilicum* L. and *Ocimum tenuiflorum* L. in vitro culture be a potential source of secondary metabolites? Food Chem 194:55–60

Bruce KK, Scott JR, David LR (2007) Characters as groups: a new approach to morphological characters in phylogenetic analysis. Taxon 56(2):479–492

Carovic K, Liber Z, Javornik B, Kolak I and Satovic Z (2006) Genetic relationships within basil (*Ocimum*) as revealed by RAPD and AFLP markers. In: XXVII International Horticultural Congress-IHC2006: II International Symposium on Plant Genetic Resources of Horticultural, pp 171–178

Carovic-Stanko K, Salinovic A, Grdisa M, Liber Z, Kolak I, Satovic Z (2011) Efficiency of morphological trait descriptors in discrimination of *Ocimum basilicum* L accessions. Plant Biosyst 145:298–305

Charles DJ, Simon JE (1990) Comparison of extraction methods for the rapid determination of essential oil content and composition of basil. J Amer Soc Hort Sci 115(3):458–462.1990

Choden D, Pokethitiyook P, Poolpak T, Kruatrachue M (2021) Phytoremediation of soil co-contaminated with zinc and crude oil using *Ocimum gratissimum* (L.) in association with pseudomonas putida MU02. Int J Phytoremediation 23:181–189

Chopra RN, Chopra IC, Handa KL, Kapoor LD (1993) Indigenous drugs of India. UN Dhar, Pvt. Ltd., Calcutta

Chowdhary K, Kumar A, Sharma S, Pathak R, Jangir M (2018) *Ocimum* sp.: source of biorational pesticides. Ind Crop Prod 122:686–701

Dada AO, Adekola FA, Odebunmi EO, Dada FE, Bello OM, Akinyemi BA, Bello OS, Umukoro OG (2020) Sustainable and low-cost *Ocimum gratissimum* for biosorption of indigo carmine dye: kinetics, isotherm, and thermodynamic studies. Int J Phytoremediation 22: 1524–1537

Devi PU (2000) Radioprotective, anticarcinogenic and antioxidant properties of Indian holy basil, *Ocimum sanctum* (Tulsi). Indian J Exp Biol 39:185–190

Dode LB, Bobrowski VL, Braga EJB, Seixas FK, Schuch MW (2003) In vitro propagation of *Ocimum basilicum* L. (Lamiaceae). Acta Sci Biol Sci 25:435–437

Giri A, Dhingra V, Giri C, Singh A, Ward OP, Narasu ML (2001) Biotransformations using plant cells, organ cultures and enzyme systems: current trends and future prospects. Biotechnol Adv 19:175–199

Grayer RG, Kite GC, Goldstone FJ, Bryan SE, Paton A, Putievsky E (1996) Infraspecific taxonomy and essential oil chemotypes in basil *Ocimum basilicum*. Phytochemistry 43:1033–1039

Gurav TP, Jayaramaiah RH, Punekar SA, Dholakia BB, Giri AP (2020) Generation of novelties in the genus *Ocimum* as a result of natural hybridization: a morphological, genetical and chemical appraisal. Ind Crop Prod 156:112859

Iijima Y, Davidovich-Rikanati R, Fridman E, Gang DR, Bar E, Lewinsohn E, Pichersky E (2004) The biochemical and molecular basis for the divergent patterns in the biosynthesis of terpenes and phenylpropenes in the peltate glands of three cultivars of basil. Plant Physiol 136: 3724–3736

Karuppusamy S (2009) A review on trends in production of secondary metabolites from higher plants by *in vitro* tissue, organ and cell cultures. J Med Plants Res 3:1222–1239

Kirtikar KR, Basu BD (1965) *Ocimum sanctum* in Indian medicinal plants. Published by LB Basu, Allahabad

Kirtikar KR and Basu BD (eds) (1984) Indian Medicinal Plants. Vol. III, Bishen Singh and Mahendra Pal Singh, Dehradun, Allahabad 1664–1666

Kitchlu S, Bhadauria R, Ram G, Bindu K, Khajuria RK, Ahuja A (2013) Chemo-divergence in essential oil composition among thirty-one core collections of *Ocimum sanctum* L. grown under sub-tropical region of Jammu. India Am J Plant Sci 4:302–308

Kothari S, Bhattacharya A, Ramesh S, Garg S, Khanuja S (2005) Volatile constituents in oil from different plant parts of methyl eugenol-rich *Ocimum tenuiflorum* L.f (syn. *O. Sanctum* L.) grown in South India. J Essent Oil Res 17:656–658. https://doi.org/10.1080/10412905.2005.9699025

Krishna H, Alizadeh M, Singh D, Singh U, Chauhan N, Eftekhari M, Sadh RK (2016) Somaclonal variations and their applications in horticultural crops improvement. 3 Biotech 6:54

Kumar A, Mishra P, Rodrigues V, Baskaran K, Verma RS, Padalia RC, Sundaresan V (2019) Delineation of *Ocimum gratissimum* L. complex combining morphological, molecular and essential oils analysis. Ind Crop Prod 139:111536

Lakshmanraj L, Gurusamy A, Gobinath MB, Chandramohan R (2009) Studies on the biosorption of hexavalent chromium from aqueous solutions by using boiled mucilaginous seeds of *Ocimum americanum*. J Hazard Mater 169:1141–1145

Lal RK (2014) Breeding for new chemotypes with stable high essential oil yield in *Ocimum*. Indust Crops Prod 59:41–49

Lal RK, Khanuja SPS, Agnihotri AK, Misra HO, Shasany AK, Naqvi AA, Dhawan OP, Kalra A, Bahl JR, Darokar MP (2003) High yielding eugenol rich oil producing variety of *Ocimum sanctum*- 'CIM-Ayu'. J Med Arom Plant Sci 25:746–747

Lal RK, Khanuja SPS, Rizavi H, Shasany AK, Ahmad R, Chandra R, Naqvi AA, Misra HO, Singh A, Singh N, Lohia RS, Bansal K, Darokar MP, Gupta AK, Kalara A, Dhawan OP, Bahl JR, Singh AK, Shankar H, Kumar D, Alam M (2008) Registration of a high yielding dark purple pigmented, variety 'CIM-Angana' of Shyam tulsi (*Ocimum sanctum* L.). J Med Arom Plant Sci 30(1):92–94

Lange BM, Ahkami A (2013) Metabolic engineering of plant monoterpenes, sesquiterpenes and diterpenes – current status and future opportunities. Plant Biotechnol J 11:169–196

Lee MSY (2006) Morphological phylogenies and the universe of useful characters. Taxon 55:5–7

Malav P, Pandey A, Bhatt KC, Krishnan S, Bisht IS (2015) Morphological variability in holy basil (*Ocimum tenuiflorum* L.) from India. Genet Resour Crop Evol 62:1245–1256

Manan AA, Taha RM, Mubarak EE, Elias H (2016) *In vitro* flowering, glandular trichomes ultrastructure, and essential oil accumulation in micropropagated *Ocimum basilicum* L. Vitro Cell Dev Biol Plant 52:303–314

Mathew R, Sankar PD (2014) Comparison of major secondary metabolites quantified in elicited cell cultures, non-elicited cell cultures, callus cultures and field grown plants of *Ocimum*. Int J Pharm Pharm Sci 6:102–106

Mauli G, Maulik N, Bhandari V, Kagan VE, Pakrashi S, Das DK (1997) Evaluation of antioxidants effectiveness of few herbal plants. Free Radic Res 27:221–228

Maurya S, Chandra M, Yadav RK, Narnoliya LK, Sangwan RS, Bansal S, Sandhu P, Singh U, Kumar D, Sangwan NS (2019) Interspecies comparative features of trichomes in *Ocimum* reveal insights for biosynthesis of specialized essential oil metabolites. Protoplasma 256:893–907

Mediratta PK, Sharma KK, Singh S (2002) Evalution of immunomodulatory potential of *Ocimum sanctum* seed oil and its possible mechanism of action. J Enthopharmacol 80:15–20

Mishra D, Awasthi A, Mishra P (2014) Phylogenetic evolution studies on different varieties of genus *Ocimum* with special reference to Rewa district of Madhya Pradesh. Sci Secure J Biotech 3(2):188–197

Mnadal S, Das DN, Dey K (1993) *Ocimum sanctum* Linn – a study on gastric ulceration and gastric secretion in rats. Indian J Physiol Pharmacol 37(91–92):25

Mondello L, Zappia G, Cotroneo A, Bonaccorsi I, Chowdhury JU, Yusuf M, Dugo G (2002) Studies on the essential oil-bearing plants of Bangladesh. Part VIII. Composition of some *Ocimum* oils *O. basilicum* L. var. *purpurascens; O. sanctum* L. green*; O. sanctum L.* purple; *O. americanum* L., citral type*; O. americanum* L., camphor type. Flavour Fragr J 17:335–340

Murthy HN, Dijkstra C, Anthony P, White DA, Davey MR, Power JB, Hahn EJ, Paek KY (2008) Establishment of *Withania somnifera* hairy root cultures for the production of withanolide A. J Integr Plant Biol 50:975–981

Nadeem M, Abbasi BH, Younas M, Ahmad W, Zahir A, Hano C (2019) LED-enhanced biosynthesis of biologically active ingredients in callus cultures of *Ocimum basilicum.* J Photochem Photobiol B 190:172–178

Namdeo A (2007) Plant cell elicitation for production of secondary metabolites: a review. Pharmacogn Rev 1:69–79

Nazim K, Ahmed M, Uzair M (2009) Growth potential of two species of basil in sandy soil of Karachi. Pakistan J Bot 41:1637–1644

Nazir M, Ullah MA, Younas M, Siddiquah A, Shah M, Giglioli-Guivarc'h N, Hano C, Abbasi BH (2020) Light-mediated biosynthesis of phenylpropanoid metabolites and antioxidant potential in callus cultures of purple basil (*Ocimum basilicum* L. var. purpurascens). Plant Cell Tissue Organ Cult 142:107–120

Nitzsche A, Tokalov SV, Gutzeit HO, Ludwig-Müller J (2004) Chemical and biological characterization of cinnamic acid derivatives from cell cultures of lavender (*Lavandula officinalis*) induced by stress and jasmonic acid. J Agric Food Chem 52:2915–2923

Pandey BP, Anita (1990) Economic botany. Chand and Company Ltd., New Delhi, p 294

Pandey H, Pandey P, Singh S, Gupta R, Banerjee S (2015) Production of anti-cancer triterpene (betulinic acid) from callus cultures of different *Ocimum* species and its elicitation. Protoplasma 252:647–655
Patel RP, Kumar RR, Singh R, Singh RR, Rao B, Singh VR, Gupta P, Lahiri R, Lal RK (2015a) Study of genetic variability pattern and their possibility of exploitation in *Ocimum* germplasm. Indust Crop Prod 66:119–122
Patel RP, Singh R, Saikia SK, Sastry KP, Rao BRR, Zaim M, Lal RK (2015b) Phenotypic characterization and stability analysis for biomass and essential oil yields of fifteen genotypes of five *Ocimum* species. Indust Crop Prod 77:21–29
Patel RP, Singh R, Rajeswara BRR, Singh RR, Srivastava A, Lal RK (2016) Differential response of genotype environment on phenology, essential oil yield and quality of natural aroma chemicals of five *Ocimum* species. Indust Crop Prod 87:210–217
Paton AJ, Springate D, Suddee S, Otieno D, Grayer RJ, Harley MM, Willis F, Simmonds MSJ, Powell MP, Savolainen V (2004) Phylogeny and evolution of basils and allies (*Ocimeae*, Labiatae) based on three plastid DNA regions. Mol Phylogenet Evol 31(1):277–299
Philip MP, Damodaran NP (1985) Chemo-types of *Ocimum sanctum* Linn. IndianPerfum 29(L-2): 49–56
Piras A, Gonçalves MJ, Alves J, Falconieri D, Porcedda S, Maxia A, Salgueiro L (2018) *Ocimum tenuiflorum* L. and *Ocimum basilicum* L., two spices of Lamiaceae family with bioactive essential oils. Ind Crop Prod 113:89–97
Prakash P, Gupta N (2005) Therapeutic uses of *Ocimum sanctum* Linn (Tulsi) with a note on eugenol and its pharmacological actions: a review. Indian J Physiol Pharmacol 49:125–131
Pushpangadan P, Bradu BL (1995) Basil. In: Chadha KL, Gupta R (eds) Advances in horticulture Vol. 11: medicinal and aromatic plants. Malhotra Publishing House, New Delhi
Rady MR, Nazif NM (2005) Rosmarinic acid content and RAPD analysis of in vitro regenerated basil (*Ocimum americanum*) plants. Fitoterapia 76:525–533
Raina AP, Misra R (2018) Chemo-divergence in essential oil composition among germplasm collection of five *Ocimum* species from eastern coastal plains of India. J Essent Oil Res 30: 47–55
Rajeshwari S (1992) Ocimum sanctum. The Indian home remedy. In: Current medical scene. Edited and published by S. Rajeshwari, Cipla Ltd., Bombay Central, Bombay
Ramírez-Sandoval M, Melchor-Partida GN, Muñiz-Hernández S, Girón-Pérez MI, Rojas-García AE, Medina-Díaz M, Robledo-Marenco ML, Velázquez-Fernández JB (2011) Phytoremediatory effect and growth of two species of *Ocimum* in endosulfan polluted soil. J Hazard Mater 192:388–392
Rana VS, Blazquez MA (2015) Essential oil composition of the aerial parts of five *Ocimum* species from Western India. J Essent Oil Bear Plants 18:1234–1241
Rastogi S, Meena S, Bhattacharya A, Ghosh S, Shukla RK, Sangwan NS, Lal RK, Gupta MM, Lavania UC, Gupta V (2014) De novo sequencing and comparative analysis of holy and sweet basil transcriptomes. BMC Genomics 15:588
Rastogi S, Kalra A, Gupta V, Khan F, Lal RK, Tripathi AK, Parameswaran S, Gopalakrishnan C, Ramaswamy G, Shasany AK (2015) Unravelling the genome of holy basil: an "incomparable" "elixir of life" of traditional Indian medicine. BMC Genomics 16(1):413. https://doi.org/10.1186/s12864-015-1640-z
Rastogi S, Shah S, Kumar R, Vashisth D, Akhtar MQ, Kumar A, Dwivedi UN, Shasany AK (2019) *Ocimum* metabolomics in response to abiotic stresses: cold, flood, drought and salinity. PLoS One 14:e0210903
Rastogi S, Shah S, Kumar R, Kumar A, Shasany AK (2020) Comparative temporal metabolomics studies to investigate interspecies variation in three *Ocimum* species. Sci Rep 10:1–15
Ray A (1995) Recent trends in stress research: focus on adaptogenesis. Proc. XXXVIIIth Conference of Indian Pharmacological Society held at Punjabi University, Patiala, p 68. 23
Santos ED, Leitão MM, Ito CN, Silva-Filho SE, Arena AC, de Souza Silva-Comar FM, Cuman RK, Oliveira RJ, Formagio AS, Kassuya CA (2021) Analgesic and anti-inflammatory articular

effects of essential oil and camphor isolated from *Ocimum kilimandscharicum* Gürke leaves. J Ethnopharmacol 269:113697

Saran PL, Tripathy V, Saha A, Kalariya KA (2017) Selection of superior *Ocimum sanctum* L. accessions for industrial application. Ind Crops Prod 108:700–707

Sarkar A, Pandey DN, Pant MC (1990) A report on the effect of *Ocimum sanctum* (Tulsi) leaves and seeds on blood and urinary uric acid, urea and urine volume in normal albino rabbits. Indian J Physiol Pharmacol 34(61–62):26

Sarkar A, Pandey DN, Pant MC (1994) Changes in the blood lipid profile level after administration of *Ocimum sanctum* (Tulsi) leaves in the normal albino rabbits. Indian J Physiol Pharmacol 38(4):311–312

Sen P (1993) Therapeutic potentials of Tulsi: from experience to facts. Drugs News Views 1(2): 15–21

Sen P, Maiti PC, Puri S, Ray A, Audulov NA, Valdman AV (1992) Mechanism of anti stress activity *of Ocimum sanctum* Linn. Eugenol, *Tinospora malabarica* in experimental animals. Indian J Exp Biol 32:592–596

Sethi J, Sood S, Seth S, Thakur A (2003) Protective effect of Tulsi (*Ocimum sanctum*) on lipid peroxidation in stress induced by anemic hypoxia in rabbits. Indian J Physiol Pharmacol 47(1): 115–119

Sharan S, Sarin N, Mukhopadhyay K (2019) Elicitor-mediated enhanced accumulation of ursolic acid and eugenol in hairy root cultures of *Ocimum tenuiflorum* L. is age, dose, and duration dependent. S Afr J Bot 124:199–210

Shazia E, Muhammad N, Shahid M, Muhammad I (2011) Genetic variation in the living repository of *Ocimum* germplasm. Pak J Agric Res 24:1–4

Singh NK, Sehgal CB (1999) Micropropagation of 'holy basil' (*Ocimum sanctum* Linn.) from young inflorescences of mature plants. Plant Growth Regul 29:161–166

Singh S, Majumdar DK, Rehan HMS (1996) Evalution of anti-inflammatory potential of *Ocimum sanctum* (holy basil) and its possible mechanism of action. J Ethnopharmacol 54:19–26

Singh S, Singh A, Singh UB, Patra DD, Khanuja SPS (2004) Intercropping of Indian basil (*Ocimum basilicum* L.) for enhancing resource utilization efficiency of aromatic grasses. J Spices Arom Crops 2:97–101

Singh E, Sharma S, Dwivedi J, Sharma S (2012) Diversified potentials of *Ocimum sanctum* Linn (Tulsi): an exhaustive survey. J Nat Prod Plant Resour 2:39–48

Singh D, Kumar Chaudhuri P, Darokar MP (2014a) New antiproliferative tricyclic sesquiterpenoid from the leaves of *Ocimum sanctum*. Helv Chim Acta 97:708–711

Singh K, Chand S, Yaseen M (2014b) Integrated nutrient management in Indian basil (*Ocimun basilicum*). Ind Crop Prod 55:225–229

Singh P, Kalunke RM, Giri AP (2015) Towards comprehension of complex chemical evolution and diversification of terpene and phenylpropanoid pathways in *Ocimum* species. RSC Adv 5: 106886–106904

Siva M, Shanmugam KR, Shanmugam B, Venkata SG, Ravi S, Sathyavelu RK, Mallikarjuna K (2016) *Ocimum sanctum*: a review on the pharmacological properties. Int J Basic Clin Pharmacol 5:558–565

Srivastava S, Srivastava AK (2007) Hairy root culture for mass-production of high-value secondary metabolites. Crit Rev Biotechnol 27:29–43

Srivastava A, Gupta AK, Sarkara S, Lal RK, Yadav A, Gupta P, Chanotiya CS (2018) Genetic and chemotypic variability in basil (*Ocimum basilicum* L.) germplasm towards future exploitation. Indust Crop Prod 112:815–820

Suddee S, Paton A, Parnell J (2005) Taxonomic revision of tribe Ocimeae Dumort. (Lamiaceae) in continental South East Asia III. Ociminae Kew Bull 60:3–75

Tilwari A, Tamrakar A, Sharma R (2013) Use of random amplified polymorphic DNA (RAPD) for assessing genetic diversity of *Ocimum sanctum* (Krishna Tulsi) from different environments of Central India. J Med Plants Res 7(24):1800–1808

Tucker AO (1986) Botanical nomenclature of culinary herbs and potherbs. In: Craker LE, Simon JE (eds) Herbs, spices and medicinal plants: recent advances 214 in botany, horticulture and pharmacology, vol 1. Binghamton, Haworth Press, Inc, pp 33–80

Upadhyay AK, Chacko AR, Gandhimathi A, Ghosh P, Harini K, Joseph AP, Joshi AG, Karpe SD, Kaushik S, Kuravadi N, Lingu CS (2015) Genome sequencing of herb Tulsi (*Ocimum tenuiflorum*) unravels key genes behind its strong medicinal properties. BMC Plant Biol 15:212

Varga F, Carovic-Stanko K, Ristic M, Grdisa M, Liber Z, Satovic Z (2017) Morphological and biochemical intraspecific characterization of *Ocimum basilicum* L. Ind Crop Prod 109:611–618. https://doi.org/10.1016/j.indcrop.2017.09.018

Vyas P, Mukhopadhyay K (2018) Elicitation of phenylpropanoids and expression analysis of PAL gene in suspension cell culture of *Ocimum tenuiflorum* L. Proc Natl Acad Sci India Sect B Biol Sci 88:1207–1217

Zárybnický T, Boušová I, Ambrož M, Skálová L (2018) Hepatotoxicity of monoterpenes and sesquiterpenes. Arch Toxicol 92:1–3

Zuccarini P (2009) Camphor: risks and benefits of a widely used natural product. J Appl Sci Environ Manag 13:69–74

Chapter 3
Ethnomedicinal Importance of Common Weeds of the Family Asteraceae in the Tribal Belt of Rajasthan, India

Supriya Kumari Sharma and Afroz Alam

Abstract The relationship between humans and their hunt for drugs in nature extends back thousands of years. Man learned to seek medications in the seeds, barks, fruits, and other parts of plants, as a result of many years of fighting against illnesses. Many commonly occurring weeds contain medicinal properties, and they have a long history in traditional medicinal systems. Asteraceae is a plant family well known for its weeds that are also grown for edible and medicinal reasons for over 3000 years. Some of the well-known weed representatives of this family are *Taraxacum officinale* F.H. Wigg. (Dandelion), *Bellis perennis* L., *Verbesina encelioides* (Cav.) Benth. & Hook. f. *ex* A. Gray. These taxa primarily contain inulin (a natural polysaccharide with significant probiotic properties), but many other bioactive compounds have also been identified in them. Almost every tribe in India, especially in Rajasthan, is aware of the therapeutic properties of various weeds of this family, although the available scientific reports on them are somewhat scattered. In Rajasthan, information on many year-round growing weeds was studied by compiling knowledge about the medicinal importance of the family Asteraceae.

Keywords Asteraceae · Ethnomedicinal · Phytochemicals · Rajasthan · Tribes · Weeds

3.1 Introduction

The link between human life and plants may most likely be traced back to the early middle era of the "Pleistocene Epoch" (2.3 million years ago) when human life first appeared in the form of "Ape" man. The climatic circumstances necessitated a high-protein diet at this time; thus, they hunted. Nevertheless, new research reveals that

---

S. K. Sharma · A. Alam (✉)
Department of Bioscience and Biotechnology, Banasthali Vidyapith, Tonk, Rajasthan, India
e-mail: aafroj@banasthali.in

Á. Máthé, I. A. Khan (eds.), *Medicinal and Aromatic Plants of India Vol. 2*,
Medicinal and Aromatic Plants of the World 9,
https://doi.org/10.1007/978-3-031-44914-7_3

they also prepared and ate plant resources. In several regions, vegetation was also limited during this time. There were a few conifers, such as pines, cypress, and yews, as well as some broadleaf trees, strewn around. On the ground, prairie grasses and members of the lily, orchid, and rose families were discovered (Arya et al. 2016). The history of human - plant relationships during that period can be gleaned from the Rigveda. It appears in Manu-Smriti. Plants were shown to have emotions and are sensitive to pain, and Smriti understands the importance of plants and forests in ensuring that the right environmental conditions are in place to proclaim that the living world is growing and developing in a balanced manner. Plants are sacred in origin, and God has designated several of them as "Abodes of Peace." Man discovered how to distinguish between poisonous and edible plants by observing animals, according to the Rigveda. There are many widely accessible and socially acceptable traditional medicinal plants. Indigenous rural populations provide an accessible and affordable healthcare system and a crucial source of income. The study of plants and the application of ancient medical knowledge has once more attracted a lot of interest. Indeed, any traditional medical system must include herbal medicine. Plants are the foundation of sophisticated traditional medical systems that have been in existence for thousands of years and are still giving different new therapies to people (Gurib-Fakim 2006). Through years of careful observation, experience, and trial-and-error experimentation, early humans learned about the therapeutic uses of plants. Traditional medicinal herbs play a significant role in both Chinese and global indigenous medical systems. Traditional medicine is any age-old, culturally based form of healthcare that deviates from modern science and is mostly passed down orally by groups of people from various ethnic backgrounds.

Ethnobotany, according to Schultes (1962), is the "study of the interaction that occurs between members of ape-like groups and the plants that surround them." In many regions of India, including the Bhoxa tribe of Bihar, the Pauri Garhwal districts, U.P. (Maheshwari and Singh 1984), the Tharus in the Kheri district, U.P., and Morni and Kalesar (Ambala-Haryana) (Jain 1984), ethnobotanical research has been carried out. Eastern Rajasthan was the subject of ethnomedical, and ethnopharmacology-statistical investigations conducted by Upadhyay et al. (2010). Herbal medication influences the patient's physical, mental, and emotional states as well as their social and cultural demands. They established that herbal medications by using conventional techniques are still active and pure, because all of the natural ingredients are preserved in their "naturally balanced form" during the lengthy grinding and combining operations. The fact that many important components exist in a "naturally balanced condition" may be the precise reason why herbal medications have so few negative effects. Since the beginning of time, they have been studied and found to have positive side effects, as compared to "dangerous" synthetic and chemical-based products. In Rajasthan, despite of the numerous ethnobotanical studies on medicinal plant resources by Kirtikar and Basu (1984), Joshi (1995), Katewa and Guria (1997), Singh and Pandey (1998), Katewa et al. (2001, 2004), Jain et al. (2004) proper documentation is still missing.

The tribes are an intrinsic part of our national life with their rich cultural heritage. In India, there are more than 500 tribal groups spread out over the nation. They have a habit of isolating themselves, which greatly aids in the preservation of their social traditions, practices, and beliefs. India is renowned for the diversity of its population, including its culture, tribes, communities, castes, and religion, all of which have been identified over the years by their clothing. These native tribes have a plethora of knowledge regarding many wild plants, which they use as traditional medicine in their day-to-day lives. Due to the diversity in their cultural, economic, political, and structural elements, tribes could not be grouped under a single description.

3.2 The Tribal Belt of Rajasthan

The term "India's tribal belt" refers to the contiguous areas where the tribal people of India resided, in contrast to other ethnic groups that coexisted widely within the Indian subcontinent. Tribal people in India are a small minority, yet they come from a wide range of ethnic groupings. The ecological conditions in which they dwell, ecological environment, population size, level of acculturation, predominant subsistence methods, level of development, language and linguistic traits, and social stratification are all different. They are scattered over the entire country while having a very unequal geographic distribution. Based on their geographic distribution, Indian tribes were divided into three zones by Hasnain (1983):

- The tribal populations of Assam, Manipur, Tripura, Eastern Kashmir, Eastern Punjab, Himachal Pradesh, and Northern Uttar Pradesh constitute the North and North-Eastern zone.
- Madhya Pradesh, Uttar Pradesh, Southern Rajasthan, Northern Maharashtra, Bihar, and Orissa constitute the Central zone.
- Andhra Pradesh, Karnataka, Cochin, Tamil Nadu, and Kerala constitute the Southern zone.

Rajasthan's tribal population has a diverse cultural environment. The Rajasthan state can be divided into three sections based on the concentration of the tribal population when it comes to geographic distribution and comprehending the demographic status of the tribes.

- Southern Region has the maximum tribal concentration about 59.9% of the total tribal population of the state.
- Western Region, with a minimum population of only 7.48% of the state's total tribal population;
- South-Eastern Region, with 32.12% of the state's total tribal population (Joshi 1995).

In India, tribal people account for 8.02% of the overall population, whereas in Rajasthan, they constitute 12.56%. According to the Census of 2001 in Rajasthan, there were a total of 70,97,706 tribal people, out of which 36,50,982 were females and 34,46,724 were males. According to the 2001 Census, Banswara district has 72.27% of the total tribal population, Dungarpur district has 65.14%, Udaipur district has 70.71%, Chittorgarh district has 55.04%, and Sirohi district has 66.55%.

Rajasthan is a part of India's tribal belt in the northwest and east. The ancestry of the tribal people in this area predates that of the Ancestral North Indians and is connected to the Ancestral South Indians. These people are believed to have roots in the Indus Valley's Harappan culture, the oldest known civilization on the Indian subcontinent, which existed between 3500 and 2500 BC. In the past, matrilineal societies predominated among the tribes of northwest India. These people's shifting fortunes and misfortunes have gradually led to the formation of a more patriarchal way of life. Although tribal communities today mainly adhere to the patrilineal principle, there are still several tribal areas where institutionalized matriarchy is still prevalent. Women are in charge of planning things like relationships and marriages, land inheritances, and money distribution (Sharma and Mishra 2021).

3.2.1 Geography of Rajasthan

Rajasthan, a desert state in India, has a diverse landscape. It is well-known for its cultural diversity and is abundant in natural resources. Rajasthan, with its undulating dunes, steep hills, freezing temperatures, intense heat, fertile plains in the east, and sparsely populated areas in the west, is the world's most diverse region.

Rajasthan is the country's largest state covering a large land area (approximately 132,139 square miles) (Fig. 3.1). It is located in India's northwestern region. Rajasthan is located in both the northern and western hemispheres of the planet, with the majority of its territory north of the Cancer Tropic. The Cancer Tropic runs through Banswara town to the south. The state's expansion lies between the North latitude and East longitude. The state has a maximum length of 826 km from north to south (Kona village in Sriganganagar district in the north to Borkund village in Banswara district in the south) and 869 km from west to east (Katra village in Jaisalmer district in the west to Silana village of Dhaulpur district in the east), presenting an uneven rhomboid shape (heterogeneous quadrilateral). The entire boundary of the state constitutes 5920 kilometers, of which 4850 kilometers are interstate borders and 1070 kilometers are international borders. The international boundary between India and Pakistan affects four major districts in the area along with its western border: Sriganganagar (210 km), Bikaner (168 km), Jaisalmer (464 km), and Barmer (228 km). This is called 'The Radcliffe Line'. The state is girdled by Punjab and Haryana states in the north, Uttar Pradesh in the east, Madhya Pradesh in the south-east, and Gujarat in the southwest (Sharma and Mishra 2021).

The Thar Desert, often referred to as the "Maru-kantar," is the largest desert in India, and it occupies a considerable portion of the arid state of Rajasthan. The state

Fig. 3.1 Map of Rajasthan depicting the different boundaries surrounding Rajasthan along with its distribution. (Source: Google maps)

is divided into two separate geographic areas by the Aravalli Range, the oldest group of fold mountains: a desert on one side and a forest belt on the other. The region's total land area is only 9.36% covered by forests. The Guru Shikhar Peak, the tallest peak in the Aravalli range at 1722 meters above sea level, is found in Mount Abu, the state's only hill station (Sharma and Mishra 2021).

3.2.2 Climate of Rajasthan

Rajasthan's climate ranges from incredibly dry to humid. Extremes in yearly temperature, low humidity, rapid wind speed, and little to no rain are the defining features of the climate in western Rajasthan. Rajasthan's eastern and southern regions experience semi-humid to humid weather. It is characterized by about the same temperature extremes but with lower wind speeds, higher humidity, and more consistent rainfall. While the eastern plain and south-eastern plateau regions are wetter with less temperature variation, the western arid plain is particularly dry and exhibits substantial temperature variability. Hyper-thermic conditions (high in temperature) describe the entire state. The average high temperature in May and June is about 45 °C, while the average low temperature is about 23 °C. The rainfall decreases from the west to the east. The Aravalli range's direction, the location of the tropic of cancer, and the continental effect all contribute to Rajasthan's "tropical desert climate."

3.2.3 *Floristic Diversity of Rajasthan*

The native vegetation is categorized as Northern Desert Thorn Forest and is found in sparse, more or less open clusters. From west to east, patches get denser and bigger, as the amount of rainfall rises. Accordingly, the forests of southern Rajasthan, particularly Mount Abu, Phulwari, Sitamata, and Kumbhalgarh wildlife sanctuaries, are known as "mega floral diversity places" in Rajasthan because of their extensive floristic diversity. These protected regions have many terrestrial orchids, tuberous plants, climbers, and lianas, as well as pteridophytic and bryophytic flora in their valleys and along the banks of their streams.

Mount Abu is home to species including bamboo, salar, and dhavand Jamun. Here, one can also find rare varieties of ferns, orchids, and wildflowers. The unique ecosystem harbors approximately 480 species of plants including 107 trees, 55 shrubs, 215 herbs, 45 climbers, 40 grass, and species of lower plants. Khair [(*Senegalia catechu* (L.f.) P.J.H.Hurter & Mabb.)], dhavado (*Anogeissus latifolia* (DC.) Wallich ex Guill. & Perr.), saledi (*Boswellia serrata* Roxb. ex Colebr), kadaya (*Sterculia urens* Roxb.), timbre (*Zanthoxylum armatum* DC.), khakhara (*Butea monosperma* (Lam.) Taub.), bor (*Ziziphus mauritiana* Lam.), desi bavalia [*Vachellia nilotica* subsp. *indica* (Benth.) Kyal. & Boatwr.], bili [*Aegle marmelos* (L.) Corrêa], dudhi (*Euphorbia hirta* L.), golar (*Ficus racemosa* L.), karanj [*Millettia pinnata* (L.) Panigrahi], karel [(*Capparis decidua* (Forssk.) Edgew.], arjun [*Terminalia arjuna* (Roxb.) Wight & Arn.],, Jamun [(*Syzygium cumini* (L.) Skeels)], behda [*Terminalia bellirica* (Gaertn.) Roxb.], khejda (*Prosopis specigera* L.), *etc.* are some local names for common species found here.

Angiosperm families are distributed widely in the different regions of Rajasthan including Burseraceae [*Protium serratum* (Wallich ex Colebr.) Engl.], Loranthaceae [*Dendrophthoe falcata* var. *coccinea* (Talb.) Sant.], Orchidaceae [*Peristylus constrictus* (Lindl.) Lindl,], Lamiaceae [*Clerodendrum serratum* (L.) Moon.], Asparagaceae (*Chlorophytum borivilianum* Santapau & R. R. Fern.), Asteraceae (*Parthenium hysterophorus* L.), Fabaceae [(*Prosopis cineraria* (L.) Druce)], etc.

The tribal people of Rajasthan have always been pastoralists. Sheep, camels, and cattle can feed on various grasses, shrubs, and low tree branches.

3.3 Tribal Populations in Rajasthan

Rajasthan's tribal community makes up 12% of the overall population (Fig. 3.2). These tribes can all be recognized by their distinctive cultures, traditions, trades, fairs, and festivals. Each and every tribe in Rajasthan has contributed with its distinctive clothes and customs, enriching the culture of the entire tribe. Their methods for building houses, celebrations, and costumes all attest to the

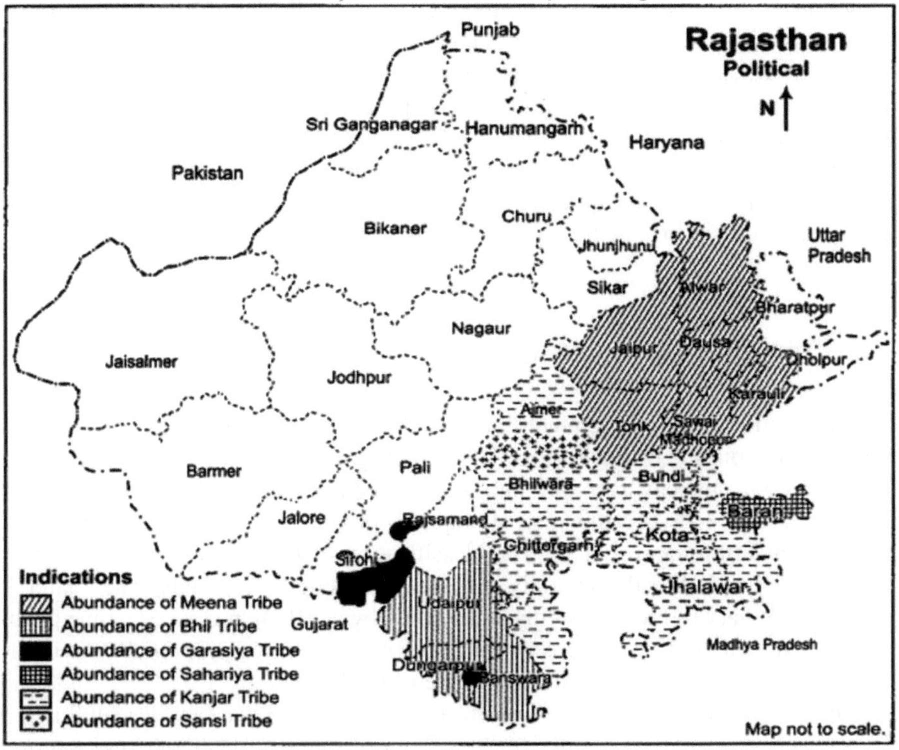

Fig. 3.2 Map showing different tribes of Rajasthan in different regions. (Source: www.rbsesolutions.com)

exquisiteness of the Rajasthani tribe's cultures. In terms of the number of scheduled tribes in India, Rajasthan comes in sixth. The isolated lifestyle of the scheduled tribes, who inhabited highlands and forests, is well known. The tribal people still avoid the plains of man even though today most of the forests have been cut off and hills may be approached by a network of transportation.

The scheduled tribes constitute around 54.75 lakh people or 12.44% of the state's total population. The biggest concentration is found in the districts of Banswara and Dungarpur, where it accounts for 73.47% and 65.84%, respectively, of the district's total population. Udaipur (46.34%), Sirohi (23.39%), Sawai Madhopur (22.47%), Bundi (20.25%), Chittorgarh (20.28%), and Baran (21.13%) are other districts with greater percentages. When examining how the state's tribal population is distributed among its various districts, one can see that the highest percentage 19.41% of the total lives in the Udaipur district, followed by 15.51% in Banswara, 10.51% in Dungarpur, 9.71% in Jaipur, and 8.10% in Sawai Madhopur. As a matter of fact, these five districts together contributed two third of the state's tribal population.

Other districts have extremely low amounts, with the district of Bikaner having the lowest at 0.04%. The districts are Ganganagar, Churu, Jhunjhunu, Sikar, Ajmer, Jaisalmer, Bikaner, Jodhpur, and Nagaur, with fewer than 1% of the state's total tribal population. The tribes' population composition is distinct from the state of the general population. Due to the tribes' isolation in their limited territory of hills and forests, there is generally low population density (Nagda 2004). The Kalbelia, Damor, Meena, Sansi, Kanjar, Bhil, Garasiya, Sahariya, and Kathodia tribes reside in the tribal belt in the southern part of Rajasthan.

3.4 Weeds of the Asteraceae Family in Rajasthan

An unintentionally growing plant that is out of place and has more negative traits than positive ones is called a weed. When they grow where they are not desirable, certain crop plants may even turn into weeds. The original occupants of the soil are referred to as weeds because man attempts to produce just the types of plants he desires, rendering them useless. The classification of plants as weeds is as old as agriculture itself. Weeds naturally emerge alongside crop plants when land is cultivated to harvest crops (Sharma and Khandelwal 2010). In contrast, some plants that are typically considered weeds may prevent erosion, provide food for wildlife and birds, as well as be a significant source of medicines.

The Asteraceae family, also known as Compositae/Aster family (Sunflowers/ Daisy family) makes up about 10% of all the flowering plants of Angiosperms. Consequently, also the weeds of this family are being widely utilized by the tribal population of this state. It has been reported by Morin et al. (2015) that the dry and semi-arid areas of subtropical and lower temperate latitudes are home to most of the species under this family. Several plant species grow wildly along roadsides, tracks of railways cultivated fields, forests as well as bare lands of Rajasthan, while species like *Ageratum conyzoides* L., *Artemisia* spp., *Blumea* spp., *Baccharis spicata* (Lam.) Baill., *Parthenium hysterophorus* L., *Verbesina encelioides* (Cav.) Benth. & Hook. f. ex A. Gray, *Xanthium strumarium* L., etc. are invasive in nature and known for their nuisance value. Other weed species, like the ragwort (*Senecio jacobeaeformis* J. Remy), are poisonous to livestock because of the presence of toxic alkaloids in their leaves. *Launaea procumbens* (Roxb.) Ramayya & Rajagopal is also a wild plant commonly used in Ayurveda due to its medicinal value. It is possible to conclude that the tribal and other demographic groups of the State of Rajasthan highly valued and employed weeds for a variety of functions, including medicinal and culinary purposes. According to Jain (1991), it is commonly recognized that weeds are undesirable, still, if they are allowed to develop properly, they may become beneficial to humanity (Jain 1991).

3.5 Data Acquisition in Tribal Studies

Human history has been shaped by our capacity to learn from one another and to share our knowledge and talents (Tomasello 1999; Pagel 2012). Long-term conservation and fast cultural change are both supported by the transfer of traditional knowledge, which also enables humans to adapt their coping mechanisms and occupy a variety of ecosystems (Guglielmino et al. 1995; Boyd et al. 2011). Traditional knowledge is transmitted in one of two ways, according to descriptions. In a process known as vertical transmission, traditional knowledge is passed down from one generation to the next, and hence from ancestral to descendant civilizations. Horizontal transmission, also known as selective diffusion or borrowing, modifies conventional knowledge, however, innovation in the absence of horizontal transmission may also alter an ethnic group's knowledge. The combination of modification and vertical and horizontal transmission establishes bodies of traditional knowledge unique to each culture, but still reflecting the traditional knowledge of their ancestors (Guglielmino et al. 1995; Pagel and Mace 2004). Herbal medicine has been used in manuscripts dating back to around 5000 years in Indian, Chinese, Egyptian, Greek, Roman, and Syrian cultures. The use of medicinal herbs is covered in India's oldest classical treatises, including the Rigveda, Atharvaveda, Charak Samhita, and Sushruta Samhita. This demonstrates how traditional medicines and herbal remedies can be traced back to ancient civilizations and scientific heritage. Herbal therapies are heavily used in all of India's officially recognized Indian medical systems, including Ayurveda, Yoga, Naturopathy, Unani, Siddha, and Homeopathy (Vaidya and Devasagayam 2007).

Our ancient writings also contain information on therapeutic plants. Except for a few Mohenjo-Daro archaeological sculptures, there is no authentic record of any kind from the pre-Vedic period in this country. However, our oldest Vedic literary sources, the Rigveda and Atharvaveda, which date from 2000 to 1000 B.C., contain valuable information about the medicinal plants used at the time. The 248 botanical drugs on the list are mostly mentioned in the Atharvaveda and Rigveda. A glossary of these therapeutic herbs is mentioned in the Charak Samhita, Sushruta, and Ashtanga Hridiyam. India has a large collection of ancient archaeological sculptures that can be used to identify the plants that were used in early civilizations. Herbaria serve as the foundation for an analytical study of a region's vegetation in order to develop methods for successful conservation and sustainable use.

Herbaria may even serve as a resource for learning the local's Materia Medica, which may have evolved over thousands of years of local practice. Identification of natural specimens for collection, use, and future research necessitates the immediate attention of herbaria where new active principles from plant resources can be viewed for recognition. Some cities in Rajasthan, such as Jodhpur, have numerous herbaria. Jaipur, Bikaner, and Sri Ganganagar have a valuable collection of plants.

Plants have evolved into a never-ending source of novel biologically active chemicals with medicinal potential. Tribals give ethnobotanists hints so that they can recommend which raw plant material can be harvested from the field. The number of reported medicinal plants and the community's claimed use of them demonstrates how deeply ingrained indigenous knowledge of medicinal plants and their uses runs.

3.6 Phytochemical Characterization of Asteraceae Species

The Asteraceae family is one of the largest flowering plant families, with over 1600 genera and 2500 species worldwide (Rolnik and Olas 2021). Despite their wide diversity, most family members share a similar chemical composition, A good example for this is the occurrence of inulin (a natural polysaccharide with prebiotic properties) in a wide range of Asteraceae species. The reported pharmacological effects can be attributed to various phytochemical compounds, including polyphenols, phenolic acids, flavonoids, acetylenes, and triterpenes. Asteraceae species are also frequent sources of sesquiterpene-lactones that are responsible for the bitter taste of many plants.

According to certain authors, approximately 25,000 phytochemicals and allelochemicals have been isolated from species of the Family Asteraceae (Bringe et al. 2006).

Most of them have antioxidant effects (Michel et al. 2020). Antioxidants, such as chlorogenic acid, gallic acid, ferulic acid, protocatechuic acid, sinapic acid, coumaric acid, caffeic acid, and caffeoylquinic acid are abundant in sunflower seeds. Vitamins, trace elements, and flavonoids like kaempferol, apigenin, quercetin, luteolin, and heliannone have been obtained from Asteraceae species, frequently regarded as weeds.

Recently, five flavonoids have been extracted from sunflower, *viz.*, the flavanol tambulin, the flavanones heliannones B and C, helianone A and chalconeskukulcanin B (Bashir et al. 2021). The sunflowers also produce colored proteins, which are the four isomers of tocopherols: alpha, beta, delta, and gamma. Sunflower seed oils include these isomers as well. Helianthinin is found as a globulin in it. Lipase is a digestive enzyme. Sunflowers were also isolated. SAP 16 is a 16 kDa protein that is isolated afterward. At high frying temperatures, sunflower seed oil remains stable. This oil is used as vegetable oil, a cosmetic component, and a lubricant. Sunflower seed extracts include dicaffeoylquinic acid, chlorogenic acid, and caffeic acid, which are isolated from the sunflower seeds' aqueous methanol extract. Sunflowers are primarily producers of carotenoids and polyphenols, which comprise flavonoids, stilbenes, and phenolic acids.

3.7 Pharmacologically Valuable Compounds of Asteraceae Species

The biologically active compounds present in plants are known as phytochemicals. According to their role in plant metabolism, these bioactive compounds are distinguished as primary phytochemicals and secondary phytochemicals. The amino acids, proteins, sugars, and chlorophylls are part of primary phytochemicals whereas terpenoids, saponins, alkaloids, phenolic compounds, tannins, and flavonoids compose the secondary phytochemicals (Hahn 1998; Sawicka et al. 2020). Phytochemical screening of *Tridax procumbens* L., a wild plant of Asteraceae has shown the existence of flavonoids, alkaloids, carotenoids, β-sitosterol, n-hexane, fumaric acid, luteolin, glucoluteolin quercetin, isoquercetin, dexamethasone, oxo ester, rutin, lauric acid, myristic, palmitic, arachidic, linoleic acid and tannin which promotes wound healing, hepatoprotective, hypotensive, anticancer and antidiabetic activities in the plant (Ankita and Jain 2012; Jayasundera et al. 2021). *Ageratum conyzoides* L. is a weed that has been known since ancient times for its curative properties and has been utilized for the treatment of various ailments such as burns and wounds. For its antimicrobial properties it was used in infectious conditions and bacterial infections, as well as a numerous health conditions, e.g.: arthrosis, headaches and dyspnea, pneumonia, analgesic, anti-inflammatory, antiasthmatic, antispasmodic and hemostatic effects, stomach ailments, gynecological diseases, leprosy, and other skin diseases. A wide range of chemical compounds isolated from this species include alkaloids, coumarins, flavonoids, chromenes, benzofurans, sterols, and terpenoids. Some of the extracts and metabolites have been found to possess also insecticidal activities (Kamboj and Saluja 2008). *Bidens pilosa* L. is an incredible source of phytochemicals that include 301 compounds that fall into the following major chemical classes: polyacetylenes, flavonoids, phenolic acids, terpenes (monoterpenes, sesquiterpenes, diterpenes, and triterpenes) pheophytins, fatty acids, phytosterols and some essential oils which are considered as the main active constituents responsible for the various pharmacological actions of the plant (Deba et al. 2008; Yang 2014). Although the weeds of this family are traditionally known to have medicinal properties and have been used as such, therefore it is only the lack of sufficient knowledge, awareness, and screening that has limited their use in the pharmaceutical sector (Jayasundera et al. 2021).

Antioxidants

Asteraceae species frequently produce powerful antioxidants including superoxide anions, hydroxyl radicals, and hydrogen peroxide, also known as free radicals. Free radicals are unguided missiles that assault healthy cells, rupturing their membranes and causing mutations and genetic damage. They react either with serum lipoprotein (LDL) to generate atheromatous plaques or with the lipid in the cell membrane to trigger polyunsaturated fatty acid peroxidation and the subsequent production of more free radicals.

Antioxidants are required by the compartments of the entire human body, including the circulatory system inside the cells, the blood-brain barrier, and the central nervous system. In processed foods, synthetic antioxidants like butylated hydroxytoluene (BHT) and butylated-hydroxy-anisole (BHA) are frequently present. However, some evidence suggests that these substances have potential negative side effects (Branen 1975; Ito et al. 1983). Additionally, it has been hypothesized that the consumption of foods high in antioxidants and the prevalence of human disease are inversely correlated (Rice-Evans et al. 1997).

Antioxidants. occur in phenolic forms, *i.e.*, tocopherols, acids derived from phenols, and flavanol derivatives. They can also occur as enzymes, such as glutathione dehydrogenase, guaiacol peroxidase, catalase, and peptides (reduced glutathione). The fat-soluble antioxidant, existing in four isomers, called Tocopherol (Vitamin E) is also frequently present. It protects the fats and oils from antioxidant rancidity. The bioactive components act as a defense against illnesses and any anxiety circumstances (Khandare 2012). The most important phenolic component of the Asteraceae is chlorogenic acid (CGA), a coumaric, cinnamic, a small amount of caffeic acid (CA), sinapic, ferulic, and hydroxy-cinnamic and consisting of small traces of vanillic, syringic, and hydroxy-benzoic acid (Weisz et al. 2009).

Different species show different antioxidant properties such as *Parthenium hysterophorus* L. a poisonous weed having phenolic and flavonoid content in their flower extracts, both of which contribute to the plant's antioxidant potential (Hamid et al. 2010). *Taraxacum officinale* F.H. Wigg. Dandelions contains beta-carotene, which is an antioxidant that helps protect cells from damage. Research shows that carotenoids, such as beta-carotene, play a vital role in reducing cell damage. *Erigeron canadensis* L. also known as horseweed possesses antioxidative properties *in vitro* protecting plasma proteins from toxicity (Saluk-Juszczak et al. 2010). *Cyanthillium cinereum* (L.) H. Rob. also known as little ironweed show potent antioxidant activity and the activity was found to be concentration dependent which may be attributed to the high flavonoid content of the plant (Suresh et al. 2015). A weed of this family, *Sphaeranthus indicus* L. treats hemicrania, a mental disease, and epileptic convulsions (Kirtikar and Basu 1987). Tribal groups of Rajasthan use *Vicoa indica* (L.) DC. as a female antifertility medication and a contraceptive. Roots are used to treat cough and jaundice, while the entire plant's infusion is used to induce abortion (Srinivasan et al. 2007). Several medical conditions can be treated with *Vernonia cinerea* (L.) Less., including cancer, abortion, diuresis, worms, pain, infections, and numerous gastrointestinal diseases. Potential sources of natural antioxidants in weeds of the Asteraceae family include the methanol extract of *Tricholepis glaberrima* DC, *Solidago gigantea* Aiton, and *Taraxacum officinale* F.H. Wigg. (Kristo et al. 2002).

3.8 Curative Uses of Weeds of Asteraceae Family

Weeds of the Asteraceae family in the Rajasthan belt have been used by the tribal people, although their utility is not always known exactly. Leaves of *Ageratum conyzoides* L. are ground and a paste used by the local people of Rajasthan is prepared from it for wound healing (Sharma and Khandelwal 2010). The root and leaves of this weed are used by the tribals also for stomach aches and healing of broken bones. *Launaea procumbens* (Roxb.) Ramayya & Rajagopal plant is used by the tribals in curing skin diseases, jaundice, Gonorrhea, and liver disorders (Bhattacharyya 1996). *Xanthium strumarium* L. leaves, seeds, and roots are used by the local people as a blood purifier and also for the treatment of fever, malarial fever, and diuretic skin diseases. Sonchus oleraceus plant parts are used by the tribes for the treatment of headaches, general pain, diarrhea, menstrual problem, fever, hepatitis, and liver infections (Sharma and Kumar 2011).

Some weed species (*Ageratum houstonianum* Mill., *Tagetes erecta* L., *Gazania rigens* L., *Chrysanthellum americanum* L., *Zinnia elegans* L., *Sanvitalia procumbens* Lam., *Bidens torta* Shreff.) are generally stored by tribal people in pots at home, in gardens, and a variety of additional locations. Their crushed leaves are used for skin care (Sharma et al. 2017; Elomaa et al. 2018). The seeds of Sunflower (occasionally found as a weed) are grown and consumed by local people due to their nutritious and valuable compounds, e.g.: unsaturated fats, vitamins (Vitamin E), fiber, proteins, copper, folate, zinc, iron, selenium, and many other compounds (Alagawany et al. 2015). Farther forms of utilization include: cooking oil, salted or roasted snacks, and confectionary nuts. Sunflower seed is used as pet feed and livestock due to its reported high content of sulphuric acid.

Local people have a great belief in herbal medicine, and women are leading the charge in implementing the formulas for herbal medicine made from these plants. Painkiller, diuretic, febrifuges, carminative, anthelmintic, anti-inflammatory, aphrodisiac, cardiotonic, tonic, stomach-ache, dyspepsia, jaundice, leprosy, cough, asthma, ulcers, vomiting, and so forth are some of the local therapeutic applications recorded in the special literature (Choudhary et al. 2000). Various research linked to the utilization of plant mixtures and their preparative home remedies has been described with the aim of exploring their additive, as well as synergic effects (Jeph and Khan 2020).

Farther weeds of the family Asteraceae growing in the Rajasthan belt such as *Erigeron canadensis* L., *Erigeron bonariensis* L., *Ageratum conyzoides* L., *Taraxacum officinale* F.H. Wigg, *Ambrosia artemisiifolia* L., *Chondrilla juncea* L., *Parthenium hysterophorus* L., etc., are known to have documented properties of pharmaceutical significance. Some farther examples are the leaves of *Erigeron canadensis* and *E. bonariensis* are used in the treatment of diarrhea, diabetes and hemorrhages (Sharma et al. 2014). *Ageratum conyzoides* has a long history of being used as remedies for conditions like arthrosis, headaches, dyspnea, common wounds, and burned wounds, as well as for treating microbes (Kamboj and Saluja

2008). *Taraxacum officinale* F.H. Wigg. (Dandelion) has been traditionally considered as a natural remedy which has been used as a diuretic, an anti-infective, and to treat bile and liver issues. In herbal medicine, dandelion is used as a mild laxative, to stimulate hunger, and as a bitter plant to aid in digestion, and the milky latex has been used to cure wart (Martinez et al. 2015). Likewise, the aerial portion of *Chondrilla juncea* have been proven to be useful for treating hyperuricemia and gout. *Parthenium hysterophorus* has many commercial applications for the manufacture of drugs to cure various diseases including relief from neuralgia, rheumatoid arthritis discomfort, diarrhoea, urinary tract infections, and dysentery. *Ambrosia artemisiifolia* pollen is an allergen extract recommended as immunotherapy for the management of allergic rhinitis caused by short ragweed pollen, with or without conjunctivitis, as determined by a positive skin test or in vitro pollen specific IgE antibody test (Parkhomenko et al. 2005).

Steam distillation of *Helichrysum italicum* (Roth) G. Don (Curry plant) is done in industries, that produce essential oil and hydrolate. Such extracts are extremely beneficial for hematomas, varicose veins, and skin care, particularly for scarred skin. *Calendula officinalis* L. (pot marigold) is most commonly used in the production of DIY skin care preparations, most notably in the form of macerated calendula oil, which is used for damaged skin and various inflammatory skin diseases.

As the practice of wild harvesting is becoming more-and-more common, the need to preserve the plants and their curative properties for future healthful usage is growing.

3.9 Medicinal Importance of Wild Plants of the Family Asteraceae

From the very beginning, wild plants of Asteraceae have been used in conventional medication as medicinal herbs, across the world (Khan 2014). Modern therapies are also using herbal drugs which are extracted from traditional herbs having a definite physiological and medicinal effect on the human body (Patwardhan et al. 2008). The study of plants used in traditional medicine in diverse cultures has produced significant medicines. The traditional knowledge system of herbal practice is rich, and it is urgent to record this knowledge also in the Asteraceae family, before it is permanently lost to the community. Figure 3.3 demonstrates the percentage of wild plants part used in medicines. Different plants and plant organs of these plants have distinct functions: such as the leaf and juice of the young stem of *Acanthospermum hispidium* DC. is used on skin scratches and wounds, with 2% calcium carbonate solution. The leaf juice of *Ageratum conyzoides* Linn. is used on skin scars and Leprosy. The whole plant pastes of *Artemisia absinthium* Linn. are used as an antiseptic and detergent agent on skin diseases. The ointment prepared from the

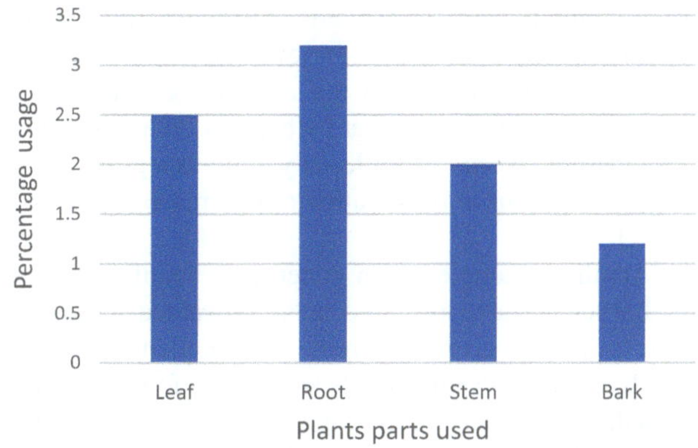

Fig. 3.3 Percentage of wild plant parts used in medicines

flowers of the corolla of *Calendula officinalis* L. is used on skin ailments: its leaf-paste is used in external ulcers and on open sores. The paste of young leaves of *Tagetes erecta* L. and its healthy flower sepals are used for wounds and injuries. The root juice of *Taraxacum officinale* F.H. Wigg. is used to wash the face to get rid of black marks (Lakshman et al. 2014).

Hence, humans have direct or indirect dependence on plants not only for food, fuel, and fodder but also for other commodities such as medicines, chemical products, fibres, *etc.* (Alam and Sharma 2012). Keeping aside all these necessities provided by plants to humans, plants also play a crucial role in the evolution of human culture all over the world. Over the last three decades, the usage of herbal medical goods and supplements has grown, with more than 80% of people globally depending on them for some aspect of primary healthcare.

3.10 From Wandering Weeds to the Pharmacy

Traditional and alternative medicine systems are regarded as a fundamental health care modalities in healthcare settings with limited resources, notwithstanding recent scientific advancements and globalization. Since the beginning of time, the herbal medicinal systems have been tested and proven via empirical observations and trial-and-error trials to promote wellness and treat illnesses and disorders. Before now, the usefulness of traditional medicinal herbs and phytotherapy was frequently overlooked and underappreciated. Currently, the public and scientific community are showing a rejuvenation and renewed interest in traditional medicinal plants. Several enormous obstacles must be eliminated to successfully and quickly promote traditional medicinal plants. The combined efforts of ethnobotanists, anthropologists, pharmacists, and physicians using cutting-edge approaches and cutting-edge scientific methods might be a practical approach to assessing and validating the use of traditional medicinal plants. Furthermore, it is essential and unavoidable to undertake clinical trials to evaluate their effectiveness and human safety. Such instances abound in the Asteraceae family, including numerous species of the *Chromolaena* genus that by being used as medicines and are thus more than just "wandering" weeds. Additionally, there is a growing interest in the current scientific use of this genus in the creation of pharmacological polyherbal medicines (Ojo et al. 2022).

Another example of an Asteraceae family weed is *Parthenium hysterophorus* L. Earlier it was considered as an unwanted weed, now its adsorbents are used in industries to successfully remove methylene blue from an aqueous solution in a batch reactor. Its root extracts are used as folk remedies, as it is applied externally on skin disorders. Decoction of the plant is often taken internally as a remedy for a wide variety of ailments (Kaur et al. 2021).

Many researchers are motivated and inspired to investigate and validate the usage of wild plants, as a result of drug development from these plants. Wild plants have now become a necessity for the production of an enormous number of drugs in various pharmaceutical industries. This has proved the evolution of plants from ethnicity to industry.

3.11 Conclusions

About 13.5% of Rajasthan's population is composed of tribal people. Tribal people have accumulated significant knowledge of the plants and their uses, particularly for medical purposes. The use of many of the medicinal plants employed by the tribal people is, however, still not known outside the restricted community.

The native tribes extensively understand the ethnobotanical uses of many weeds. In Table 3.1, the ethnomedicinal importance of Asteraceae species, including weeds, that are found in Rajasthan is compiled and explained in a tabular form. Globally, there is a resurgence of the traditional system of medicine because of its user-friendly nature and because of the intrinsic side effects of modern medicines. But the knowledge of traditional healing practices mainly through the use of wild plants is now fast disappearing because of modernization, globalization, and the tendency to change traditional lifestyles.

It is important to properly record and utilize the knowledge of herbal remedies held by ethnic communities. There is an increasing necessity for ethnobotanical research, to catalog therapeutic plants and related indigenous knowledge. These investigations are very helpful in identifying endangered medicinal plant species and in determining the appropriate conservation measures in the near future.

Table 3.1 Species of the family Asteraceae that are common among the tribes of Rajasthan with their medicinal uses

S. No	Botanical name of plant and growth habit	Vernacular name	Compounds isolated	Tribes	Plant parts	Medicinal uses	References
1.	*Ageratum conyzoides* L. (Herb)	Jangli pudina, Visadodi, Semandulu, Ghabuti, Bhakumbar, Mukhada	Flavonoids, Alkaloids, Chromene, Terpenoids, Coumarins, Sterols, Caryophyllene Fumaric acid	Bhil, Kathodia, Garasia	Whole plant	Used to treat fever, rheumatism, headache, colic, wounds caused by burns, dyspepsia, eye problem, uterine disorders, and pneumonia. It has anti-asthmatic, antispasmodic, and hemostatic properties in traditional medicine.	Sharma and Sharma (2001), Okunade (2002)
2.	*Ambrosia artemisiifolia* L. (Herb)	Bitterweed	Phenol carboxylic acids, Coumarins, Flavonoids	Bhil, Garasia	LeavesRoots	The leaves are astringent, febrifuge, and emetic. Internally, they are used as a tea to treat fevers, pneumonia, nausea, intestinal cramps, diarrhea, and mucous discharges. They are administered topically to bug bites, rheumatoid joints, and numerous skin ailments. The disinfecting juice from the withered leaves is applied to infected toes. Menstrual difficulties and stroke are treated using a tea produced from the roots.	Boericke (2002), Sharma and Kumar et al. (2011)
3.	*Ambrosia confertiflora* DC. (Herb)	Estafiate, Istafiate, Chi'ichibo, Chibchibo	Santamarine, Reynosin, 1,10-Epoxyparthenolide	Garasia, Kathodia	Leaves, Flowers, Root	Intestinal parasites, stomachache, fever, lack of appetite, menstrual symptoms	Rastogi et al. (1990), Robles-Zepeda et al. (2013)
4.	*Artemisia dranuculus* L. (Herb)	NaagadaunaTerragon, Tarkhun	Flavonoid compounds, Phenolic acids, Coumarins, Alkamides	Garasia	Root	Carminative, digestive, anti-inflammatory, antipyretic, antiseptic, antispasmodic, antiparasitic, antimicrobial, anthelmintic, and fungicidal effects	Chopra et al. (1956), Duke (2001)

(continued)

Table 3.1 (continued)

S. No	Botanical name of plant and growth habit	Vernacular name	Compounds isolated	Tribes	Plant parts	Medicinal uses	References
5.	*Bellis perennis* L. (Herb)	Daisy, Gowlan, Luckin Gowan	Flavonoids, Phenols, Saponins	Sehariya	Whole plant	Bruises, fractured bones, and wounds have all been treated with it. Traditional medicine has been used to treat headaches, colds, stomachaches, eye illnesses, eczema, skin boils, gastritis, diarrhea, bleeding, rheumatism, inflammation, and infections of the upper respiratory tract.	Rao et al. (2011)
6.	*Chondrilla juncea* L. (Herb)	Rush skeleton-weed, Lampri	Phenols, Sesquiterpenes lactones, Caffeoyltartonic acid, coumarins	Garasia	Leaves, Stem	Lowers blood pressure, helps cure insomnia, purifies the blood, and hyper-uricmia heals wounds ad gout	Rawat et al. (2016)
7.	*Chrysanthemum*× *grandiflorum* Ramat. (Herb/ Shrub)	Guldaudi	Flavonoids, Alkaloids, Phenolic compounds, Triterpene	Bhil	Flowers	Used to treat chest pain (angina), high blood pressure, type 2 diabetes, fever, cold, headache, dizziness, and swelling	Gurung et al. (2019), Shahrajabian et al. (2019)
8.	*Chrysanthemum* × *morifolium* Ramat. (Herb/ Shrub)	Mums, Guldaudi	Flavonoids, Alkaloids, Phenolic compounds, Triterpene	Bhil, Damor	Flowers	Inflammation, hypertension, and respiratory diseases Although it is most commonly used in tea preparations, it is also used in tinctures, creams, and lotions.	Choi et al. (2009), Toppo et al. (2015)
9.	*Cyanthillium cinereum* (L.) H. Rob. (Herb)	Sahadevi, Sadodi, Puvamkuruntal	β-amyrin, taraxasterol, lupeol, betulin	Kathodia, Kalbelia	Leaf, Stem, Flower	*It* has therapeutic potential against asthma, cancer, cholera, colic pain, cough, diarrhea, dysentery, impotency, and night blindness. The seeds are employed as an alternative to leprosy and chronic skin illness, as well as a source of alexipharmic and anthelmintic medications.	Nadkarni (1954), Johnson (1998), Hsu (1967), Lin (2005)

10.	*Erigeron bonariensis* L. (Herb)	Goojugha, Gulava, Mrichbooti	Flavonoids, Triterpenes, Caffeoylic derivatives, Steroids	Bhil	Leaves	Treat gastrointestinal problems, roots are used to cure diarrhea and dysentery treatment	Kumar et al. (2011), Bukhari et al. (2013)
11.	*Erigeron canadensis* L. (Herb)	Makshikavish, Jarayupriya	Saponins, Diterpenoids, Terpenoids, Glycosides, Tannin, Anthraquinone, Steroids, Flavonoids	Meena, Sehariya, Kathodia	Leaves	Treat allergic diarrhea, stomatitis, otitis media, conjunctivitis, and acute toothache	Haq and Singh (2020)
12.	*Gaillardia aristata* Pursh (Herb)	Blanket flower	Pinene, Limonene, myrcene	Kalbelia, Bhil	Root	Treat wounds, cure painful urination, diuretic and settle fevers	Singh (1970)
13.	*Gaillardia pinnatifida* Torr. (Herb)	Red dome blanket flower	Flavonoids	Kalbelia, Bhil	Leaves	Gastroenteritis is treated with root tea, and skin diseases and painful eyes are treated with chewed powdered root.	Singh (1970)
14.	*Gnaphalium coarctatum* Willd. (Herb)	Cudweed, Gordolobo	Flavonoids, Sesquiterpenes, Diterpenes, Triterpenes, Phytosterols, Anthraquinones, Caffeoylquinic acid derivatives	Meena, Garasia	Leaves	Treat various respiratory diseases, such as grippe, fever, asthma, cough, cold, bronchitis, expectorating, and bronchial affections relief of stomach diseases, swelling, wounds, prostatism, lumbago, neuritis, and angina ache, also used for the lowering of blood pressure, or as a diuretic	Singh and Sharma (1988)
15.	*Helichrysum luteoalbum* L. (Herb)	Balraksha	Limonene, Pinene, Linalool	Minas, Meena	Leaves	Leaves are used as an astringent, cholagogue, diuretic, febrifuge, and hemostatic. Used as a tonic and for the treatment of tumors, gout, and dermatitis. To repair damaged bones, plants are mixed with dried fish and administered as a poultice. The Kavirajes of Chalna employed it as a tonic for tumors, gout, and dermatitis. It was utilized in the treatment of cancer in Belgium (breast)	Gupta et al. (2008)

(continued)

Table 3.1 (continued)

S. No	Botanical name of plant and growth habit	Vernacular name	Compounds isolated	Tribes	Plant parts	Medicinal uses	References
16.	*Helianthus annus* L. (Herb)	Surajmukhi Suryakanti, Suryamukhi, Sooryamukhiphool	Sesquiterpenes, Lactones, Diterpenes, Flavonoids	Kathodia, Garasia	Seeds, Leaves, Flowers	The tincture is used to treat laryngitis, lung infections, bronchitis, coughs, and colds, as well as whooping cough. It is also used to treat malarial fever. Typically, leaves are laid on a bed and covered with a towel that has been wet with warm milk before the patient is curled up in it. In the therapy of bronchiectasis, a tincture of the flowers and leaves is combined with balsamic vinegar.	Subashini and Rakshitha (2012), Suo and Yang (2014)
17.	*Parthenium hysterophorus* L. (Herb)	Gajarghas Vayaribhama	Parthenin	Bhil	Whole plant	Remedy for skin inflammation, rheumatic pain, diarrhea, urinary tract infections, dysentery, malaria, and neuralgia	Patel (2011)
18.	*Rhanterium eppanosum* Desf. (Shrub)	Arfaj	Camphene, Myrcene, Limonene, Pinene	Bhil, Meena, Gadolia	Leaves	Hydrophobia, rabies, fevers, diarrhea, cancer, rheumatoid arthritis	Romero et al. (2015), Bhatt et al. (2021)
19.	*Sonchus asper* (L.) Hill (Herb)	Doodhi	Saponins, Flavonoids, Phenols	Bhil, Garasia	Young shoot, seeds, whole plant	Wounds and burns, cough, bronchitis and asthma, gastrointestinal infection, inflammation, diabetes, and cardiac kidney and liver disorders (Rivera and a reproductive disorder like impotence (erectile dysfunction) in humans' jaundice and cancer.	Koche et al. (2008)
20.	*Stevia rebaudiana* Cav. (Herb)	Madhu Patrika	Flavonoids, Phenols	Meena	Whole plant	Strengthening the heart, and the circulatory system and regulating blood pressure.	Savita et al. (2004), Marcinek and Krejpcio (2016)

21.	*Symphyotrichum Novi-belgii* (L.) G. L. Nesom	Michaelmas daisy	Saponins, Flavonoids	Garasia	Flowers, Roots, Stem	Fever, cold, tonsillitis, snake bite, and bee sting	Shao et al. (1995), Mukherjee (2006)
22.	*Tagetes erecta* L. (herb)	Genda, Hajara	Carotenoids, flavonoids, Monoterpenoids	Garasia, Meena	Leaves, flowers	Leaves are used as antiseptic and in kidney troubles, muscular pain, piles and applied to boils and carbuncles. The leaves are used as an antiseptic and in the treatment of renal problems, muscle discomfort, piles, boils, and carbuncles. Fevers, epileptic fits (Ayurveda), astringent, carminative, stomachic, scabies, and liver ailments are all treated with the flower, which is also used to treat eye disorders.	Gopi et al. (2012).
23.	*Tagetes patula* L. (herb)	Genda	Flavonoids, Glycosides, Limonene, Volatile oils, Sesquiterpenes, Saponins, Ocimenones, Spathulenol	Garasia	Leaves, Flowers	Lowers blood sugar, treatment of pain, inflammation, and cuts and wounds	Faizi et al. (2011), Singh et al. (2016)
24.	*Taraxacum officinale* F.H. Wigg (Herb)	Dudhali, Dudal, Dudh batthal, Dudhal	Sesquiterpene, Lactones, Monoterpene, Phytosterol, terpenes, Coumarin,	Garasia, Bhil	Whole plant	The root possesses choleretic, cholagogic, tonic, anti-rheumatic, digestive-stimulant, alternative, and depurative qualities, while the leaf has diuretic, choleretic, and anti-inflammatory activities. Hypertension, diabetes, dyspepsia, irritable bowel syndrome, and ovarian androgen excess	Bhatia et al. (2014), Bokelmann (2022)
25.	*Verbesina encelioides* (Cav.) Benth. & Hook. f. Ex A. Gray (Herb)	Jungli surajmukhi	Alkaloids, Flavonoids, Tannin, Saponin	Bhil, Meena	Flowers, Leaves	Spider bites, as well as gum sores and hemorrhoid therapy	Mehal (2021)

Acknowledgments The authors would like to express their gratitude to Prof. Ina Aditya Shastri, Vice-Chancellor, Banasthali Vidyapith, Rajasthan for her support and encouragement. We also thank DST for providing networking support through the FIST program at the Department of Bioscience and Biotechnology, Banasthali as well as the Bioinformatics Center, Banasthali Vidyapith, funded by DBT.

References

Alagawany M, Farag MR, Abd El-Hack ME, Dhama K (2015) The practical application of sunflower meal in poultry nutrition. Adv Anim Vet Sci 3:634–648. https://doi.org/10.14737/journal.aavs/2015/3.12.634.648

Alam A, Sharma V (2012) Textbook of economic botany. Pointer Publishers, Jaipur, p 317

Ankita J, Jain A (2012) *Tridax procumbens* (L.): a weed with immense medicinal importance: a review. Int J Pharma Bio Sci 3(1):544–552

Arya R, Singh DC, Tiwari RC, Tripathi BM (2016) An overview of ethnomedicine and future aspect of ethnomedicinal plants. Int J Ayur Pharma Res 4:29–33

Bashir T, Zia-Ur-Rehman Mashwani KZ, Haider S, Shaista Tabassum M (2021) 02. Chemistry, pharmacology and ethnomedicinal uses of *Helianthus annuus* (sunflower): a review. Pure Appl Biol 4(2):226–235

Bhatia H, Sharma YP, Manhas RK, Kumar K (2014) Ethnomedicinal plants used by the villagers of district Udhampur, J&K, India. J Ethnopharmacol 151(2):1005–1018. https://doi.org/10.1016/j.jep.2013.12.017

Bhatt A, Caron MM, Gallacher D, Souza-Filho PRDM (2021) Storage duration, light, temperature, and salinity exposure influence germination of the glycophyte *Rhanterium eppanosum*. Botany 99(5):261–267. https://doi.org/10.1139/cjb-2020-0153

Bhattacharyya G (1996) Medico-ethnobotanical value of Saurashtra weeds. J Econ Taxon Bot Addit Ser 12:166–168

Boericke W (2002) New Manual of Homoeopathic Materia Medica & Repertory [with Relationship of Remedies], Second Re-Augmented & Revised Edition Based on Ninth Edition, Reprint Edition, B. Jain Publishers, New Delhi

Bokelmann JM (2022) 39 – dandelion (*Taraxacum officinale*): leaf and root. In: Jean M (ed) Bokelmann, medicinal herbs in primary care, Elsevier, pp 303–307. https://doi.org/10.1016/B978-0-323-84676-9.00039

Boyd R, Richerson PJ, Henrich J (2011) The cultural niche: why social learning is essential for human adaptation. Proc Natl Acad Sci U S A 108(Suppl. 2):10918–10925. https://doi.org/10.1073/pnas.1100290108

Branen AL (1975) Toxicology and biochemistry of butylated hydroxyanisole and butylated hydroxytoluene. JAOCS 52:59–63. https://doi.org/10.12691/jfnr-1-3-3

Bringe K, Schumacher CF, Schmitz-Eiberger M, Steiner U, Oerke EC (2006) Ontogenetic variation in chemical and physical characteristics of the adaxial apple leaf surface. Phytochemistry 67: 161–170. https://doi.org/10.1016/j.phytochem.2005.10.018

Bukhari IA, Shah AJ, Khan RA, Meo SA, Khan A, Gilani AH (2013) Gut modulator effects of *Conyza bonariensis* explain its traditional use in constipation and diarrhea. Eur Rev Med Pharmacol Sci 17:552–558

Choi G, Yoon T, Cheon MS, Choo BK, Kim HK (2009) Anti-inflammatory activity of *Chrysanthemum indicum* extract in acute and chronic cutaneous inflammation. J Ethnopharmacol 123(1):149–154. https://doi.org/10.1016/j.jep.2009.02.009

Chopra RN, Nayar SL, Chopra IC, Varma BS (1956) Glossary of Indian medicinal plants. Council of Scientific and Industrial Research, New Delhi, p 174

Choudhary M, Ahmad S, Ali H, Sher MS (2000) Technical report: a market study of medicinal herbs in Malakand, Peshawar, Lahore, and Karachi. SDC- Inter Co-operation, Peshawar

Deba F, Xuan TD, Yasuda M, Tawata S (2008) Chemical composition and antioxidant, antibacterial and antifungal activities of the essential oils from *Bidens pilosa* Linn. Var. Radiata Food Control 19(4):346–352. https://doi.org/10.1016/j.foodcont.2007.04.011

Duke JA (2001) Handbook of medicinal plants. CRC Press, London, p 68. https://doi.org/10.1201/9781420043174

Elomaa P, Zhao Y, Zhang T (2018) Flower heads in Asteraceae-recruitment of conserved developmental regulators to control the flower-like inflorescence architecture. Hortic Res 5(1). https://doi.org/10.1038/s41438-018-0056-8

Faizi SA, Dar H, Siddiqi S, Naqvi A, Naz S, Bano L (2011) Bioassay-guided isolation of antioxidant agents with analgesic properties from flowers of *Tagetes patula*. Pharm Biol 49: 516–525. https://doi.org/10.3109/13880209.2010.523006

Gopi G, Elumalai A, Jayasri P (2012) A concise review on *Tagetes erecta*. Int J Phytopharm Res 3(1):16–19

Guglielmino CR, Viganotti C, Hewlett B, Cavalli-Sforza LL (1995) Cultural variation in Africa: role of mechanisms of transmission and adaptation. Proc Natl Acad Sci U S A 92:7585–7589. https://doi.org/10.1073/pnas.92.16.7585

Gupta A, Joshi SP, Manhas RK (2008) Multivariate analysis of diversity and composition of weed communities of wheat fields in Doon Valley, India. Trop Ecol 49(2):103

Gurib-Fakim A (2006) Medicinal plants: traditions of yesterday and drugs of tomorrow. Mol Asp Med 27:1–93. https://doi.org/10.1016/j.mam.2005.07.008

Gurung A, Vanlalneihi B, Bennurmath P (2019) Breeding for abiotic stress tolerance in chrysanthemum (*Dendranthema x grandiflora* Tzvelev). Int J Farm Sci 9(4):99–105. https://doi.org/10.5958/2250-0499.2019.00103.4

Hahn NI (1998) Are phytoestrogens nature's cure for what ails us? A look at the research. J Am Diet Assoc 98(9):974–977. https://doi.org/10.1016/s0002-8223(98)00223-5

Hamid AA, Aiyelaagbe OO, Usman LA, Ameen OM, Lawal A (2010) Antioxidants: its medicinal and pharmacological applications. AJPAC 4:142–151

Haq SM, Singh B (2020) Ethnobotany as a science of preserving traditional knowledge: traditional uses of wild medicinal plants from district Reasi, J&K (northwestern Himalaya), India. In: Botanical leads for drug discovery. Springer, Singapore, pp 277–293. https://doi.org/10.1007/978-981-15-5917-4_13

Hasnain N (1983) Tribal India Today. Harun Publications, New Delhi

Hsu YT (1967) Study on the Chinese drugs used as cancer remedy. J South Asian Res 3:63

Ito N, Fukushima S, Hassegawa A, Shibata M, Ogiso T (1983) Carcinogenicity of butylated hydroxyl anisole in F 344 rats. J Natl Cancer Inst 70:343–347. https://doi.org/10.1093/jnci/70.2.343

Jain SP (1984) Ethnobotany of Morni and Kalesar (Ambala-Haryana). J. Econ. Tax. Bot. 5:809–813

Jain SK (1991) Dictionary of Indian folk medicine and ethnobotany. Deep Publications, New Delhi. https://doi.org/10.12691/plant-5-2-1

Jain A, Katewa SS, Choudhary BL, Galav P (2004) Folk herbal medicines used in birth control and sexual diseases by tribals of southern Rajasthan. India J Ethnopharmacol 90:171–177. https://doi.org/10.1016/j.jep.2003.09.041

Jayasundera M, Florentine S, Tennakoon KU, Chauhan BS (2021) Medicinal value of three agricultural weed species of the Asteraceae family: a review. Pharm J 13(1):264–277. https://doi.org/10.5530/pj.2021.13.36

Jeph A, Khan JB (2020) Ethnomedicinal study in the reserve forest area of Jhunjhunu District, Rajasthan, India. J Soc Trop Plant Res 7(2):379–387. https://doi.org/10.22271/tpr.2020.v7.i2.044

Johnson T (1998) CRC ethnobotany desk reference. CRC Press, Boca Raton. https://doi.org/10.1201/9781351070942

Joshi P (1995) Ethnobotany of the primitive tribes in Rajasthan. Rupa Books Pvt Ltd., Jaipur

Kamboj A, Saluja AK (2008) *Ageratum conyzoides* L.: a review on its phytochemical and pharmacological profile. Int J Green Pharm 2. https://doi.org/10.4103/0973-8258.41171

Katewa SS, Guria BD (1997) Ethnobotanical observations on certain wild plants from southern Aravalli Hills of Rajasthan. Vasundhara 2:85–86

Katewa SS, Guria BD, Jain A (2001) Ethnomedicinal and obnoxious grasses of Rajasthan, India. J Ethnopharmacol 76:293–297. https://doi.org/10.1016/S0378-8741(01)00233-1

Katewa SS, Chaudhary BL, Jain A (2004) Folk herbal medicines from tribal area of Rajasthan, India. J Ethnopharmacol 92:41–46. https://doi.org/10.1016/j.jep.2004.01.011

Kaur L, Malhi DS, Cooper R, Kaur M, Sohal HS, Mutreja V, Sharma A (2021) A comprehensive review on ethnobotanical uses, phytochemistry, biological potential and toxicology of *Parthenium hysterophorus* L.: a journey from noxious weed to a therapeutic medicinal plant. J Ethnopharmacol 281:114525. https://doi.org/10.1016/j.jep.2021.114525

Khan H (2014) Medicinal plants in light of history and recognized therapeutic modality. J Evid-Based Integr Med 19:216–219. https://doi.org/10.1177/2156587214533346

Khandare NA (2012) Qualitative Phytochemical analysis of ethnomedicinally important plant *Capparis phyla* Roth (Capparidaceae) from Akola District, Maharashtra, India. Int Res J Pharm 3(4):206–207

Kirtikar KR, Basu BD (1984) Indian medicinal plants. Lalit Mohan, Allahabad

Kirtikar KR, Basu BD (1987) Indian medicinal plants, 2nd edn. International Book Distributors, Dehradun

Koche DK, Shirsat RP, Imran S, Nafees M, Zingare AK, Donode KA (2008) Ethnomedicinal survey of Nigeria wildlife sanctuary, district Gondia (M.S.). India- part II. Ethno Leaflets 12: 532–537. https://doi.org/10.12692/ijb/10.1.327-334

Kristo ST, Ganzler K, Apati P, Szőke E, Kéry A (2002) Analysis of antioxidant flavonoids from Asteraceae and Moraceae plants by capillary electrophoresis. Chromatographia 56:121–126. https://doi.org/10.1007/BF02494124

Kumar S, Joseph L, George M, Sharma A (2011) A review on anticoagulant/antithrombotic activity of natural plants used in traditional medicine. Int J Pharm Sci Rev Res 8(1):70–74

Lakshman HC, Yeasmin T, Gabriel KP (2014) Herbs of Asteraceae and their ethano-medicinal uses in dermatological problems. J BioSci 22:127–129. https://doi.org/10.3329/jbs.v22i0.30016

Lin KW (2005) Ethnobotanical study of medicinal plants used by the Jah Hut peoples in Malaysia. Indian J Med Sci 59(4):156–161

Maheshwari JK, Singh JP (1984) Contribution to the ethnobotany of Bhoxa tribe of Bihar and Pauri Garhwal districts, U.P. J Econ Tax Bot 5:251–259

Marcinek K, Krejpcio Z (2016) *Stevia rebaudiana* Bertoni: health-promoting properties and therapeutic applications. J Verbr Lebensm 11:3–8. https://doi.org/10.1007/s00003-015-0968-2

Martinez M, Poirrier P, Chamy R, Prufer D, Schulze-Gronover C, Jorquera L, Ruiz G (2015) *Taraxacum officinale* and related species – an ethnopharmacological review and its potential as a commercial medicinal plant. J Ethnopharmacol 169:244–262. https://doi.org/10.1016/j.jep.2015.03.067

Mehal KK (2021) *Verbesina encelioides*: a fast spreading weed in semi-arid regions of North-Western India. Is climate change responsible? J Sci Res 13(1):275–282. https://doi.org/10.3389/fphar.2020.00852

Michel J, Abd Rani NZ, Husain K (2020) A review on the potential use of medicinal plants from Asteraceae and Lamiaceae plant family in cardiovascular diseases. Front Pharmacol 11:852. https://doi.org/10.3389/fphar.2020.00852

Morin NR, Brouillet L, Levin GA (2015) Flora of North America north of Mexico. Rodriguésia 66: 973–981. https://doi.org/10.1590/2175-7860201566416

Mukherjee SK (2006) Medicinal plants of Asteraceae in India and their uses. In: Gupta, SK, Mitra, BR (eds) Proceedings of National Seminar, pp 43–49

Nadkarni KM (1954) Indian Materia Medica with Ayurvedic Unani-Tibbi, siddha, allopathic, homeopathic, naturopathic & home remedies. Popular Prakashan, Mumbai

Nagda BL (2004) Tribal population and health in Rajasthan. Stud Tribes Tribals 2(1):1–8. https://doi.org/10.1080/0972639X.2004.11886496

Ojo O, Mphahlele MP, Oladeji OS, Mmutlane EM, Ndinteh DT (2022) From wandering weeds to pharmacy: an insight into traditional uses, phytochemicals, and pharmacology of genus *Chromolaena* (Asteraceae). J Ethnopharmacol 291:115155. https://doi.org/10.1016/j.jep.2022.115155

Okunade AL (2002) *Ageratum conyzoides* L. Asteraceae. Fitoterapia 73:1–16

Pagel M (2012) Evolution: adapted to culture. Nature 482:297–299. https://doi.org/10.1038/482297a

Pagel M, Mace R (2004) The cultural wealth of nations. Nature 428:275–278. https://doi.org/10.1038/428275a

Parkhomenko AY, Andreeva OA, Oganesyan ET, Ivashev MN (2005) *Ambrosia artemisiifolia* as a source of biologically active substances. Pharm Chem J 39(3):149–153. https://doi.org/10.1007/s11094-005-0106-z

Patel S (2011) Harmful and beneficial aspects of *Parthenium hysterophorus*: an update. 3 Biotech 1:1–9

Patwardhan B, Vaidya A, Chorghade M, Joshi S (2008) Reverse pharmacology and systems approach for drug discovery and development. Curr Bioact Compd 4:201–212

Rao PP, Subramanian P, Reddy P (2011) Pharmacognostic and physico-chemical evaluation of *Bellis perennis* L.–A Homeopathic Drug. J Res Educ Indian Med 17(3–4):65–73

Rastogi RP, Mehrotra BN, Sinha S, Pant P, Seth R (1990) Compendium of Indian medicinal plants: 1985–1989, vol 4. Central Drug Research Institute and Publications & Information Directorate, New Delhi

Rawat P, Saroj LM, Kumar A, Singh TD, Tewari SK, Pal M (2016) Phytochemicals and cytotoxicity of *Launaea procumbens* on human cancer cell lines. Pharmacogn Mag 12(Suppl 4):S431–S435. https://doi.org/10.4103/0973-1296.191452

Rice-Evans CA, Sampson J, Bramley PM, Hollowa DE (1997) Why do we expect carotenoids to be antioxidants in vivo. Free Rad Res 26:381–389. https://doi.org/10.3109/10715769709097818

Robles-Zepeda RE, Coronado-Aceves EW, Velázquez-Contreras CA, Ruiz-Bustos E, Navarro-Navarro M, Garibay-Escobar A (2013) In vitro anti-mycobacterial activity of nine medicinal plants used by ethnic groups in Sonora. Mexico BMC Complement Altern Med 13:329. https://doi.org/10.1186/1472-6882-13-329

Rolnik A, Olas B (2021) The plants of the Asteraceae Family as agents in the protection of human health. Biomed Pharmacother 22(6):1–10

Romero M, Zanuy M, Rosell E, Cascante M, Piulats J, Font-Bardia M et al (2015) Optimization of xanthatin extraction from *Xanthium spinosum* L. and its cytotoxic, anti-angiogenesis and antiviral properties. Eur J Med Chem 90:491–496. https://doi.org/10.1016/j.ejmech.2014.11.060

Saluk-Juszczak J, Olas B, Nowak P, Wachowicz B, Bald E, Głowacki R, Gancarz R (2010) Extract from *Conyza canadensis* as a modulator of plasma protein oxidation induced by peroxynitrite in vitro. Cent Eur J Biol 5(6):800–807. https://doi.org/10.2478/s11535-010-0065-6

Savita SM, Sheela K, Sunanda S, Shankar AG, Ramakrishna P, Sakey S (2004) Health implications of *Stevia rebaudiana*. J Hum Ecol 15(3):191–194. https://doi.org/10.1080/09709274.2004.11905691

Sawicka B, Skiba D, Pszczolkowski P, Aslan I, Sharifi-Rad J, Krochmal-Marczak B (2020) Jerusalem artichoke (Helianthus tuberosus L.) as a medicinal plant and its natural products. Nat Rev Mol Cell Biol 66:1–10. https://doi.org/10.14715/cmb/2020.66.4.20

Schultes RE (1962) The role of ethnobotanist in search for new medicinal plants. Lloydia 25(4): 257–266

Shahrajabian MH, Sun W, Zandi P, Qi C (2019) A review of chrysanthemum, the eastern queen in traditional Chinese medicine with healing power in modern pharmaceutical sciences. Appl Ecol Environ Res 17:13355–13369

Shao Y, Zhou BN, Ma K, Wu HM (1995) New triterpenoid saponins, asterbatanoside D and E, from *Aster batangensis*. Planta Med 61(03):246–249. https://doi.org/10.1055/s-2006-958065
Sharma L, Khandelwal S (2010) Weeds of Rajasthan and their ethno-botanical importance. Stud Ethno-Med 4:75–79. https://doi.org/10.1080/09735070.2010.11886363
Sharma H, Kumar A (2011) Ethnobotanical studies on medicinal plants of Rajasthan (India). A review. J Med Plants Res 5(7):1107–1112
Sharma PK, Mishra P (2021) Geography of Rajasthan. Pareek Publication, Jaipur
Sharma K, Sharma OP (2001) Analysis of precocenes in the essential oil of *ageratum* spp. by reverse-phase high-performance liquid chromatography. Phytochem Anal 12(4):263–265. https://doi.org/10.1002/pca.587
Sharma RK, Verma N, Jha KK, Singh NK, Kumar B (2014) Phytochemistry, pharmacological activity, traditional and medicinal uses of *Erigeron* species: a review. IJARI 2:379–383
Sharma G, Sahu NP, Shukla N (2017) Effect of bio-organic and inorganic nutrient sources on growth and flower production of African marigold. Horticulture 3:11. https://doi.org/10.3390/horticulturae3010011
Singh R (1970) *Gaillardia aristata* Pursh., a naturally occurring new host of cucumber mosaic virus from India. Phytopathol Mediterr 9(2–3):191–192
Singh V, Pandey RP (1998) Ethnobotany of Rajasthan, India. Scientific Publishers, Jodhpur
Singh H, Sharma M (1988) *Gnaphalium coarctatum* Willd-a south American taxon naturalised in Chamba District (Himachal Pradesh), India. Nelumbo 30(1–4):181–184
Singh P, Krishna A, Kumar V, Krishna S, Singh K, Gupta M, Singh S (2016) Chemistry and biology of industrial crop *Tagetes* species: a review. J Essent Oil Res 28(1):1–14. https://doi.org/10.1080/10412905.2015.1076740
Srinivasan K, Natarajan D, Mohanasundari C, Venkatakrishnan C, Nagamurugan N (2007) Antibacterial, preliminary phytochemical and pharmacognostical screening on the leaves of *Vicoa indica* (L.) DC. Iran J Pharmacol Ther 6:109–113
Subashini R, Rakshitha SU (2012) Phytochemical screening, antimicrobial activity and in vitro antioxidant investigation of methanolic extract of seeds from *Helianthus annuus* L. Chem Sci Rev Lett 1(1):30–34
Suo M, Yang J (2014) Ceramides isolated from *Helianthus annuus* L. Helv Chim Acta 97(3): 355–360. https://doi.org/10.1002/hlca.201300194
Suresh SN, Varsha V, Prejeena V (2015) Phytochemical screening of *Cyanthellium cinereum* leaf extracts. Int J Med Pharm Res 3(6):1238–1241
Tomasello M (1999) The human adaptation for culture. Annu Rev Anthropol 28:509–529. https://doi.org/10.1146/annurev.anthro.28.1.509
Toppo KI, Gupta S, Karkun D, Kumar A (2015) Study of antimicrobial effect of *Chrysanthemum morifolium* Ramat. (Asteraceae) against some human pathogens. Int J Pharmacol Biol Sci 9(2)
Upadhyay B, Dhaker AK, Singh KP, Kumar A (2010) Phytochemical analysis and influence of edaphic factors on Lawsone content of *Lawsonia inermis* L. J Phytol Phytochem 2:47–54
Vaidya AD, Devasagayam TP (2007) Current status of herbal drugs in India: an overview. J Clin Biochem Nutr 41:1–11
Weisz GM, Kammerer DR, Carle R (2009) Identification and quantification of phenolic compounds from sunflower (*Helianthus annuus* L.) kernels and shells by HPLC-DAD/ESI-MSn. Food Chem 115(2):758–765. https://doi.org/10.1016/j.foodchem.2008.12.074
Yang WC (2014) Review article botanical, pharmacological, phytochemical, and toxicological aspects of the antidiabetic plant *Bidens pilosa* L. Evid-Based Complem Altern Med 2014:1–14. https://doi.org/10.1155/2014/698617

Chapter 4
Cydonia oblonga Mill.: Wound Healing Properties

Elhan Khan and Iffat Zareen Ahmad

Cydonia oblonga Mill. (Photo by: Iffat Zareen Ahmad, Srinagar, India)

Abstract Traditional and indigenous medicines make significant use of natural raw materials and derivatives of natural products. *Cydonia oblonga* Mill, commonly known as 'Behi', in India or 'Quince' in English, belongs to the family of Rosaceae. Quince is one of the plants that has been traditionally cultivated for its medicinal, nutritional, and decorative properties. In India, quince is typically cultivated in backyards and on fence corners in Jammu and Kashmir and certain portions of Himachal Pradesh. Quince seed mucilage (QSM) is reported to have a significant

E. Khan · I. Z. Ahmad (✉)
Natural Products Laboratory, Department of Bioengineering, Integral University, Lucknow, Uttar Pradesh, India
e-mail: iffat@iul.ac.in

Á. Máthé, I. A. Khan (eds.), *Medicinal and Aromatic Plants of India Vol. 2*,
Medicinal and Aromatic Plants of the World 9,
https://doi.org/10.1007/978-3-031-44914-7_4

wound healing potential, as it is believed to be the best for biomedical functions as it contains biocompatible, inexpensive, water-soluble glucuronoxylan and glucuronic acid-based biomaterials. QSM is utilized to make creams, nano-bandages, and dermal patches for the treatment of both two-degree burn wounds and inflammation caused by microbial infections. Nanoparticles of QSM (silver and zinc oxide) are also reported to attain synergistic wound healing activity with increased efficacy. There is still a need for clinical trials and research to evaluate quince-based extracts and QSM for their efficacy and safety.

Keywords *Cydonia oblonga* · Quince · Wound healing · Burns · Inflammations · Quince seed mucilage · Dermal patches

4.1 Introduction

Abrasion, laceration, contusion, or hematoma are some of the problems associated with the normal tissue structure and function arising from the loss of epithelial integrity due to splitting or incision in the skin caused by physical injuries (Erickson and Echeverri 2018). The process of wound healing commences at the very moment the injury happens, and the rate of progression toward complete healing depends on the extent of the injury, which determines the integrity and appearance of the healed tissues. Hemostasis, inflammation, proliferation, and remodeling or resolution of tissues are the four highly interwoven and overlapping stages involved in the wound healing progression (Wilkinson and Hardman 2020). The regeneration process is disrupted by metabolic abnormalities and diseased circumstances, which causes the restorative mechanism to be delayed. As a result, it places a significant economic strain on both developed and emerging markets. As a consequence, cost-effective and alternative therapies based on indigenous plant-based remedies have been investigated (Tottoli et al. 2020). China and India, Asia's two largest nations, are known to have a diverse range of certified and relatively eminent therapeutical plants. As a result of the immense variety of codified and tribal knowledge about pharmaceutical plants in India, as well as the vast variances in climatic conditions, rainfall, and attitudes, this country is a veritable treasure trove for herbal medicines. As the subcontinent of India is recognized for its diverse forest commodities and practices that are useful for age-old healthcare practices, it is critical to reinforce these conventional values in both domestic and international perceptions while considering current development trends in therapeutical medicine knowledge (Kala et al. 2006).

Traditional Indian medicine, notably Ayurveda, is well-engrained in a folk medical system that has acquired overwhelming acceptance. From the report of the World Health Organization (WHO), around 80% of the global population uses natural remedies for primary healthcare. Only 15% of the world's 300,000 plant species have been screened for potent therapeutic applications (De Luca et al. 2012). However, a ministry established by the government of India for the environment and forests documented more than 9500 species of plants depending on their relevance in the pharmaceutical business (Maiti and Geetha 2013). Since ancient times, the seeds of *Cydonia oblonga*, commonly known as quince, have been recognized to offer a significant activity of healing wounds (Derakhshanfar et al. 2019).

Cydonia oblonga Mill., in the Rosaceae family, is a lone species and belongs to the monotypic genus Cydonia. Cydonia is derived from the term κυδονία, originates from the ancient city of the Greek island of Crete, "Kydo", and also belongs to Dioscorides. While the term "oblong epithet" is derived from the Latin word "oblong", which represents the shape of the quince fruit (Rather et al. 2020). It has been farmed for a long period of time throughout the Mediterranean region as a native plant. Commercial production seems to be mostly in eastern Europe and Asia Minor. Nowadays, it is extended to all over the world, above all in Iraq, Iran, Afghanistan, Syria, Algeria, Tunis, countries of southern Europe, France, and Portugal (Ashraf et al. 2016). In India, quince is mostly grown in Jammu and Kashmir and some parts of Himachal Pradesh, most frequently in backyards and fence corners.

4.2 Taxonomic Characteristics

Cydonia oblonga Mill., is a member of the Rosaceae family and was first mentioned by Philip Miller in 1768. It is known by a variety of common names in different parts of the globe, in addition to the English term quince. It is recognized by several names in the Indian Ayurveda (Sanskrit) system, including Pataalaa, Amritphala, and Imbitikaa. Other vernacular names include: Behi (Urdu/Hindi), Bamsutu (Kashmiri), Cognaisser/Coing (French), Al Sefarjal (Arabic), Wen po (Chinese), Marumero (Japanese), Quitte/Quittenbaum (German), Membrillero (Spanish), Ajva (Russian), Ayva (Turkey), Marmaelo (Portuguese), Shul (Persian), and Kvitten (Swedish) (Lopes et al. 2018).

4.3 Crude Drug Used

The ripe quince fruits have been utilized for culinary purposes like jams, jellies, candy, cookies, and marmalades, which are the most common products of quince fruit. Furthermore, the various parts of quince, such as leaves, seeds, and fruit, are used in the preparation of various solvent-based extracts to treat various illnesses in traditional medicine. The extract of various parts (pulp, peel, seed, and leaf) of quince has been continuously used to prepare various decoctions and formulations for prophylactic and therapeutic applications (Karimi et al. 2017).

4.4 Morphological Description

Cydonia oblonga, commonly known as "Behi" in India, is a small deciduous shrub tree, rather bushy, cultivated under warm temperatures in gardens or farmlands, growing to a height of eight meters and a width of about four meters. The quince leaves are simple with an ovate and stipule oblique arrangement, while the flowers of the quince are normally pink or white in color with a large solitary look. The yellow-juicy fruit of *Cydonia oblonga* is categorized into three varieties based on its shape, namely Lusitanica (Portugal quince with larger sized fruits), Maliformis (fruits having an apple shape) and Pyriformis (pear-shaped), having a sweet fragrance along with the five carpellary chambers of the fruit, which contain a huge number of red or black colored mucilage-coated plano-convex seeds tightly packed in two vertical rows of each chamber (Postman 2009).

4.5 Geographic Distribution

Cydonia is indigenous to Persia but is now cultivated and distributed throughout the globe. Normally native to Armenia, Azerbaijan, Uzbekistan, Iran, Tajikistan, Georgia, Afghanistan, Pakistan, and India, the Behi plant is now cultivated and harvested in Afghanistan's north-west Frontier Province, Poland, Albania, Greece, the Republic of Macedonia, Bosnia, Slovenia, Syria, Croatia, Turkey, Lebanon, Cyprus, Romania, Moldova, Bulgaria, Ukraine, and Serbia (Postman 2009).

In India, *Cydonia oblonga* is not a primary fruit crop but an essential temperate fruit. The temperate conditions that are required for its cultivation are provided in a few regions of India, as its fruit demands repetitive cold temperature periods for growth of about seven degrees Celsius for blossoming as well as a moderately mild environment for the rest of the year (Sharma and Sumbali 2000). Jammu and Kashmir are the major cultivar regions, while small production is provided by the regions of Himachal Pradesh. More than 470 hectares of quince are reported to be farmed in the Kashmir division, especially in the Anantnag, Baramulla, Budgam, and Pulwama districts. These trees have a considerable level of genetic variability, which means they have a lot of possibilities for developing or producing refined cultivars with superior fruit quality and quantity (Ahmad et al. 2008).

4.6 Major Chemical Constituents and Bioactive Compounds

Quince is a reported nutraceutical fruit due to its reservoir of many health-promoting phytochemicals. The existence of polyphenols along with certain polysaccharide fractions such as flavanols and their derivatives, tannins, ionone glycosides, and tetracyclic sesterterpenes (Pirvu et al. 2018). Rasheed et al. (2018) reported the antioxidant activity of ascorbic acid present in quince fruit. A chemical profiling analysis has reported about 13 major phenolic compounds to be present in quince fruits, including the derivatives of caffeoylquinic acid (such as 5-O-caffeoylquinic, 4-O-caffeoylquinic, 3-O-caffeoylquinic and 3,5-dicaffeoylquinic), rutin, and quercitin-3-galactoside, and also contains p-coumaric acid, acetylated quercetin glycosides and acetylated glycosides of kaempferol (Rasheed et al. 2018). The HPLC analysis of fruit jams prepared from both unpeeled and peeled quince fruits was reported, and the analysis revealed that jams produced from unpeeled quince fruits (19%) contain a higher concentration of flavonoids than those prepared from peeled quince fruits (3%) (Silva et al. 2002). One of the other reported HPAEC studies confirms the presence of different sugar derivatives such as ribose, maltose, fructose, rhamnose, isomaltose, sucrose, D-glucose, D-trehalose, D-galactitol and D-sorbitol, along with the presence of a similar percentage of amino acids. *C. oblonga* fruit is majorly utilized for the treatment of irritable bowel syndrome (IBD), as it is a rich source of pectin (Silva et al. 2004). In addition, carotenoid pigments present in *C. oblonga* fruit and concentrated in the epidermal cells, such as zeaxanthin, lycopene, β-cryptoxanthin, β-carotene and lutein, have been shown to possess high antioxidant activity. Mass spectrometric methods such as gas chromatography (GC) and nuclear magnetic resonance (NMR) combined with mass analyzers were also used to identify various carotenoid derivatives (Lopes et al. 2018).

Sut et al. (2019) used LC-MS combined with diode array detection to assess the content and quantity of secondary metabolites in both pulp and peel, which they compared to a few apple cultivars. A significant amount of quinic acid and shikimic acid derivatives, as well as procyanidins and flavonoids, were reported to be present in the fruit. 3-O-caffeoylquinic acid, on the other hand, was found to be the most predominant component in fruit pulp, according to the profiling. Organic acids that are reported by chemical profiling are malic, oxalic, quinic, ascorbic, shikimic, fumaric and citric acids (Silva et al. 2004). Researchers revealed that the metabolite 5-p-coumaroylquinic acid had substantial antioxidant properties (Baroni et al. 2018). During the sample evaluation, four hydroxycinnamic acids and quinic acid (an organic acid) were discovered. Flavanols found in a combination of kaempferol and a mixture of glycosylated quercetin and aglycone may be responsible for the antioxidant action. The flavanol quercetin-3-O-rutinoside was found in a significant percentage. Catechins and procyanidins were the most commonly found flavanols, with epicatechin being the most abundant in all of the samples (Baroni et al. 2018). Quince leaves were reported to possess the highest percentage of 5-O-caffeoylquinic acid (36.2%), followed by quercitin-3-O-rutinoside and kaempferol-3-O-rutinoside (Tsuneya et al. 1983). A study on fatty and organic acid composition reported by Dzhan (2016) showed the presence of quinic acid with the highest concentration of about 72.7%, along with the presence of other organic acids like oxalic acid, citric acid, shikimic acid, fumaric acids and malic acid (Dzhan 2016). Quince leaves also contained 40 different essential oils, fatty acids, oxygenated monoterpene, aromatic aldehydes and sequesterpene hydrocarbons ((E)-ionone, benzaldehyde, linalool, hexadecanoic acid, and germacrene D). Essential oils extracted from quince leaves have been found to contain (Z)-3-hexenal, (Z)-benzaldehyde, germacrene D, farnesene, and (E) phytol (Winterhalter and Schreier 1988). One of the important components discovered in quince leaves is 2-methylbutanoate (Velièkoviæ et al. 2016).

Flavones make up the majority of the phytochemical constituents of quince seeds (63–66%), with caffeoylquinic acids and their derivatives, along with isoschaftoside, being the most prominent. The seed phytochemical profiling has shown the presence of organic acids such as fumaric, malic, ascorbic, D (−)-quinic, citric, and L-shikimic acids, while lacking oxalic acid from the seed extract. In addition, carotenoid pigments present in *C. oblonga* fruit and concentrated in the epidermal cells, such as zeaxanthin, lycopene, β-cryptoxanthin, β-carotene and lutein, have been shown to possess high antioxidant activity. Mass spectrometric methods such as gas chromatography (GC) and nuclear magnetic resonance (NMR) combined with mass analyzers were also used to identify various carotenoid derivatives (Lopes et al. 2018).

4.7 Ecological Requirements

The plant has a good tolerance against cold temperatures but is sensitive to drought owing to its root structure. In the months of August and September, the flowering and fruiting of the plant take place. Quince is a resilient, drought-tolerant shrub that thrives well in a variety of relatively low to moderate soil pH conditions. It can withstand both shade and sunlight, although sunlight is necessary for large flower blooming and the ripening of the fruit. It is a resilient plant that tolerates seasons without pruning or any pest or disease challenges. To blossom successfully, it demands a colder season, with temperatures below 7 °C. Although the tree self-pollinates, cross-pollination results in higher yields (Cumo 2013).

4.8 Traditional Use and Common Knowledge

In the food processing sector, the fruits of quince are currently used to make certain beverages, jellies, marmalades, and jams. It is also used in canning and distillation for aromatic products. Since it contains a wide spectrum of nutritional and therapeutic properties, *C. oblonga* has been extensively exploited by the folk medical system, particularly by Iranian traditional medicine practitioners. The newlyweds would often eat a quince during a Roman wedding since the quince is seen as a fertility symbol in Roman culture and is thought to offer essence. Quince leaves have historically been used as an antiseptic and astringent. Quince root was employed as an amulet towards Scrofula, while the leaves were applied to make a decoction to cure hypertension, pains of the stomach, diarrhea and hyperglycemia. For swollen eyes, the flowers, whether fresh or dried, have responded significantly (Tita et al. 2009). Ashraf et al. (2016) explain that pectin, which is abundant in these fruits, protects the colon from damage by promoting colon cell proliferation and growth in colitis (Ashraf et al. 2016). The polyphenols present in quince fruits are held accountable for anti-inflammatory and antioxidant actions, such as flavonoids and chlorogenic acids like rutin, kaempferol and quercetin, which also promote the healing of IBD-induced colon damage (Minaiyan et al. 2012).

The fruits are constantly utilized for ailments like leucorrhea, liver conditions, uterine hemorrhages, haemoptysis, dysentery and diarrhea, owing to their potential as antiseptic, liver-protective, anti-inflammatory, and astringent (Fazeenah and Quamri 2016). The seeds have been utilized since ancient times for curing gastrointestinal-associated problems (constipation, canker sores, intestinal colic, diarrhea, and gum concerns), emollient, demulcent, and for respiratory system disorders like rhinitis, sore throats, bronchitis, and asthma (Khan and Ahmad 2021). A concoction prepared by the mixture of lemon juice and an extract of quince (aqueous), namely, Gencydo®, a remedy for asthma and rhinitis (Gründemann et al. 2011).

4.9 Modern Medicine Based on Its Traditional Medicine Uses

Human colon and kidney cancer cells were shown to be resistant to development and proliferation when treated with quince polar extracts of flowers, fruit (seeds, pulp and peel), and vegetative leaves. By stimulating the apoptotic signaling pathway, the well-known polyphenols isolated from quince fruit extract were reported to synergistically suppress the growth of human adenocarcinoma LS17 cells. These chemicals may also account for the extract's demonstrated anti-inflammatory effects against human-derived THP-1 macrophages that are stimulated by LPS (Riahi-Chebbi et al. 2016). An extract was subjected to LC-MS analysis, which revealed that neochlorogenic acid was the primary phenolic present, and that it inhibited basophil degranulation while also significantly reducing the production of tumor necrosis factor (TNF) and interleukin IL-8 by mast cells (Huber et al. 2012). Furthermore, the presence of polyphenols in quince, which have hypoglycemic effects, may potentially be a contributing factor to the prevention and reduction of diabetes consequences (Tang et al. 2016).

The presence of catechins and flavones is believed to be the most dominant and effective in defending against ROS. Suppression of lipid peroxidation and the ensuing damage to cellular membranes, accompanied by a succession of detrimental events that culminate in the death of cells, seems to be the most likely mechanism. Aside from attracting free radicals, they also attract a variety of inflammatory mediators, which raise the overall level of inflammation and cell damage (Nijveldt et al. 2001). Because of their capacity to trap and quench ROS, phenolics, notably caffeoylquinic derivatives, along with quercetin, flavonoids, quercetin, astragalin, and kaempferol derivatives in quince leaves, have cardioprotective potential (Pirvu et al. 2018). Several studies have shown that extracts from leaves and fruits that are high in caffeoylquinic acid percentage, as well as kaempferol glycosides and quercetin, are effective in lowering lipid concentrations and alleviating the symptoms of progressive atherosclerotic disorders (Khademi et al. 2013). The antioxidant behavior of polyphenols, as well as their cooperation with the mucilage fractions, may be linked to the skin-healing and antiallergic properties of quince-based products (Shinomiya et al. 2009). Some of the pharmacological activities reported are described in Table 4.1.

Table 4.1 Pharmacological activities of *Cydonia oblonga* Mill

Activities	Parts of quince	Results of the study	References
Aphrodisiac activity	Hydroalcoholic extract of the fruits.	An *in-vivo* study on Wistar rats demonstrated that the rats' mounting frequency and mating performance rose significantly.	Aslam and Sial (2014)
Protective activity	Ethanolic extract of the leaves	The possible preventive benefits of hypercholesterolemia-induced kidney damage in rabbits might be related to both antioxidants and lipid-lowering activities.	Jouyban et al. (2011)
Antimicrobial activity	Acetone and ethanolic and acetone extracts of the fruits	In comparison to aqueous and acetone extracts, quince seed ethanolic extracts were the most effective against infections of bacteria.	Alizadeh et al. (2013)
Anti-allergic properties	Quince fruit extract prepared by hot water.	The studies concluded that hot water extract treated mice showed decreased immunoglobulin E (IgE) levels as compared to the control group.	Sabir et al. (2015)
Antidiabetic properties	Aqueous extract of the fruits	In streptozotocin-induced diabetic rats, they have hypolipidemic effects and ameliorate some of the symptoms of diabetes.	Mirmohammadlu et al. (2015)
Anti-inflammatory property	Extracts of the leaves	In rats, the treatment reduced edema by inhibiting TNF-α, lipid peroxidation, nitric oxide, and IL-6.	Ahmad and Bastawy (2014)
Anti-cancer potential	Aqueous quince fruit extract	DEN-induced hepatocellular carcinoma in rats when treated with quince shows a reduced level of serum biomarkers.	Adiban et al. (2019)
Antioxidant activity	Cookies containing quince fruit	When compared to control cookies, they were shown to have a larger quantity of volatile chemicals, which provides them with significant radical scavenging activities.	Antoniewska et al. (2019)
Ultraviolet protective effects	Methanolic extract of the leaves	This research reveals that the extract can protect catfish red blood cells from UVA-induced damage and hematotoxin stress.	Sayed et al. (2013)

4.9.1 Wound Healing Potential of **Cydonia oblonga** *Mill*

One of the biological macromolecules with therapeutic potential is quince seed, which has been extensively utilized in Iranian ancient medicine for the alleviation of inflammation and pain, the healing of wounds and skin re-epithelialization (Ghafourian et al. 2015). Quince-seed is gaining popularity in the research field due to its health-associated benefits, primarily owing to its attributes of phenolic composition, anti-infection, antioxidant potential, anti-inflammatory and antibacterial capabilities of seed mucilage (Hussain et al. 2019). The gum and mucilage extracted from the seeds of quince have been the most prominent forms utilized for wound healing studies. It's worth noting that extracting gums or isolating secondary metabolites of plants necessitates greater ethanol levels of 70–90%, while mucilage is produced when seeds are soaked in either cold or hot water and promptly generate a large amount of translucent gelatinous substance. Water is among the excellent solvents for conventional medical applications in terms of pharmacology. As a result, mucilage is an excellent choice for a wide spectrum of biomedical functions. Unlike a polysaccharide called chitosan, which is retained in an acidic medium, quince seed mucilage (QSM) is made up of glucuronoxylan and glucuronic acid-based biomaterials that are water-soluble, biocompatible, and inexpensive polysaccharides that inflate in water, which is attributed to the existence of appropriate hydrophilic functional groups like amide (–CONH2), carboxylic acid (–COOH), hydroxyl (–OH), and amino (–NH2) (Szymańska and Winnicka 2015).

4.9.2 Quince Seed Mucilage (QSM) for Dermal Treatments

An investigation was undertaken to assess the efficacy of QSM towards wound healing when it was formulated as creams comprising 5%, 10%, and 20% QSM in an Eucerin base. The statistical evaluation identified a significant difference in the contraction of the wound between creams containing 10% and 20% QSM and the control groups ($P < 0.05$). Further examinations revealed that the cream containing 20% QSM had the best potency, as further examination revealed better tissue resistance, increased growth factors in wound fluids, and an increased content of hydroxyproline, culminating in complete healing in 13 days (Tamri et al. 2014). A 34-patient double-blind clinical trial of benign lesions for excisional biopsy was conducted. The use of a 10% quince mucilage ointment aided wound healing and was reported in 10.72 days. The results showed significant results in comparison to Eucerin groups and untreated groups, suggesting an elaborated assessment of the quince mucilage impact and its potential applicability for chronic ulcer treatment (Moosavi et al. 2006). A chemical analysis was conducted on quince mucilage to determine its dry weight (95.62%), 4.38% of moisture (4.38%) and ash content of about 8.24%, with a yield of 10.9% (Fekri et al. 2008). A study was designed to evaluate the antioxidant potential of an ethanolic extract prepared from *Cydonia oblonga* seeds to treat second-degree wounds caused by burns. Treatment with one percentage of seed extract showed a significant wound healing capability of about 99.50% recovery in comparison to groups of controls treated with sulfadiazine (92.97%) with a shorter healing time (comparatively) (Tajoddini et al. 2013).

Similarly, another study was conducted by Hemmati et al. (2012) to determine the wound healing potential of QSM on the rabbits' damaged skin. The animals were subsequently classified into five experimental groups. Poison-treated rabbits (positive controls) were placed in Group 1, while negative controls treated with Eucerin were placed in Group 2, respectively. The remaining were classified based on the treatment of QSM at different concentrations of 5%, 10%, and 15%. The skin of the rabbits was damaged by consequently administering methanolic T-2 toxin solution twice at a 24-h interval. After 8 days of evaluation, it was seen that rabbits treated with 10% and 15% QSM showed complete wound healing and retained normal hair growth. In contrast to the positive and negative controls that were only receiving 5% QSM treatment, they developed erythema and inflammation. This whole study was evaluated in detail to determine the plausible mechanisms accountable for the QSM efficacy against wound healing (represented in Fig. 4.1) and those were: (a) functioning as a barrier between skin and T-2 toxin and also preventing water evaporation; (b) mitigating disrupted T-2 toxin-induced protein synthesis; (c) it behaves as both a growth factor and an antioxidant; (d) the amplification of the blood-flow process and the formation of granulated tissue; and (e) enhancing fibroblast activity and prompting collagen formation (Hemmati et al. 2012). Recently, various quince seed extracts, especially QSM, have been utilized for developing

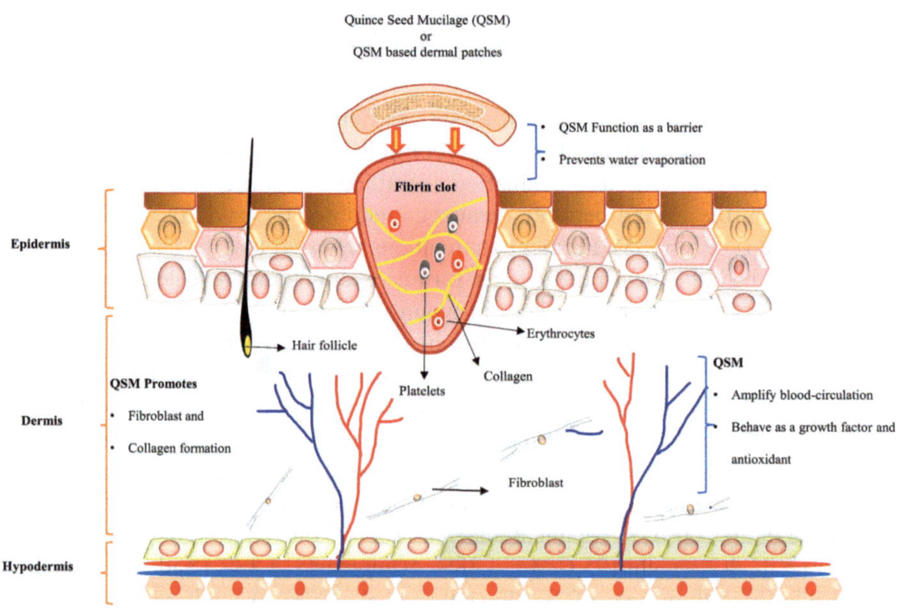

Fig. 4.1 The plausible action responsible for quince seed mucilage efficacy in wound healing. (Image: Khan E)

dermal patches or/and dermal films for the treatment of wounds and burn injuries as they contain the polysaccharide glucuronoxylan. However, it still has to be evaluated for quality and safety (Ashraf et al. 2016).

4.9.3 Quince Incorporation with Nanotechnology for Wound Healing

Recent discoveries enable the manufacture of wound dressings that transfer active chemicals and/or medications to the sites of wounds (Negut et al. 2018). A substantial quantity of scientific evidence has been gathered to support the use of biocompatible silver nanoparticles as a nanostructured material (Burduşel et al. 2018). In their in vivo investigations, Mohseni et al. (2019) examined antimicrobial dressings for chronic wound healing conjugated with silver nanoparticles (AgNPs) and silver sulfadiazine (SSD). Although both SSD and AgNPs had excellent antimicrobial potential towards *S. aureus* at equivalent concentrations, AgNPs had better biocompatibility, accelerated healing, epithelization, and skin regeneration. In wound healing, the scientists determined that nano-silver was more potent than silver sulfadiazine (Mohseni et al. 2019). A leaf extract of *C. oblonga* (methanolic and acetonic) along with AgNPs of QSM were also evaluated against the infected wounds caused by *Staphylococcus aureus*. The findings demonstrated by another study that extracts of quince leaf and silver nanoparticles may both suppress the growth of *Aspergillus niger* represent their activity against *A. niger*, and a synergistic effect between an ethanolic extract of quince's leaf and AgNPs was revealed. The combination of an ethanolic extract of quince's leaf and silver nanoparticles has synergistic antifungal activity against *A. niger* and has the potential to be beneficial in the management of this pathogen (Alizadeh et al. 2014). One of the studies reported the preparation of nano-badges utilizing a green method. This method involves the preparation of zinc oxide nanoparticles using QSM (as a biopolymer) and walnut leaf extract against two types of severe burns and antibacterial infections. The nano-bandages containing the nanoparticles showed a significant healing effect, with no infections detected. In an animal study, the wound showed complete healing in 21 days with no burn evidence on the skin. The proposed nano-bandage had a substantially better impact on wound healing than the control or without a nanoparticle nano-bandage (Darvishi et al. 2021).

4.10 Conclusion

There are 15 agroclimatic zones in India, with 17,000–18,000 different varieties of flowering plants, of which 6000–7000 is thought to have medicinal characteristics. The usage of these therapeutic plants may be found in many Indian communities and is documented in Indian medical systems such as Ayurveda, Unani, Siddha, Sowa-rigpa, and Homeopathy. Researchers have been interested in *Cydonia oblonga* for the past two decades, not only because of its nutritional value but also because of its inherent chemical and physical properties, which make it a potential candidate for a variety of applications, including those in the pharmaceutical industry. According to research, QSM is a reported natural wound healing agent with no hostile properties. In fact, QSM works as a moisture and oxygen barrier, speeding wound closure via collagen formation (Hemmati and Mohammadian 2000).

According to the published research, Quince seed mucilage (QSM) used with creams for wound healing promotes cell development, which enhances wound burn treatments. As it enhances a variety of physical qualities, QSM is a very critical component in a variety of cosmetics. Purified QSM may be employed at an industrial level, notably in food packaging and as a food additive, since it has been shown to be an excellent emulsifying, stabilizing, and film-forming agent. Thorough investigations into the therapeutic benefits of QSM on keratinocytes linked with skin inflammation should precede its commercial production (Hussain et al. 2019).

References

Adiban H, Shirazi FH, Gholami S, Kamalinejad M, Hosseini SH, Noubarani M, Eskandari MR (2019) Chemopreventive effect of quince (*Cydonia oblonga* Mill.) fruit extract on hepatocellular carcinoma induced by diethylnitrosamine in rats. Int Pharm Acta 2(1):1–12

Ahmad MF, Wani N, Hassan GI, Zafar G, Khan IA (2008) Variability in quince (*Cydonia oblonga* Mill) from Baramulla district of Kashmir valley. Indian J Genet 68(2):215–218

Ahmed MM, Bastawy S (2014) Evaluation of anti-inflammatory properties and possible mechanism of action of Egyptian quince (*Cydonia oblonga*) leaf. Egypt J Biochem Mol Biol 32(2): 190–205

Alizadeh H, Rahnema M, Semnani SN, Hajizadeh N (2013) Detection of compounds and antibacterial effect of quince (*Cydonia oblonga* Miller) extracts *in vitro* and *in vivo*. J Biol Act Prod Nat 3:303–309

Alizadeh H, Rahnema M, Semnani SN, Ajalli M (2014) Synergistic antifungal effects of quince leaf's extracts and silver nanoparticles on *aspergillus Niger*. J Appl Biol Sci 8(3):10–13

Antoniewska A, Rutkowska J, Pineda MM (2019) Antioxidative, sensory and volatile profiles of cookies enriched with freeze-dried Japanese quince (*Chaenomeles japonica*) fruits. Food Chem 286:376–387

Ashraf MU, Muhammad G, Hussain MA, Bukhari SN (2016) *Cydonia oblonga* M., a medicinal plant rich in phytonutrients for pharmaceuticals. Front Pharmacol 7:163

Aslam M, Sial AA (2014) Effect of Hydroalcoholic extract of *Cydonia oblonga* Miller (quince) on sexual behaviour of Wistar rats. Adv Pharmacol Sci:1–6

Baroni MV, Gastaminza J, Podio NS, Lingua MS, Wunderlin DA, Rovasio JL, Dotti R, Rosso JC, Ghione S, Ribotta PD (2018) Changes in the antioxidant properties of quince fruit (*Cydonia oblonga* Miller) during jam production at industrial scale. J Food Qual 2018:1–9

Burduşel AC, Gherasim O, Grumezescu AM, Mogoantă L, Ficai A, Andronescu E (2018) Biomedical applications of silver nanoparticles: an up-to-date overview. Nano 8(9):681

Cumo C (ed) (2013) Encyclopedia of cultivated plants: from *Acacia* to *Zinnia* [3 volumes]: from acacia to Zinnia. ABC-CLIO

Darvishi E, Kahrizi D, Arkan E, Hosseinabadi S, Nematpour N (2021) Preparation of bio-nano bandage from quince seed mucilage/ZnO nanoparticles and its application for the treatment of burn. J Mol Liq 339:116598
De Luca V, Salim V, Atsumi SM, Yu F (2012) Mining the biodiversity of plants: a revolution in the making. Science 336(6089):1658–1661
Derakhshanfar A, Moayedi J, Derakhshanfar G, Poostforoosh Fard A (2019) The role of Iranian medicinal plants in experimental surgical skin wound healing: an integrative review. Iran J Basic Med Sci 22(6):590–600
Dzhan T (2016) The study of fatty and organic acids composition in quince leaves and fruits (*Cydonia oblonga* Mill). EUREKA Life Sci 5:39–44
Erickson JR, Echeverri K (2018) Learning from regeneration research organisms: the circuitous road to scar free wound healing. Dev Biol 433(2):144–154
Fazeenah AHA, Quamri MA (2016) Behidana (*Cydonia oblonga* Miller.)- a review. World J Pharm Res 5(11):79–94
Fekri N, Khayami M, Heidari R, Jamee R (2008) Chemical analysis of flaxseed, sweet basil, dragon head and quince seed mucilages. Res J Biol Sci 3(2):166–170
Ghafourian M, Tamri P, Hemmati AA (2015) Enhancement of human skin fibroblasts proliferation as a result of treating with quince seed mucilage. Jundishapur J Nat Pharm Prod 10:e18820
Gründemann C, Papagiannopoulos M, Lamy E, Mersch-Sundermann V, Huber R (2011) Immunomodulatory properties of a lemon-quince preparation (Gencydo®) as an indicator of anti-allergic potency. Phytomed: Int J Phytother Phytopharmacol 18(8–9):760–768
Hemmati AA, Mohammadian F (2000) An investigation into the effects of mucilage of quince seeds on wound healing in rabbit. Int J Geogr Inf Syst 7(4):41–46
Hemmati AA, Kalantari H, Rezai S, Zadeh H (2012) Healing effect of quince seed mucilage on T-2 toxin-induced dermal toxicity in rabbit. Exp Toxicol Pathol 64:181–186
Huber R, Stintzing FC, Briemle D, Beckmann C, Meyer U, Gründemann C (2012) *In vitro* antiallergic effects of aqueous fermented preparations from citrus and *Cydonia* fruits. Planta Med 78(4):334–340
Hussain MA, Muhammad G, Haseeb MT, Tahir MN (2019) Quince seed mucilage: a stimuli-responsive/smart biopolymer. Functional biopolymers. Polymers and polymeric composites: a reference series. Springer, Cham, p 127
Jouyban A, Shoja MM, Ardalan MR, Khoubnasabjafari M, Sadighi A, Tubbs RS, Agutter PS, Ghabili K (2011) The effect of quince leaf decoction on renal injury induced by hypercholesterolemia in rabbits: a pilot study. J Med Plants Res 5(21):5291–5295
Kala CP, Dhyani PP, Sajwan BS (2006) Developing the medicinal plants sector in northern India: challenges and opportunities. J Ethnobiol Ethnomed 2(1):1–5
Karimi A, Movahhed M, Haji Mehdipoor H, Allahyari F (2017) A review on *Cydonia oblonga* Miller as an herbal medicine. INDO Am J Pharm Sci 4:4370–4386
Khademi F, Danesh B, Nejad DM, Rad JS (2013) The comparative effects of atorvastatin and quince leaf extract on atherosclerosis. Iran Red Crescent Med J 15(8):639–643
Khan E, Ahmad IZ (2021) An insight into the prophylactic and therapeutic activities of golden apple (*Cydonia oblonga* Mill.) for the future cancer care and prevention: a review. Ann Phytomed 10(2):22–35
Lopes MA, Sanches AG, de Souza KO, Silva EO (2018) Quince – *Cydonia oblonga*. In: Rodrirues SEO, Brito ES (eds) Exotic fruits. Academic, Amsterdam, p 363
Maiti S, Geetha KA (2013). Country status report on Medicinal and Aromatic Plants in India. In Expert consultation on promotion of medicinal and aromatic plants in the Asia-Pacific region: proceedings, pp 101–23
Minaiyan M, Ghannadi A, Etemad M, Mahzouni P (2012) A study of the effects of *Cydonia oblonga* Miller (quince) on TNBS-induced ulcerative colitis in rats. Res Pharm Sci 7:103–110
Mirmohammadlu M, Hosseini SH, Kamalinejad M, Gavgani ME, Noubarani M, Eskandari MR (2015) Hypolipidemic, hepatoprotective and renoprotective effects of *Cydonia oblonga* Mill. Fruit in streptozotocin-induced diabetic rats. Iran J Pharm Res 14:1207–1214

Mohseni M, Shamloo A, Aghababaie Z, Afjoul H, Abdi S, Moravvej H, Vossoughi M (2019) A comparative study of wound dressings loaded with silver sulfadiazine and silver nanoparticles: in vitro and in vivo evaluation. Int J Pharm 564:350–358

Moosavi Z, Meshki M, Hemmati AA, Veisi M, Rafiee R (2006) Evaluation of the efficacy of quince mucilage on wound healing. J Derm Dis 9:260–263

Negut I, Grumezescu V, Grumezescu AM (2018) Treatment strategies for infected wounds. Molecules 23(9):2392

Nijveldt RJ, van Nood E, van Hoorn DE, Boelens PG, van Norren K, van Leeuwen PA (2001) Flavonoids: a review of probable mechanisms of action and potential applications. Am J Clin Nutr 74(4):418–425

Pirvu L, Stefaniu A, Neagu G, Albu B, Pintilie L (2018) In vitro cytotoxic and antiproliferative activity of *Cydonia oblonga* flower petals, leaf and fruit pellet ethanolic extracts. Docking simulation of the active flavonoids on the anti-apototic protein Bcl-2. Open Chem 16(1): 591–604

Postman J (2009) *Cydonia oblonga*: the unappreciated quince. Arnoldia 67:2–9

Rasheed M, Hussain I, Rafiq S, Hayat I, Qayyum A, Ishaq S, Awan MS (2018) Chemical composition and antioxidant activity of quince fruit pulp collected from different locations. Int J Food Prop 21(1):2320–2327

Rather GA, Bhat MY, Sana SS, Ali A, Gul MZ, Nanda A, Hassan M (2020) Quince. In: Antioxidants in fruits: properties and health benefits. Springer, Singapore, p 397

Riahi-Chebbi I, Haoues M, Essafi M, Zakraoui O, Fattouch S, Karoui H, Essafi-Benkhadir K (2016) Quince peel polyphenolic extract blocks human colon adenocarcinoma LS174 cell growth and potentiates 5- fluorouracil efficacy. Cancer Cell Int 16:1

Sabir S, Qureshi R, Arshad M, Amjad MS, Fatima S, Masood M, Saboon CSK (2015) Pharmacognostic and clinical aspects of *Cydonia oblonga*: a review. Asian Pac Trop Dis 5(11):850–855

Sayed AH, Abdel-Tawab HS, Abdel Hakeem SS, Mekkawy IA (2013) The protective role of quince leaf extract against the adverse impacts of ultraviolet: a radiation on some tissues of *Clarias gariepinus* (Burchell, 1822). J Photochem Photobiol B Biol 119:09–14

Sharma Y, Sumbali G (2000) Impact of different storage conditions, systems and associated mycoflora on the ascorbic acid content of dried fruit slices of quinces (*Cydonia oblonga* Mill). J Ind Bot Soc 79(1–4):1–211

Shinomiya F, Hamauzu Y, Kawahara T (2009) Anti-allergic effect of a hot-water extract of quince (*Cydonia oblonga*). Biosci Biotechnol Biochem 73(8):1773–1778

Silva BM, Andrade PB, Ferreres F, Domingues AL, Seabra RM, Ferreira MA (2002) Phenolic profile of quince fruit (*Cydonia oblonga* Miller) (pulp and peel). J Agric Food Chem 50: 4615–4618

Silva BM, Andrade PB, Gonçalves A, Seabra R, Oliveira M, Ferreira M (2004) Influence of jam processing upon the contents of phenolics, organic acids and free amino acids in quince fruit (*Cydonia oblonga* Miller). Eur Food Res Technol 218:385–389

Sut S, Dall'Acqua S, Poloniato G, Maggi F, Malagoli M (2019) Preliminary evaluation of quince (*Cydonia oblonga* Mill.) fruit as extraction source of antioxidant phytoconstituents for nutraceutical and functional food applications. J Sci Food Agric 99:1046–1054

Szymańska E, Winnicka K (2015) Stability of chitosan – a challenge for pharmaceutical and biomedical applications. Mar Drugs 13(4):1819–1846

Tajoddini A, Rafieian-kopaei M, Namjoo AR, Sedeh M, Ansari R, Shahinfard N (2013) Effect of Ethanolic extract of *Cydonia oblonga* seed on the healing of second-degree burn wounds. Armaghan-eDanesh 17(6):494–501

Tamri P, Hemmati A, Boroujerdnia MG (2014) Wound healing properties of quince seed mucilage: *in vivo* evaluation in rabbit full-thickness wound model. Int J Surg 12:843–847

Tang D, Xie L, Xin X, Aisa HA (2016) Antioxidant action of *Cydonia oblonga* seed extract: improvement of glucose metabolism via activation PI3K/AKT signaling pathway research and reviews. J Pharmacog Phytochem 4(2):1–13

Tita I, Mogosanu G, Tita M (2009) Ethnobotanical inventory of medicinal plants from the south-west of Romania. Farmacia 57:141–156

Tottoli EM, Dorati R, Genta I, Chiesa E, Pisani S, Conti B (2020) Skin wound healing process and new emerging Technologies for Skin Wound Care and Regeneration. Pharmaceutics 12(8):735

Tsuneya T, Ishihara M, Shiota H, Shiga M (1983) Volatile components of quince fruit (*Cydonia oblonga* Mill.). Agric Biol Chem 47:2495–2502

Velièkoviæ DT, Ristiæ MS, Milosavljeviæ NP, Davidoviæ DN, Milenoviæ DM, Velièkoviæ AS (2016) Volatiles of quince fruit and leaf (*Cydonia oblonga* Mill.) from Serbia. BiologicaNyssana 7(2):145–149

Wilkinson HN, Hardman MJ (2020) Wound healing: cellular mechanisms and pathological outcomes. Open Biol 10(9):200223

Winterhalter P, Schreier P (1988) Free and bound C13 norisoprenoids in quince (*Cydonia oblonga*, Mill.) fruit. J Agric Food Chem 36:1251–1256

Chapter 5
Artemisia annua L.: Comprehensive Review of Pharmacological Properties

Nafish Ahamd, Riyazuddine Khan, and Iffat Zareen Ahmad

Artemisia annua L. (Photos: by Nafish Ahmad, in Integral university, Lucknow, India)

Abstract *Artemisia annua* L. has been diagnosed as a potent medicinal herb and has been well documented in the historical pharmacopoeias of distinctive Asian and European nations. The WHO has suggested *Artemisia annua* as a potential pharmacological agent and as an antimalaria drug. The entire plant has been recognized to possess antipyretic, antihelminthic, antispasmodic, antiseptic, stimulant, carminative, and stomachic properties. In African, *Artemisia annua* brew has been used to deal with malaria. *A. annua* has a key ingredient known as artemisinin, which serves as the chemical basis for the world's antimalarial programmes and combinatorial

N. Ahamd · R. Khan · I. Z. Ahmad (✉)
Natural Products Laboratory, Department of Bioengineering, Integral University, Lucknow, Uttar Pradesh, India
e-mail: iffat@iul.ac.in

Á. Máthé, I. A. Khan (eds.), *Medicinal and Aromatic Plants of India Vol. 2*,
Medicinal and Aromatic Plants of the World 9,
https://doi.org/10.1007/978-3-031-44914-7_5

curative therapy. Current research indicates that artemisinin is active in killing human breast cancer cells. As a result, the characterization and separation of artemisinin from *Artemisia annua* L. has received augmented global interest.

Keywords Artemisia · Pharmacology · Antimalarial · Artemisinin

5.1 Introduction

Artemisia annua L., often recognized as “candy Annie”, “sugar wormwood”, “candy sage wort”, and “annual wormwood”, is a member of the Asteraceae family having significant medicinal and economic value. It is a definite, wind-pollinated worldwide species that is particularly abundant in temperate regions of the northern hemisphere’s mid-to high latitudes, especially in dry and semiarid habitats, and has just a few specimens in the southern hemisphere. *Artemisia annua* is the only species of the genus Artemisia that flowers annually (Willcox et al. 2004). It is referred to as Qing Hao in ancient Chinese, which translates as “green herb.” There are two mutually conflicting explanations for the origin of the name. Artemisia is a group of plants named after the Greek goddess Artemis, whose name means “she who heals disease.” (Willcox et al. 2004) People in the area used to believe that she cured diseases and fought evil.

In India, through the initiatives of CSIR-CIMAP, this crop is currently increasing in popularity with improved varieties among several farmers as a beneficial endeavor. The plant is currently ecologically farmed to improve rural livelihoods and to generate employment. Furthermore, it integrates well with the farmers’ existing agricultural strategy (Kumar et al. 2015). In collaboration with the CSIR-CDRI, Lucknow, the CSIR-CIMAP discovered the derivative of artemisinin (α,-β-arteether) for the management of the complicated cerebral malaria patients. The epimeric combination of α- and β-arteether has the benefit of having better oil solubility and being economically produced on a massive scale (Varkey 2015). Furthermore, IPCA dominates the Indian antimalarial market for all dosage types, which accounts for around 18% of the company’s revenue, or about 60 million US dollars (Shukla et al. 2018) (Fig. 5.1).

5.2 Taxonomic Characteristics

Wormwood is the prevalent moniker for the genus Artemisia. It is a reference name for *Artemisia absinthium* L., among the most widely prevalent and well-known species of this genus. *Artemisia vulgaris* is another type species, in addition to *Artemisia annua* L., including *Artemisia afra*, *Artemisia abrotanum* and *Artemisia absinthium*. *Artemisia verlotiorum* Lamotte, has been introduced to Mascarene

Fig. 5.1 *Artemisia annua* L. (Photo: Nafish Ahmad)

Islands but is now commercially cultivated in different parts of the world including India.

Vernacular Names: Sweet sage wort, annual wormwood, candy wormwood, candy Annie (English); armoise annually (French); Gae-tong-sook, Hwang-hwa-ho, Chui-ho (Korean); Kusoninjin (Japanese); Cao Haozi, Cao Qinghao, Caohao, Haozi, Chou Qinghao, Chouhao, Kuhao, Jiu Bingcao, Xiang Qinghao, Xiyehao, Xianghao, San Gengcao, Xiang Sicao (Chinese); Thanh cao hoa vàng (Vietnamese).

Habitat: Terrestrial/Wetlands.

5.3 Crude Drug Used

The leaves of *Artemisia annua* are extensively used for the preparation of various herbal concoctions and extractions of metabolites of interest, especially artemisinin-compounds. The leaves of Artemisia are reported to be infused in tea to treat malarial fever (Weathers et al. 2014).

5.4 Major Chemical Constituents and Bioactive Compounds

The leading development of the antimalarial drug, i.e., artemisinin, in *A. annua* has emerged as a source of intense phytochemicals. The leaves and flowers contain a high proportion of protein, crude fat, and digestible fraction. Manganese and copper are abundant in plant tissue. In addition, this plant has a lot of amino acids and nutrients, which makes it a good herb to eat (Das 2012). *Artemisia annua* is an efficient antimalarial herb that is used in folk medicine, homoeopathy, and Ayurvedic medication. Numerous chemicals have been identified by phytochemical analysis, together with phenolics, coumarins, steroids, purines, triterpenoids, flavonoids, aliphatic compounds, lipids, monoterpenoids, alkaloids, essential oils, and glycosides (Zanjani et al. 2012).

According to the special literature, aqueous administration of the dried plant was originally used in China to treat fever, malaria, skin ailments, jaundice, and haemorrhoids. Significant terpene derivatives have been found, including artemisinic alcohol, artemisia ketone, myrcene hydroperoxide, and arteannuin B (Tellez et al. 1999). Several of these are also present in essential oils (Brown 2010). Each essential oil contains non-volatile and volatile ingredients such as artemisia ketone, camphene hydrate, 1,8-cineole, culminal (Das 2012), germacrene D, beta-caryophyllene, and alpha-pinene. Sesquiterpenoids, flavonoids, and coumarins, as well as β-glucosidase, β-galactosidase, stigma-sterol, and β-sitosterol, comprise the non-volatile components of vital oil (Das 2012). It comprises camphor (7.25%), erythritol (50.30%), pinocarveol (4.13%), as well as diethoxy ethane (2.18%) (Haghighian et al. 2008). Scopoletin is a member of the coumarins identified in extracts of *A. annua* (Tzeng et al. 2007). Scopoletin (coumarin), scopoline (coumarin glycoside), chrysosplenol-D (flavonoid), domesticoside (phloroacetophenone), and nonanoic acid (bisnor-cadinane) are all necessary phytonutrients (Yang et al. 2009). In 1972, artemisinin (sesquiterpene lactone) was isolated from *A. annua* for the first time (Ogwang et al. 2012). Though artemisias number around 400 species (Ferreira 2004), artemisinin and imperative oil concentrations in the *A. annua* leaves varied between 0.01–1.4% and 0.04–1.9%, respectively (Damtew et al. 2011). *A. annua* leaves are used as a natural artemisinin source as well as for other important secondary metabolites that can be used to synthesize derivatives with pharmacological value (Cafferata et al. 2010). The World Health Organization recommends excessive use of the vigorous ingredient Artemisinin and its organic variants; hence, they are widely utilized as an antimalarial medication on a global scale. Artemisinin is a sesquiterpene lactone (Fig. 5.2) derived from sweet wormwood, and this plant is utilized as an antimalarial in the treatment of multi-drug resistant *falciparum* malaria infections. It functions also as a plant metabolite. Chemically, it is an organic peroxide and a sesquiterpene lactone.

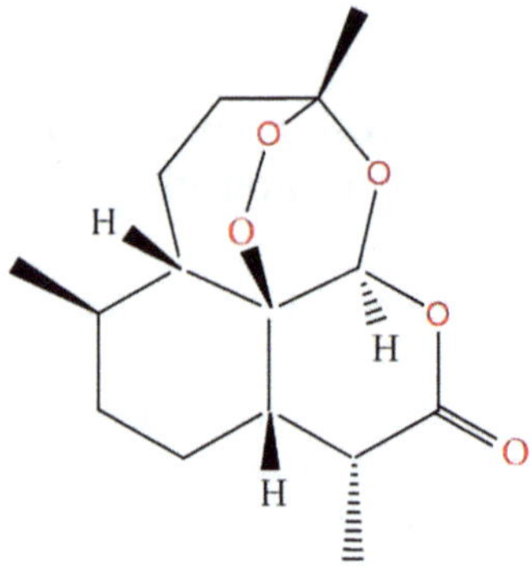

Fig. 5.2 The primary 2D structure of Artemisinin, the major component of *Artemisia annua*

5.5 Morphological Description of *Artemisia annua* L.

Artemisia annua is a large herb that frequently exceeds two meters in height. It is normally single-stemmed with alternative branches. The fragrant leaves are highly separated and range in length from 2.5 to 5 cm. Each 10-celled biseriate trichome and five telephone filamentous trichomes compose the leaves and plant life. Typically, variation occurs in the leaf and aerial portions. The borders of the leaves are no longer complete, but the base is uneven (Fig. 5.1). Both the external and internal surfaces are magnificent. On each surface, glandular and non-glandular trichomes are present. Spongy parenchyma is composed of four to six layers of loosely arranged cells. Artemisinin is synthesized, stored, and secreted by *Artemisia annua* trichomes.

5.6 Geographical Distribution

Artemisia annua occurs naturally in the northern sections of China's Sui yuan and Chahar provinces as a phase of desert vegetation, between 1000 and 1500 m above sea level, and has naturalized in a number of countries (Fig. 5.3).

It is extensively cultivated in the regions of the subtropical, temperate, and cool temperate zones of the world, most commonly in Asian countries, especially in the south and south-west areas. Other countries include China, some parts of Europe, North Africa, and the Mediterranean area. Small-scale growing has emerged in India as well as many worldwide locations in Africa, South Europe, and South America. It is now extensively farmed in Kenya, Romania, Vietnam, Tanzania, France, Bulgaria, Argentina, Hungary, Spain, the United States, Italy, and Yugoslavia (Khosravi et al. 2011; Lestari et al. 2011). It is cultivated contractually by the

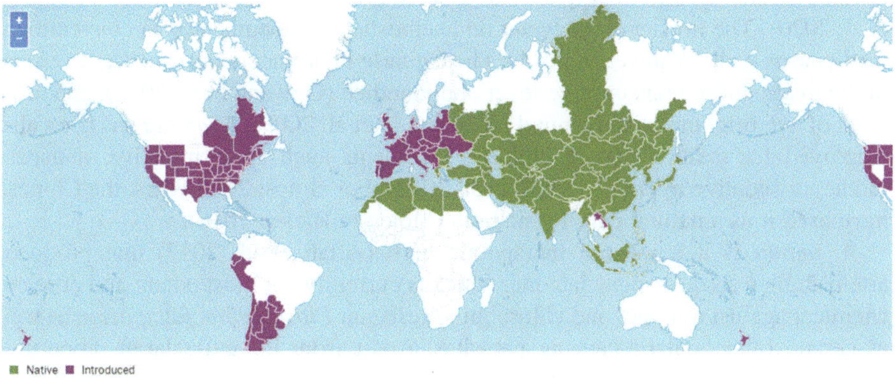

Fig. 5.3 Global geographic distribution of *Artemisia annua* L. (Source: https://powo.science.kew.org/taxon/urn:lsid:ipni.org:names:304416-2#distributions)

industry in several states of India, including Gujrat, Madhya Pradesh, Bihar, Uttar Pradesh (UP), and Uttarakhand. Currently, Uttar Pradesh is the largest producer, followed by Gujrat, Uttarakhand, and then other states (Kumar et al. 2015).

5.7 Ecological Requirements

Artemisia annua has high ecological adaptability, still it prefers sunny environments for better growth. *A. annua* cultivation is prevalent on hillsides but can also thrive in wastelands and at the borders of woods. It can grow at various elevations: in China, even at an altitude of 600–800 m. Humid tropical monsoon, with an average temperature of 17.6–28.4 °C throughout the growth period and a yearly amount of rainfall is 1150–1350 mm provides the most favorable conditions for growth of *A. annua*. It can thrive in a variety of soil types with a pH in between 4.5 and 8.5, the deep topsoil and there should be excellent drainage conditions (WHO Monograph 2006). Sunlight and well-aerated soils are required for an increased rate of seed germination, and hence seeds are initially germinated in a greenhouse to enhance the plant growth rate. The optimum period of exposure to light is required to optimize the production of active principles. Highest concentrations of artemisinin contents were observed during the blooming period (Ekiert et al. 2021).

5.8 Traditional Use and Common Knowledge

Historically, *Artemisia annua* has been referred to a flavouring ingredient. Due to its powerful aroma, it may eventually become a source of important oils for the perfume industry. Fifty-two "prescriptions" (dating all the way back to 168 BC) refer to *A. annua* as a therapeutic plant. It is part of traditional medicine. *A. annua* decoctions have also been used to treat diarrhoea as an antihemorrhagic (Septembre-Malaterre et al. 2020). The impact of *A. annua* on haemostasis is documented in conventional medicine as well (Wang et al. 2011). The Chinese Pharmacopoeia specifies its usage in the treatment of consumptive fever and jaundice (Ogwang et al. 2012), enhancement of eye brightness, and wound healing (Liu et al. 2009). *A. annua* has been also employed for treating certain children's ailments such as carminative, antispasmodic, or sedative/hypnotic (Aftab et al. 2014). For almost 2000 years, the Chinese have used it as a natural cure for malaria (Meier zu Biesen 2010).

A. annua is a renowned therapeutic herb (Aftab et al. 2017) that produces artemisinin, a sesquiterpene lactone, which is currently the most potent and efficient chemical against quinine- and chloroquine-resistant *Plasmodium falciparum* as well as certain other malaria-causing parasites. Apart from its antimalarial properties, *A. annua* also exhibits anti-inflammatory, antibacterial, and anticancer properties. In fact, it is the only Chinese medicinal plant that has been discovered and developed to meet the standards of western medicinal drug research and is renowned for its efficacy and minimal toxicity in the treatment of ague (Wang et al. 2011). At present, the phytoconstituents are being extensively investigated for the treatment of malaria that is drug-resistant (Misra et al. 2013).

5.9 Modern Medicine Based on Its Traditional Medicine Uses

5.9.1 Antimicrobial Activity

Numerous studies have been conducted to determine the antibacterial activity of *A. annua* essential oils. Fundamental oil was shown to be antibacterial against an extensive variety of gram-negative and positive bacteria and fungus in experiments. The oil exhibited significant inhibitory activity against bacterial species like *Escherichia coli*, *Staphylococcus aureus*, as well as *Enterococcus hirae*. *Pseudomonas aeruginosa* was shown to be unresponsive to essential oils. In vitro antifungal activity of *A. annua* extracts against fungi, particularly *Candida albicans* and *Saccharomyces cerevisiae* (Juteau et al. 2002). Additionally, this research demonstrated that essential oils had a greater effect on fungal cells than on gram-positive bacterial lines (Verdian-rizi et al. 2008). Chemical comparisons of extracts of plants demonstrated that phytoconstituents impart this antibacterial activity. The most critical molecules that have been investigated for this activity are scopoletin (Tzeng et al. 2007), artemisinin and a range of different by-product compounds. Chemical comparisons of plant extracts demonstrated that phytoconstituents impart this antibacterial activity. The most critical molecules that have been investigated for this activity are the scopoleti motion mechanisms. In *Mycobacterium TB*, *Mycobacterium smegmatis*, and *Escherichia coli*, compounds have been examined at the molecular level. Artemisinic acid is a common chemical that is used in the semi-synthesis of artemisinin. Like artemisinin, it has been studied for its antibacterial effects (Huang et al. 2010).

5.9.2 Anti-inflammatory Activity

The anti-inflammatory action of aqueous methanolic extract of plants was investigated for chronic and acute irritation using a number of inflammatory models. The aqueous extract of the plant demonstrated a dose-dependent anti-inflammatory effect, resulting in claimed anti-edema action. Phytochemical examination determines the existence of essential chemical groups, including triterpenoids, flavonoids, coumarin, and polyphenols. As a result, these chemicals operate synergistically and confer inhibitory effects on the edoema in both acute and chronic forms (Das 2012). Additionally, another search assessment discusses various anti-inflammatory substances such as scopoletin, dihydroartemisinin, and artemisinin. *In vivo* studies show that these drugs pointedly decrease humeral responses at high concentrations. However, recent research indicates that pure chemicals may have lost their enormous potency in eliciting a chronic hypersensitive action (Bhakuni et al. 2001). Additional investigational research on murine macrophages such as RAW 264.7 cells validated scopoletin's dose-dependent effect. As a result, many research studies say scopoletin could be a good anti-inflammatory remedy (Tzeng et al. 2007).

5.9.3 Antimalarial Activity

To address this condition, a coordinated approach is required that includes preventative initiatives, therapeutic medications, and patient healing. As a result, the mining of artemisinin from *A. annua* has paved the way for the development of novel and strikingly good substitutes (Kim et al. 2015). This plant has been correctly recognised around the world (Willcox 2009), and it is being used in over 50 nations as a potent treatment alternative against malaria, mainly chloroquine-resistant malaria (Ferreira 2007). Artemin, chrysoplenetin, casticin, chrysoplenol-D, eupatorine, and cirsilineol are all flavonoids that have been shown to be effective against malaria (Castilho et al. 2008). The action of methoxylated flavonoids is connected to artemisinin initiation, which clarifies their critical significance in enhancing artemisinin's interaction with plasmodial haemoglobin via the catabolic route that creates artemisinin peroxide. Additionally, artemisinin peroxide constrains heme polymerization, which results in antimalarial activity against the protozoan *Falciparum vivax*. Another mechanism for flavonoids is that they inhibit hypoxanthine uptake through Plasmodium. The antiplasmodial effects of artemisinin are caused by the alkylation of malaria-specific proteins (Bhakuni et al. 2001). However, certain flavonoids that don't have their own antiplasmodial properties can boost the antiplasmodial effects of artemisin (Ferreira et al. 2010).

There are findings in which over 3000 malaria patients were clinically treated with artemisinin and its derivatives. These findings support the therapeutic efficacy of artemisinin compounds, particularly against drug-resistant *Plasmodium falciparum* (Weathers and Towler 2012). A relative medical study has been shown to determine the effectiveness of *A. annua* 's complete herb and chloroquine. Chloroquine is not as good at treating malaria as organic extracts of *A. annua*, which are faster acting, much more effective, and less toxic (Tayebe et al. 2012). It suggestsively decreases parasitaemia and enhances the immunological action by boosting macrophage phagocytic action. The intake of a whole plant extract is more recommended owing to the presence of a range of phytoconstituents with synergistic antimalarial activity. So, it is clear that cutting-edge combinatorial methods may also be re-presenting combinations of phytoconstituents that work together just like in nature (Donno et al. 2012).

5.9.4 Antioxidant Activity

Plant is an excellent cause of concentrated antioxidants and nutritional components (Das 2012). Crude natural extracts of aerial parts have an extreme antioxidant potential, which is likely owing to the leaf's increase flavonoid variety and concentration, including the recently identified C-glycosyl flavonoid as a potential antioxidant. Antioxidant properties are imparted by essential oil and flavonoids content materials found in *A. annua*. As a result of this research, *A. annua* was listed among the top medicinal flora on the list, principally because of their ideal antioxidant manageability (Kim et al. 2015). Significant sources of polymethoxylated and hydroxylated flavonoids have been recognized, including eupatin, cirsilineol, chrysosplenol-D, cirsilineol, chrysop lenetin, artemetin and casticin (Ferreira et al. 2010).

5.9.5 Immunosuppressive Activity

Plants are an excellent source of concentrated antioxidants and nutritional components (Das 2012). Crude natural extracts of aerial parts have an extremely high antioxidant potential, which is likely owing to the leaf's increased flavonoid variety and concentration, including the recently identified C-glycosyl flavonoid as a potential antioxidant. Antioxidant properties are provided by the essential oil and flavonoids content materials found in *A. annua*. As a result of this research, it was listed among the top medicinal flora on the list, principally because of its ideal antioxidant manageability (Kim et al. 2015). It has been recognized as a significant source of polymethoxylated and hydroxylated flavonoids, including eupatin, cirsilineol, chrysosplenol-D, cirsilineol, chrysop lenetin, artemetin, and casticin (Ferreira et al. 2010).

5.9.6 Anti-hypertensive Activity

The antihypertensive efficacy of leaf extract of artemisia from numerous species was investigated *in vivo* by administering 100–390 mg kg^{-1} to diabetic rabbits and rats for 2–4 weeks. The outcomes demonstrated the beneficial effects of aqueous leaf extracts by demonstrating a decrease in blood levels. As a result, this motion inhibits the degree of glycosylation of haemoglobin from increasing and results in hypoliposis. Additionally, it explains the shielding effect on body weight reduction in animals with diabetes. It also significantly reduces the contraction caused by phenylephrine and increases the rat's aortic ring relaxation in response to endothelium stimulation (Das 2012).

5.9.7 Anti-arthritic Activity

An experimental study has demonstrated that the artemisinin by-product derived from *A. annua* (SM905) inhibits inflammatory responses, resulting in an increase in collagen-induced arthritis. This study examined collagen-induced arthritis in DBA/1 mice using a type II bovine collagen mannequin and the artemisinin derivative SM905. The prevalence and severity of disorders have been tracked on a regular basis. Additionally, the expression of genes and the stage of T helper (Th) 17/Th1/Th2 cytokine creation were investigated. The findings of this study indicate that the SM905 molecule has a crucial action in delaying the development of illness, hence lowering the prevalence of arthritis. Additionally, it inhibits the overexpression of a range of pro-inflammatory chemokines and cytokines (Das 2012).

5.9.8 Antiparasitic Activity

Artemisinin tablets may be effective against *Trypanosoma babesia*, *Leishmania*, *coccidiosis,* or *Eimeria*, and trematodal blood flukes (Septembre-Malaterre et al. 2020). Research was conducted in response to *Neospora canum*. For 14 days, artemisinin was added to either Vero cells or mouse peritoneal macrophages. After 11 days, most microscopic emphases of N. caninum were totally eradicated at concentrations of 10 or 20 lg/mL, and similar outcomes were found at concentrations of 0.1 lg/mL. As a result, artemisinin has been shown to be effective in reducing the intracellular replication of *N. caninum* tachyzoites. Previously, the effect of artemether was evaluated in comparison to *Schistosoma mansoni* larval ranges. It was shown that after artemether therapy, animals did not develop schistosomiasis.

5.9.9 Antiviral Activity

The antiviral potential of *A. annua* tea distillations against HIV were investigated for the first time in a scientific study. Toxicological research has been conducted on two unrelated cell structures at a relatively modest concentration (2.0 lg/mL). The *A. annua* tea brew exhibits a very broad spectrum of activity. However, artemisinin was previously proven to be ineffective at increased concentrations (25 lg/mL). Likewise, no cytotoxic activities on cells have been seen as a result of tea infusion. Thus, this *in vitro* research shows that artemisinin exerts a limited effect and may possibly operate synergistically with anti-HIV activity (Castilho et al. 2008). Presently, artemisinin and its derivatives are the subject of scientific research to decide their efficacy in contrast to an extensive variety of viruses (Ferreira 2007), with the goal of developing improved antiviral amalgamation therapy approaches (Weathers and Towler 2012).

5.9.10 Anticancer Activity

A. annua is renowned for its pharmacological properties in well-known treatments, and it is now undergoing investigation to identify a cure for most malignancies (Cafferata et al. 2010). Numerous natural extracts of *A. annua* have been investigated for their anticancer activity by evaluating their cytotoxic practicable in TC221 cells and HeLa malignancy cells. These opinions set up that methanol extracts are a great deal more cytotoxic than dichloromethane extracts (Efferth et al. 2011). Cytotoxicity research on artemisinin and quercetagetin-6, 7, 3, 4, tetramethyl ether against a variety of tumour cells, including A-549, P-388, MCF-7, and Ht-29 cells, demonstrated their effectiveness on a large scale. *In vitro* as well as *in vivo* anticancer trials have demonstrated artemisinin's well-known promising properties, and comparable research has revealed its method of action, which provides insight into its constitutional property. Artemisinin is an endoperoxide derivative that possesses anticancer properties. Artemisinin, like several other chemicals like hydrogen peroxide, interacts with ferrous iron to form free radical species. Anticancer activity is triggered by these free radicals. Further extensive research reveals that these anticancer actions become even more apparent when iron complexes are added to cell culture. Because artemisinin forms a covalent bond with transferrin (human iron transport protein), the artemisinin as well as transferrin conjugate are vigorously transported into most cancer cells by the transferrin receptor (TfR)-mediated endocytosis pathway, resulting in the cited anticancer endeavor *in vitro* cell cultures. This additionally explains the function of iron metabolism in improving artemisinin's anticancer potential. Additionally, artemisinin as well as its derivatives induce programmed cell mortality of cancer cells by activating the cytochrome C-mediated pathway leading to apoptosis (Ferreira et al. 2010). Additionally, its chemical and structural properties suggest that it might serve as a lead ingredient, serving as the cornerstone for medication expansion (Bhakuni et al. 2001). Research has also found that several other important molecules have anticancer properties, like artemisinin, scopoletin, and their derivatives (Tzeng et al. 2007; Kumar et al. 2003).

5.9.11 Antifeedant Properties

The research was conducted using a variety of parameters for evaluating antifeedant recreation for crude extracts of *A. annua*. Moreover, its precise antifeedant properties, ovicidal potential, and increased regulatory impact strongly suggest it as a precise antifeedant herb (Haghighian et al. 2008), as well as an anti-Helminthes and anti-insecticidal agent (Khosravi et al. 2011; Vici domini 2011). According to certain research, crude extracts of *A. annua* include artemisinin as well as its derivatives, which serve as insecticides in their natural state (Weathers et al. 2011).

5.10 Conclusions

Artemisia annua has multiple medicinal properties, due to which it is commercially used as a potent drug. *A. annua* has been comprehensively evaluated for its therapeutical properties against numerous disorders. These are attributed to the presence of a wide spectrum of biological compounds such as certain phenolic acids, coumarins, flavonoids, and sesquiterpene lactones. The plant is widely known for its content of artemisinin, a compound that has antiprotozoal activity. Apart from the antimalarial effects, it has a range of different therapeutical properties, such as anti-tumor, anti-inflammatory, antibacterial, cytokinin-like and angiotensin-changing enzyme inhibitory activities. The current review is centered on *Artemisia annua*, with a focus on the investigation of its anticancer and antiviral outcomes, particularly for HIV/AIDS. The studies conducted in India have been equally focused on fundamental and applied aspects of the plant. It represents a practical illustration of "lab-to-land" and "farm-to-pharma" systems.

References

Aftab T, Ferreira JF, Khan MM, Naeem M (eds) (2014) *Artemisia annua* – pharmacology and biotechnology. Springer, Berlin, Heidelberg

Aftab T, Naeem M, Khan MM (eds) (2017) *Artemisia annua*: prospects, applications and therapeutic uses. CRC Press

Bhakuni RS, Jain DC, Sharma RP, Kumar S (2001) Secondary metabolites of *Artemisia annua* and their biological activity. Curr Sci 80(1):35–48

Brown GD (2010) The biosynthesis of artemisinin (Qinghaosu) and the phytochemistry of *Artemisia annua* L. (Qinghao). Molecules 15:7603–7698

Cafferata LFR, Gatti WO, Mijailosky S (2010) Secondary gaseous metabolites analyses of wild *Artemisia annua* L. Mol Med Chem 21:48–52

Castilho PC, Gouveia SC, Rodrigues AI (2008) Quantification of artemisinin in *Artemisia annua* extracts by 1H-NMR. Phytochem Anal 9(4):329–334

Damtew Z, Tesfaye B, Bisrat D (2011) Leaf, essential oil and artemisinin yield of artemisia (*Artemisia annua* L.) as influenced by harvesting age and plant population density. World J Agri Sci 7(4):404–412

Das S (2012) *Artemisia annua* (Qinghao): a pharmacological review. Int J Pharmac Sci Res 3(12): 4573–4577

Donno AD, Grassi T, Idolo A, Guido M, Papadia P, Caccioppola A, Villanova L, Merendino A, Bagordo F, Fanizzi FP (2012) First-time comparison of the in vitro antimalarial activity of *Artemisia annua* herbal tea and artemisinin. Trans R Soc Trop Med Hyg 106(11):696–700

Efferth T, Herrmann F, Tahrani A, Wink M (2011) Cytotoxic activity of secondary metabolites derived from *Artemisia annua* L. towards cancer cells in comparison to its designated active constituent artemisinin. Phytomedicine 18(11):959–969

Ekiert H, Świątkowska J, Klin P, Rzepiela A, Szopa A (2021) *Artemisia annua*–importance in traditional medicine and current state of knowledge on the chemistry, biological activity and possible applications. Planta Med 87(08):584–599

Ferreira JFS (2004) *Artemisia annua* L. the hope against malaria and cancer. In: Proceedings of medicinal and aromatic plants: production, business and applications. Mountain State University, Beckley, pp 15–17

Ferreira JFS (2007) Nutrient deficiency in the production of artemisinin, dihydroartemisinic acid, and artemisinic acid in *Artemisia annua* L. J Agric Food Chem 55(5):1686–1694
Ferreira JFS, Luthria DL, Sasaki T, Heyerick A (2010) Flavonoids from *Artemisia annua* L. as antioxidants and their potential synergism with artemisinin against malaria and cancer. Molecules 15:3135–3170
Haghighian F, Sendi JJ, Aliakbar A, Javaherdashti M (2008) The growth regulatory, deterrency and ovicidal activity of worm wood (*Artemisia annua* L.) on Tribolium confusum duv. And identification of its chemical constituents by GC-MS. Pestycydy 1(2):51–59
Huang L, Xie C, Duan B, Chen S (2010) Mapping the potential distribution of high artemisininyielding *Artemisia annua* L. (Qinghao) in China with a geographic information system. Chin Med 5:18
Juteau F, Masotti V, Bessie're JM, Dherbomez M, Viano J (2002) Antibacterial and antioxidant activities of *Artemisia annua* essential oil. Fitoterapia 73:532–535
Khosravi R, Sendi JJ, Ghadamyari M, Yezdani E (2011) Effect of sweet wormwood *Artemisia annua* crude leaf extracts on some biological and physiological characteristics of the lesser mulberry pyralid. Glyphodes Pyloalis J Insect Sci 11:156
Kim WS, Choi WJ, Lee S, Kim WJ, Lee DC, Sohn UD, Shin HS, Kim W (2015) Anti-inflammatory antioxidant and antimicrobial effects of artemisinin extracts from *Artemisia annua* L. Korean J Physiol Pharmacol 19(1):21–27
Kumar S, Gupta SK, Singh P, Bajpai P, Gupta MM, Singh D, Gupta AK, Ram G, Shasany AK, Sharma S (2003) High yields of artemisinin by multi-harvest of *Artemisia annua* crops. Ind Crop Prod 19:77–90
Kumar S, Suresh R, Verma DK, Dangesh A, Tomar VKS (2015) A study of *Artemisia annua* in Uttar Pradesh. Curr Sci 109:1237–1239
Lestari EG, Syukur M, Purnamaningsih R, Yunita R, Firdaus R (2011) Evaluation and selection of mutative artemisia (*Artemisia annua* L.) according to the altitude variants. HAYATI. J Biosci 18(1):16–20
Liu S, Tian N, Li J, Huang J, Liu Z (2009) Simple and rapid micro-scale quantification of artemisinin in living *Artemisia annua* L. by improved gas chromatography with electroncapture detection. Biomed Chromatogr 23:1101–1107
Misra H, Mehta D, Mehta BK, Jain DC (2013) Microwave-assisted extraction studies of target analyte artemisinin from dried leaves of *Artemisia annua* L. Org Chem Int 6
Ogwang PE, Ogwal JO, Kasasa S, Olila D, Ejobi F, Kabasa D, Obua C (2012) *Artemisia annua* L. infusion consumed once a week reduces risk of multiple episodes of malaria: a randomised trial in a ugandan community. Trop J Pharmac Res 11(3):445–453
Septembre-Malaterre A, Lalarizo Rakoto M, Marodon C, Bedoui Y, Nakab J, Simon E, Hoarau L, Savriama S, Strasberg D, Guiraud P, Selambarom J (2020) *Artemisia annua*, a traditional plant brought to light. Int J Mol Sci 21(14):4986
Shukla AK, Shasany AK, Khanuja SP (2018) Research and Development on *Artemisia annua* in India. New Age Herbals Springer, Singapore, pp 15–27
Tayebe S, Mehrnaz K, Khosro P, Tahere H (2012) Morphological evaluation of hairy roots induced in *Artemisia annua* L. and investigating elicitation effects on the hairy roots' biomass production. Int J Agric: Res Rev 2:1005–1013
Tellez MR, Canel C, Rimando AM, Duke SO (1999) Differential accumulation of isoprenoids in glanded and glandless *Artemisia annua* L. Phytochemistry 52(6):1035–1040
Tzeng TC, Lin YL, Jong TT, Chang CMJ (2007) Ethanol modified supercritical fluids extraction of scopoletin and artemisinin from *Artemisia annua* L. Sep Purif Technol 56(1):18–24
Varkey MJ (2015) Nobel prize for artemisinin research: Indian side of the story. Curr Sci 109(12): 2172
Verdian-rizi MR, Sadat-Ebrahimi E, Hadjiakhoondi A, Fazeli MR, Hamedani PM (2008) Chemical composition and antimicrobial activity of *Artemisia annua* L. essential oil from Iran. J Med Plants 7(4):58–62

Wang B, Sui J, Yu Z, Zhu L (2011) Screening the hemostatic active fraction of *Artemisia annua* L. *in-vitro*. Iranian J Pharm Res 10(1):57–62

Weathers PJ, Towler MJ (2012) The flavonoids casticin and artemetin are poorly extracted and are unstable in an *Artemisia annua* tea infusion. Planta Med 78(10):1024–1026

Weathers PJ, Arsenault PR, Covello PS, McMickle A, Teoh KH, Reed DW (2011) Artemisinin production in *Artemisia annua*: studies in planta and results of a novel delivery method for treating malaria and other neglected diseases. Phytochem Rev 10(2):173–183

Weathers PJ, Towler M, Hassanali A, Lutgen P, Engeu PO (2014) Dried-leaf *Artemisia annua*: a practical malaria therapeutic for developing countries? World J Pharmacol 4:39–55

WHO (2005) Monograph for cultivation of herbal antimalarial: *Artemisia annua*. WHO Drug Inf 19(3):215

World Health Organization (2006) WHO monograph on good agricultural and collection practices (GACP) for *Artemisia annua* L. Geneva: World Health Organization 300. https://apps.who.int/iris/bitstream/handle/10665/43509/9241594438_eng.pdf;jsessionid=A927467AFDCFEA879E90A88074766735?sequence=1. Accessed 25 Sep 2022

Willcox M (2009) Artemisia species: from traditional medicines to modern antimalarials and back again. J Altern Complement Med 15(2):101–109

Willcox M, Bodeker G, Bourdy G, Dhingra V, Falquet J, Ferreira JFS, Graz B, Hirt HM, Hsu E, Melillo de Magalhães P, Provendier D, Wright CW (2004) *Artemisia annua* as a traditional herbal antimalarial. In: Wilcox ML, Bodeker G, Rasoanaivo P (eds) Traditional medicinal plants and malaria, vol 4. CRC Press, Boca Raton, pp 43–59

Yang GE, Bao L, Zhang XQ, Wang Y, Li Q, Zhang WK, Ye WC (2009) Studies on flavonoids and their antioxidant activities of *Artemisia annua*. J Chin Med Mater 32(11):1683–1686

Zanjani KE, Rad AS, Bitarafan Z, Aghdam AM, Taherkhani T, Khalili P (2012) Physiological response of sweet wormwood to salt stress under salicylic acid application and non-application conditions. Life Sci J 9(4):1097–8135

zu Biesen CM (2010) The rise to prominence of *Artemisia annua* L. the transformation of a Chinese plant to a global pharmaceutical. Afr Sociol Rev 14(2):24–46

Chapter 6
Mesua ferrea L.: Ethnobotany, Phytochemistry and Pharmacology

Mushfa Khatoon and Amita Dubey

Mesua ferrea L. image is adapted from Chahar et al. (2012) – (i) Tree, (ii) Fruit, (iii) Leaves, (iv) Flower and (v) Seeds

Abstract The plant kingdom has plenty of plants with herbal activities. Amongst them *Mesua ferrea*, also known as "Nagakesar", is a species with several medicinal properties. *M. ferrea* is a rare plant species, typically found in the tropical region.

M. Khatoon · A. Dubey (✉)
Plant Molecular Biology Laboratory, Department of Biosciences, Integral University, Lucknow, Uttar Pradesh, India
e-mail: amita@iul.ac.in

Á. Máthé, I. A. Khan (eds.), *Medicinal and Aromatic Plants of India Vol. 2*,
Medicinal and Aromatic Plants of the World 9,
https://doi.org/10.1007/978-3-031-44914-7_6

Possesses various medicinal properties such as anti-inflammatory, anti-arthritic, analgesic, anti-diabetic, anti-cancer, cardioprotective etc. Due to its pharmacological potential, *Mesua ferrea* is traditionally used by tribal people. This chapter summarizes knowledge about its traditional use, phytoconstituents and pharmacological properties.

Keywords *Mesua ferrea* · Pharmacology · Anticancer activity · Anti-inflammatory activity · Traditional uses · Medicinal plant

6.1 Introduction

Mesua ferrea Linn, also popularly called "Nagakesara" in hindi, is from the Calophyllaceae family. The English name for *Mesua* is "Ceylon iron wood" (Sharma et al. 2017). Although *Mesua* is a large genus with approximately 48 species, only *M. ferrea* has been studied by most of the researchers (Chahar et al. 2012). It is a medium to large sized evergreen tree with a small short trunk. *Mesua* trees are ornamental in nature and are produced for their very attractive flowers. The leaves and flowers of the plant possess high medicinal properties and act as an anti-venom agent for fatal snake bites (Kritikar and Basu 1956). The oil extracted from seeds of *Mesua* is used for curing itches. The most important medicinal use of *Mesua* flowers is in treating the burning of feet, as well as in curing bleeding piles by applying a paste of its flowers mixed with sugar and butter (Sharma et al. 2017). The plant is also usually utilized for its antimicrobial, antiprotozoal and antibacterial properties (Mazumder et al. 2003; Verotta et al. 2004).

Mesua ferrea has been an important component in Ayurvedic medicines such as Naga-keshara yaga and Nagakeshara-adi-churna for treating piles and bacillary dysentery, respectively. It is also used as an essential ingredient in Unani drugs, Jawarish Shehryaran (a liver tonic) and an appetizer, Hab Pachuluna (Sharma et al. 2017). It also cures seizures and acts as an anti-convulsant agent (Smith and Bleck 1991). *M. ferrea* flowers usually possess germacrene-D, as well as α-copaene. Some of its important biologically active components include flavanoids, glycosides, xanthones, resins and triglycerides. *Mesua ferrea* also contains some steroids, essential oils, fatty acids, proteins, tannins and saponins, as well (Choudhury et al. 1998).

6.2 Taxonomic Characteristics

Mesua ferrea commonly known as "Nagkesar" is a plant of Calophyllaceae family. Its English name is "Ceylon wood" (Sharma et al. 2017). The wood of *Mesua ferrea* is very hard. It is also known as ironwood that is another morphologically different species in the same genus. *Mesua ferrea* and the closely related genus *Kayea* make up the taxa that form the currently proposed genus *Mesua*. According to the number of seeds present in the fruit, combined *Kayea* under genus *Mesua*. Hence, the taxa which were prior known as *Kayea* were renamed as *Mesua*. Several scholars strongly supported this taxonomic classification of *Kayea/Mesua* (Zakaria et al. 2007).

6.3 Crude Drug Used

Crude drug used: Bark, root, flowers, leaves, fruits, stamens and seeds. Its flowers and leaves are used as anti-venom for scorpion sting and bite of snake. The essential oil is utilized for treating sores, skin infection, wounds, rheumatism and scabies and flowers of this plant are also used as an astringent, stomachic and expectorant. The decoction of roots and bark are used to treat snake bites (Santamaría 1978) and acts as a bitter tonic; used in treating bronchitis and gastritis (Sahni 1998; Husain et al. 1992; Joy et al. 1998; Nadkarni 1976).

On the basis of literature, the major traditional uses of different organs of *Mesua ferrea* have been compiled in Table 6.1.

6.4 Major Chemical Constituents and Bioactive Compounds

Phytochemical studies have found *Mesua ferrea* to contain secondary metabolites of various classes including xanthones, phenyl coumarins and triterpenoides (Chow and Quon 1968; Bandaranayake et al. 1975; Raju et al. 1976). Dichloromethane stem extract consists of Friedelin (composed of α- amyrin and β-amyrin), β-sitosterol and lupeol (Keawsa-ard et al. 2015). The heartwood of *Mesua ferrea* contains

Table 6.1 Traditional uses of *Mesua ferrea*

S. no.	Plant part	Traditional uses	References
1.	Leaves	Cure cold, joint pain and burning sensation in feet and hands	Sharkar et al. (2013)
2.	Kernal	Skin outbreak condition	Sharma et al. (2017)
3.	Root	Used in treating bronchitis and gastritis	Sahni (1998)
4	Flower	Anti-venom for scorpion stings and bite of snake	Santamaría (1978)
5.	Essential oil	Skin infection, wounds, scabies and rheumatism	Sahni (1998)
6.	Whole plant	Used as an antipyretic, cardiotonic and carminative agent, also curing cold, fever and asthma	Sharma et al. (2017)
7.	Bark	Cure dysentery, sore throat, cough and vomiting	Keawsa-ard and Kongtaweelert (2012)

ferrxanthone, which can be chemically characterized as 1, 3- dimethoxy-5,6-dihydroxyxanthone (Walia and Mukerjee 1984). 1,5-dihydroxyxanthone (II), β-sitosterol, mesuaxanthone-A, euxanthone 7-methyl ether (IV) and mesuaxanthone-B were isolated from heartwood by several researchers (Govindchari et al. 1967; Chow and Quon 1968). Govindachari et al. later isolated an alkylcoumarin, Ferrol-A from the *M. ferrea* trunk bark (Govindchari et al. 1967). Gunasekera and his colleagues isolated several different types of xanthones such as 4-hydroxy-, 1- hydroxy-5-methoxy-, 3-hydroxy-4- methoxy, 1-hydroxy-7-methoxy-, 2-Hydroxy-, 2-methoxy-, 1,5-dihydroxy-, and 1,5,6-trihydroxyxanthone (Gunasekera et al. 1975). The main oil extracted from mature and young leaves consists of β caryophyllene (26.0% and 18.8%) and α-copaene (9.9% and 19.3%) respectively, while the oil extracted from bark consists of α-selinene (12.2%) and (E)-α-bisabolene (31.3%). The oil extracted from the flowers and bud of the plant mainly contains germacrene D (16.1% and 19.0%) and α-copaene (20.2% and 28.7%) (Choudhury et al. 1998). A study revealed that the root bark of *Mesua* consists of macluraxanthone, β-sitosterol, betulinic acid, caloxanthone C and friedelin, a new xanthone was isolated from root bark, mesuaferrin C (Ee et al. 2012). Another group studied root bark and isolated Mesuaferrin-A and -B, 1,8-dihydro-3-methoxy-6- methyl anthraquinone, caloxanthone C, friedelin, betulinic acid and β-sitosterol (Teh et al. 2012). A combination of amyrins were isolated from stem and stem bark of *Mesua* along with calophyllin-B, euxanthone, ferruol A, friedelin, mesuaxanthone-A, β-sitosterol, dehydrocycloguanandin, euxanthone 7-methyl ether (IV), ferrxanthone, lupeol, mesuaxanthone-B, stigmasterol, 6-desoxy jacareubin, 1,5-dihydroxyxanthone (II) and jacareubin (Gunasekera et al. 1975; Keawsa-ard et al. 2015; Lim 2012). Rajesh et al. extracted *M. ferrea* in chloroform and methanol and its HPLC analysis revealed that it contains several types of antioxidants such as ellagic acid, kaempferol, rutin, vanillic acid, coumaric acid, gallic acid, myricetin and quercetin (Rajesh et al. 2013). Alakh et al. isolated two essential oil components from the bark oil i. e. α-selinene and (E)-α-bisabolene (Alakh et al. 2014). The primary phytochemical studies of leaves extract of *M. ferrea* revealed that it consists of total tannin of about 11.25 mg/g of dry weight extract, total flavonoid content of about 30 mg/g of dry weight extract, total phenolic content of about 14.72 mg/g of dry weight extract and total flavanol content of about 3.60 mg/g of dry weight extract respectively (Sahu Alakh et al. 2013). In the stem bark of *M. nagassarium*, the presence of 3β friedelanol, 3-oxo-betulin, spinasterol, friedelin and lupeol was reported (Islam et al. 2014).

6.5 Morphological Description

Mesua ferrea is an evergreen tree with a height ranging between 20 and 30 m usually pinkish or creamy white in colour. Its base is grooved having very hard, bitter, dark red coloured heart wood. The bark surface of the plant is smooth (Orwa et al. 2009). *Mesua sp.* possesses entire and simple and oppositely located leaves. The flowers are dioecious- polygamous, attractive, red, white or yellow in colour. *M. ferrea* flowers consist decussated or imbricated two to six sepals. It also consist two to six imbricated petals with several stamens which have golden colour. Because of the presence of various secondary veins reaching to the margin of the leaves, they appear to be glossy. The fruit of the plant is conically pointed consist of globose to ovoid shape. They are striate having one to four seeds and one to ten locules (Sharma et al. 2017). In dry season, *Mesua* flowers and produces leaves after it, in the monsoon. The flowering time of bisexual flowers is between 3 a.m. and 4 a.m. with closure around the sunset (Orwa et al. 2009). The mature leaves are blue grey to dark green in colour, whereas the young leaves appear reddish yellow in colour with a length of 7–15 cm. The length of the fruit is about 2.5–5 cm (Dassanayake 1980).

6.6 Geographic Distribution

The distribution of *M. ferrea* is vast; it expands to the southern part of Konkon, eastern Himalaya's dense mountains, vast forests of Western Ghats to Travancore via southern Kanara, Andamans, Tenasserim Burma, Bengal and Assam (Kritikar and Basu 1981). *M. ferrea* is most profusely found in the Himalayas to an altitude of about 1500 m, in North India, from eastward Nepal, the Andaman Island and Deccan peninsula (Sharma et al. 2017).

6.7 Ecological Requirements

Mesua ferrea tolerates shade, thrives in dense or moist fertile soils, and has important mycorrhizal relationships for nitrogen fixation (Mitra et al. 2021). It grows on fertile loamy soil as well as well drained soils. The usual pH values for the cultivation of *Mesua sp.* range between 5 and 5.5, it has been reported to tolerate pH-values between 4.3 and 6.9 (Fern 2022).

6.8 Traditional Use Part(s) Used and Common Knowledge

The edible parts of *Mesua ferrea* are fruits, seeds, leaves and flowers. Local people in India use *M. ferrea* for boosting immunity. In Thailand, people eat flowers which treat numerous diseases. Although seeds are edible after cooking, they do not taste good. When the ripe fruit is eaten, they taste like chestnut. The leaves are eaten raw despite their astringent-sour taste (Lim 2012).

Various species of *Mesua* genus have been traditionally utilized by the people of Asian countries for curing illnesses such as renal disease, cough, fever, nausea, asthma, dyspepsia and itchiness etc. *Mesua* species possesses several pharmacological properties like antimicrobial, antitumor, immunomodulatory, antioxidant and antiviral, which already have been proved (Teh et al. 2012; Asif et al. 2016). *M. ferrea* is traditionally utilized as an antimicrobial, carminative, diuretic, antipyretic, anticancer, cardiotonic and expectorant (Chahar et al. 2012; Rahman et al. 2008). Traditionally, the barks are used for curing dysentery, sore throat, cough and vomiting (Keawsa-ard and Kongtaweelert 2012). The local communities of Bangladesh use powder of dried leaves and fruits, mixing it with ghee for curing the cold, joint pain and burning sensation in feet and hands (Sharkar et al. 2013). *Mesua ferrea* is used in septic conditions and in cases of inflammation (Rai et al. 2000). The local people from tribals of Assam use Mesua for its purgative, worm-control, antiseptic tonic and blood purifier properties (Parukutty and Chandra 1984). *Mesua ferrea* is used as herbal medicine for treating several ailments such as cough, dysentery, headache, itching, scabies, small tumors, bleeding piles, cardiovascular diseases, dehydration, hiccup, sweating, skin diseases and vomiting (Roshy Joseph et al. 2010; Lim 2012). The ashes of leaves of *Mesua* are used for curing sore eye. The plant is utilized as an antipyretic, cardiotonic and carminative agent, also curing cold, fever and asthma. The skin outbreak condition is cured by *Mesua ferrea* kernels (Sharma et al. 2017). Skin infection, wounds, scabies and rheumatism is treated by using *M. ferrea* oil (Sahni 1998).

6.9 Modern Medicine Based on Its Traditional Medicine Uses

6.9.1 Analgesic Activity

A significant amount of analgesic activity was shown by ethyl acetate, n-hexane and methanol extract of leaves of *M. ferrea* (at 125 and 250 mg/kg) in writhing response induced by acetic acid in mouse. For higher dosage, the writhing response was reduced by 19.63%, 42.21% and 17.06% in ethyl acetate, n- Hexane and methanol extract respectively while in lower dosage it was 16.33%, 36.08% and 10.21% respectively (Hassan et al. 2006).

6.9.2 Anti-inflammatory and Anti-arthritis Activity

The xanthones from *M. ferrea* such as Mesuaxanthone A(MXA) and Mesuaxanthone B (MXB) were tested in albino rats through cotton pellet implantation, carrageenan induced hind paw oedema and granuloma pouch tests. These xanthones were applied at the dosage level of 50 mg/kg in the above methods. By the oral intake of *Mesua* xanthone i. e. MXA and MXB showed 37% and 49% reduction in comparison to control group in carrageenan induced hind paw oedema.

The xanthones, MXA and MXB showed substantial anti-inflammatory activity in normal rats and in adrenalectomized rats also. The reduction in inflammation in comparison to control group in MXA is 38% and MXB is 22%. In cotton pellets granuloma experiment, the reduction of 47% was reported in inflammation. The xanthones MXA and MXB reported 46% and 49% reduction in inflammation respectively in granuloma pouch tests. Thus, significant anti-inflammation potential was reported in xanthones of *Mesua* in this study (Gopalakrishnan et al. 1980). An ayurvedic medicine, Shirishavaleha which consists of *Mesua* along with several other herbs has showed inhibition of oedema development present in carrageenan-induced paw oedema model (Yadav et al. 2010).

Two separate *in vivo* models i.e. CFA (Complete Freund's Adjuvant) and FI (Formaldehyde induced) were examined in rats for the anti-arthritic potential of seed extracts of *Mesua ferrea*. Significant reduction in the arthritic lesions was shown in CFA injected paw by observing the swelling volume by the seed extracts of *Mesua* in treated animals. Similarly, in FI model; the swelling in formaldehyde injected paw was reduced in treated animals in comparison to the control by the *M. ferrea* seed extract.

The *M. Ferrea* treated rats showed significant weight gain when compared to control in the final stage of treatment where as in untreated rats, a loss of weight was detected (Jalalpure et al. 2001). Another group of researchers reported the anti-inflammatory potential in several *in vitro* bioassays of 80% ethanol extract of *Mesua ferrea* stem bark.

When compared to the standard drug available which is Indomethacin in concentration of 100 μg/mL, reported significant anti-inflammatory potential in the 80% ethanol extract at the conc. of 100, 200 and 500 μg/mL (Ranganathaiah et al. 2016).

6.9.3 *Diuretic Properties*

Kaliuretic, diuretic and natriuretic properties were induced in the albino rats by the application of combination of herbs i. e. Draksharishta-M and -T and its formulation available in the market which consists of *M. ferrea* stamens for a period of 5 h at the dosage of 2.0 ml/Kg in comparison to the control group (Tiwari and Patel 2011).

6.9.4 *Anti-hemorrhoidal Activity*

In initial clinical research which involved 22 participants, a polyhedral formulation containing *M. ferrea* was investigated for its efficiency in treating bleeding piles. The study reported that 16 patients showed improvement out of 22 patients, with negligible harmful effects with reduction in bleeding (Paranjpe et al. 2000). *M. ferrea* has been reported to be effective in the standard herbal formulations i. e. Roidosanal® and Daflon® by studying the improvement in Grade I and grade II patients of anorectal conditions. The pain and bleeding were reduced in the hemorrhoid patients by the above medications (Aggrawal et al. 2014).

6.9.5 CNS Depressant and Anticonvulsant Activities

The CNS depressive potential of *M. ferrea* xanthone which are mesuaxanthone-B, dehydrocycloguanandin, jacareubin, mesuaxanthone-A, calophyllin-B, euxanthone and 6-desoxy jacareubin were studied in both rat and mouse models. Usual CNS depressive effects, such as ptosis, muscular tone reduction, drowsiness, and decreased spontaneous muscular movement, were detected in xanthone-treated rats. Furthermore, rats treated with xanthone, the anaesthetic potential of phenobarbitone and ether-induced sleeping time was enhanced (Gopalakrishnan et al. 1980; Lim 2012).

The *M. ferrea* flowers extracted in ethanol was tested for its anticonvulsant property in albino mice using the Maximum electroshock seizure (MES) test at three distinct dosing levels which are 200, 400 and 600 mg/kg. In comparison to the MES model, the extract shortened the duration of hind limb tonic extension (HLTE) in a dose-dependent manner. *M. ferrea* ethanolic extract prevented MES-induced convulsions. The inhibition percentage of the extract at the dosage of 200, 400, and 600 mg/kg were 100% ($p < 0.01$), 60% ($p < 0.01$) and 100% ($p < 0.001$), respectively. According to the findings of this investigation, *M. ferrea* flowers significantly enhanced the start time and decreased the duration of seizures induced by electroconvulsive shock (Tiwari and Patel 2012).

6.9.6 Immunomodulatory and Hormone Balancing Activities

When irradiated animals were compared to drug-treated or normal animals, there was no substantial difference in their haemoglobin content. Also, no significant change was shown by ACII in the ratio of lymphocyte-neutrophil. The cellularity of bone marrow was greatly improved, as were the -esterase positive cells. The mass of the thymus increased in ACII-treated mice in comparison to irradiated animals (Tharakan et al. 2006). Furthermore, the immunomodulatory effect by ACII was reported in animals treated with cyclophosphamide (Tharakan et al. 2003) an in normal animals as well (Tharakan et al. 2004).

The activity of mesuol extracted from *M. ferrea* seed oil on the immune system was investigated utilising both humoral and cellular immunological models. The mesuol in humoral immunological models in rats substantially increased the antibody titer values. These antibodies were already tested and immunized by introducing SRBCs (sheep red blood cells) which is then immunosuppressed by cyclophosphamide. Moreover, mesuol stimulated T-cells and produced cellular immunological reaction in immunosuppressant rats induced by cyclophosphamide.

When rats treated with mesuol, were exposed to SRBCs which is an irritant, the thickness of their foot pads increased (Chahar et al. 2012). The *Mesua* ferrea flower extract were reported to show effects similar to progesterone and oestrogen. It was proposed for helping in the balancing of menstrual diseases (Lim 2012).

6.9.7 Antidiabetic Activity

In diabetic mice induced by streptozotocin, the M. ferrea leaves extracted in methanol showed good antidiabetic action. It has been proposed that extract can boost insulin release from pancreatic β -cells. In terms of increasing insulin secretion, the leaf extract lowered blood glucose levels and restored body weight in diabetic rats when compared to untreated animals. In vitro investigations employing a MIN6-β-cells (mouse insulinoma pancreatic β-cell line) revealed an increase in insulin levels depending upon dosage, as a result of treatment by methanol extract, with the effects being more pronounced in hyperglycemic settings in comparisons to normal cell culture settings (Balekari and Veeresham 2015).

6.9.8 Hepatoprotective Activity

In male Wistar rats inoculated with Staphylococcus aureus, the hepatoprotective properties of *M. ferrea* flowers extracted in methanol were tested in vivo. After 1 week of treatment with methanol extract at dosage of 50, 100 and 200 mg/kg demonstrated increment in the liver enzyme levels such SOD, GR, CAT and GPx with decrement in the level of enzymes such as AST and AAT At a dose of 100 mg/kg of methanol extract, significant effects were detected (Garg et al. 2009). In another experiment, the hepatoprotective properties of several stamen extracts were assessed utilising an in vitro carbon tetrachloride-induced oxidative stress liver slice culture model. Among the extracts tested, n-hexane and ethanol extracts of stamens preserved cultured liver slice cells from oxidative stress caused by carbon tetrachloride. The effective extracts also exhibited improved antioxidant properties in various in vitro free radical scavenging models, including DPPH, ABTS+, SOD, and NO (Rajopadhye and Upadhye 2012).

6.9.9 Cardioprotective Activities

In the albino rat model, “Ashwagandharishta” (a polyherbal combination) and its commercial preparation incorporating *M. ferrea* stamens were demonstrated to defend from isoproterenol-induced myocardial infarction. Herbal medicine treatment also substantially reduced changes induced by the isoproterenol in serum marker enzyme like aspartate aminotransferase, lactate dehydrogenase, alanine aminotransferase and creatine kinase resulting in the serum lipid profile improvement. Furthermore, pre-treatment with a herbal formulation in animals resulted in a decrease in malondialdehyde (MDA) levels and a considerable rise in glutathione (GSH) levels.

Thus, it was postulated that the cardioprotective property of herbal medicine in the treated rats may be attributed to an increment in in vivo antioxidant levels such as GSH and decrement in lipid peroxidation of cardiac membranes (Tiwari and Patel 2012).

6.9.10 *Anti-cancer Activities*

The extract of oleo-gum resin was demonstrated ROS-mediated apoptotic pathways to trigger apoptosis in HCT 116 cells. Unexpectedly, the extract of oleo-gum resin did not cause toxicity in CCD-18co (normal colon cells) (Asif et al. 2016). Furthermore, Asif et al. found that terpene-rich stem bark extract exhibits broad-spectrum anti-cancerous properties in their previous work. The sensitivity order towards F-3 of cancer cell lines from high to low was HCT 116 > MNK–74 > PC–3 > T-47D > MIA PaCa–2 > HT– 29 > PANC–1 > MCF–7 > Capan-1 > EA. hy926 > 3 T3-L1 > CCD- 18co (Asif et al. 2016). There is one study that indicates the in vivo efficacy of *M. ferrea* flowers extracted in ethyl acetate and chloroform against Ehrlich ascites cancer in Swiss albino mice, in addition to a number of in vitro anticancer investigations. The percentage decrement of cancer in rats treated with ethyl acetate and chloroform was 41.7% and 54.8%, respectively (Rana et al. 2004).

6.10 Conclusions

Mesua ferrea is an important medicinal plant belonging to the family Calophyllaceae. It contains various bioactive compounds that are responsible for its important anti-inflammatory, anti-arthritic, analgesic, anti-diabetic, anti-cancer, cardioprotective. Each plant organ contains several phytoconstituents. Tribal people in India have been using its crude drug, as *Mesua* is known for its numerous medicinal properties. Despite of it being used not only in India but also throughout the world, the validation of its pharmacological properties remains to be made. There is a need for more research to reveal its true medicinal potentials.

References

Aggrawal K et al (2014) Efficacy of a standardized herbal preparation (Roidosanal®) in the treatment of hemorrhoids: a randomized, controlled, open-label multicentre study. J Ayurveda Integr Med 5:117–124

Alakh NS, Hemalatha S, Sairam K (2014) Phyto-pharmacological review of *Mesua ferrea* Linn. Int J Phytopharmacol 5:6–14

Asif M et al (2016) Isoledene from *Mesua ferrea* oleo-gum resin induces apoptosis in HCT 116 cells through ROS-mediated modulation of multiple proteins in the apoptotic pathways: a mechanistic study. Tox Lett 257:84–96

Balekari U, Veeresham C (2015) Insulinotropic activity of Methanolic extract of *Mesua ferrea* Linn. J Basic Appl Sci 11:410–417

Bandaranayake WM et al (1975) Xanthones and 4-phenylcoumarins of *Mesua thwaitesii*. Phytochemistry 14:265

Chahar M et al (2012) In-vivo antioxidant and immunomodulatory activity of mesuol isolated from *Mesua ferrea* L. seed oil. Int Immunopharmacol 13:386–391

Choudhury S, Ahmed R, Barthel A, Leclercq PA (1998) Volatile oils of *Mesua ferrea*(L.). J Essent Oil Res 10(5):497–501. https://doi.org/10.1080/10412905.1998.9700955

Chow YL, Quon HH (1968) Chemical constituents of the heartwood of *Mesua ferrea*. Phytochemistry 7:1871

Dassanayake MD (1980) Revised handbook of the Flora of Ceylon, vol I. Oxonian Press Pvt. Ltd, Faridabad, p 105

Ee GCL, Teh SS, Rahmani M, Taufiq-Yap YH, Go R, Mah SH (2012) A new furanoxanthone from the root bark of *Mesua ferrea*. Lett Org Chem 9:457–459

Fern K Tropical Plants Database. (2022) Tropical.theferns.info. 2022-10-21

Garg S, Kameshwar S, Rajeev R, Pankaj A, Parshuram M (2009) *In vivo* antioxidant activity and hepatoprotective effects of methanolic extracts of *Mesua ferrea* L. Int J Pharmatechnol Res 1: 1692–1696

Gopalakrishnan C et al (1980) Anti-inflammatory and C.N.S. Depressant activities of xanthones from *Calophyllum inophyllum* and *Mesua ferrea*. Indian J Pharmacol 12:181–191

Govindchari TR, Pai BR, Suramaniam PS, Ramdas Rao U, Muthukumarswamy N (1967) Constituents of *Mesua ferrea* L.-I Mesuaxanthone a and Mesuaxanthone B. Tetrahedron 23:243–248

Gunasekera SP, Ramachandran S, Selliah S, Sultanbawa MUS (1975) Chemical investigation of ceylonese plants. Part XVII. Isolation and structures of the xanthones in the extractives of *Mesua ferrea* L. (form M. Salicina Pl. and Tr.) (Guttiferae). J Chem Soc Perkin Trans 1:2447–2450

Hassan MT et al (2006) Analgesic activity of *Mesua ferrea* Linn. Dhaka Univ J Pharm Sci 5:73–75

Husain A, Virmani OP, Popli SP, Misra LN, Gupta MM, Srivastava GN, Abraham Z, Singh AK (1992) Dictionary of Indian medicinal plants. CIMAP, Lucknow, p 546

Islam R, Ahmed I, Sikder AA, Haque MR, Al-Mansur A, Ahmed M, Rasheed M, Rashid MA (2014) Chemical investigation of *Mesua nagassarium* (Burm. f.) Kosterm. J Basic Appl Sci 10:124–128

Jalalpure SS et al (2001) Antiarthritic activity of various extracts of *Mesua ferrea* Linn. seed. J Ethnopharmacol 138:700–704

Joy PP, Thomas J, Mathew S, Skaria BP (1998) Kerala agricultural university. Arom Med Plant Res:106–107

Keawsa-ard S, Kongtaweelert S (2012) Antioxidant, antibacterial, anticancer activities and chemical constituents of the essential oil from *Mesua ferrea* leaves. Chiang Mai J Sci 39:455–463

Keawsa-ard S, Liawruangrath B, Kongtaweelert S (2015) Bioactive compounds from *Mesua ferrea* stems. Chiang Mai J Sci 42:185–195

Kritikar KR, Basu BD (1956) Indian medicinal plants, vol 171, 2nd edn, Allahabad, pp 267–270

Kritikar KR, Basu BD (1981) Indian medicinal plants, vol 1, 2nd edn. International book distributors, Dehradun. p 274

Lim TK (2012) Edible medicinal and non-medicinal plants. Springer, New York

Mazumder R et al (2003) Emergence of *Mesua ferrea* Linn. Leaf extract as a potent bactericide. Ancient Sci Life 22:160–165

Mitra S, Ghose A, Gujre N, Senthilkumar S, Borah P, Paul A, Rangan L (2021) A review on environmental and socioeconomic perspectives of three promising biofuel plants *Jatropha curcas*, *Pongamia pinnata* and *Mesua ferrea*. Biomass Bioenergy 151:106173

Nadkarni KM (1976) Indian Materia Medica. Sangam Books Ltd., London, p 1319

Orwa C et al (2009) Agroforestree: a tree reference and selection guide version 4.0; http://www.worldagroforestry.org/sites/treedatabase.asp

Paranjpe P et al (2000) Efficacy of an indigenous formulation in patients with bleeding piles: a preliminary clinical study. Fitoterapia 71:41–45

Parukutty B, Chandra GS (1984) Studies on the medicinal uses of plants by the Boro tribals of Assam-II. J Econ Taxon Bot 5:599–604

Rahman SMM, Shabnom S, Quader MA, Hossain MA (2008) Phytochemical study on the ethylacetate extract of the leaves of *Mesua ferrea* Linn. Indo J Chem 8:242–244

Rai LK, Pankaj P, Sharma E (2000) Conservation threats to some important medicinal plants of the Sikkim Himalaya. Biol Conserv 93:27–33

Rajesh KP, Manjunatha H, Krishna V, Kumara Swamy BE (2013) Potential *in vitro* antioxidant and protective effects of *Mesua ferrea* Linn. bark extracts on induced oxidative damage. Ind Crop Prod 47:186–198

Rajopadhye AA, Upadhye AS (2012) Hepatoprotective effect of stamen extracts of *Mesua ferrea* L. against oxidative stress induced by CCl4 in liver slice culture model. Nat Prod Sci 18:76–82

Raju MS et al (1976) Structure of Mesuaferrone-B a new biflavanone from the stamens of *Mesua ferrea* Linn. Tetrahedron Lett 49:4509

Rana AYKMM, Khanam JA, Asad-Ud-Daula M (2004) Antineoplastic screening of some medicinal plants against Ehrlich ascites carcinoma in mice. Int J Med Sci 4:142–145

Ranganathaiah P et al (2016) Evaluation of *in vitro* anti-inflammatory activity of stem bark extracts of *Mesua ferrea* Linn. Int J Pharm Pharm Sci 8:173–177

Roshy Joseph C, Iianchezhian R, Biswajyoti P, Harish CR (2010) Pharmacognostical study of nagakeshara (*Mesua ferrea* Linn)-an ingredient in Vyaghrihareetaki Avaleha. Int J Res Ayur Pharm 1:264–272

Sahni KC (1998) The book of Indian trees. Bombay Natural History Society, Mumbai

Sahu Alakh N, Hemalatha S, Sairam K (2013) Quantitative phytochemical and heavy metal estimation of *Mesua ferrea* flowers and *Argyreia speciosa* leaves. Int J Pharm Sci Rev Res 22:276–278

Santamaría FJ (1978) Diccionario de Mejicanismos, 3rd edn. Mejico, Editorial Porrua

Sharkar P, Rahman MM, Haque Masum GZ, Nayeem MA, Hossen MM, Azad AK (2013) Ethnomedicinal importance of the plants in villages in kushtia sador and mirpur upozila, Bangladesh. Int J Geogr Inf Syst 19(4):401–417

Sharma A et al (2017) *Mesua ferrea* Linn.: a review of the Indian medical herb. Syst Rev Pharm 8(1):19

Smith MC, Bleck TP (1991) Convulsive disorders: toxicity of anticonvulsants. Clin Neuropharmacol 14:97–115

Teh SS et al (2012) *Mesua beccariana* (Clusiaceae), a source of potential anti-cancer Lead compounds in drug discovery. Molecules 17:10791–10800

Tharakan TS, Kesavan M, Kuttan G, Kuttan R (2003) Immunomodulatory and toxicity study of NCV I AND ACII drugs useful against human immunodeficiency virus. Amala Res Bull 2:64

Tharakan TS, Kuttan G, Kesavan M Sr, Austin RK, Kuttan R (2004) Effect of NCV I and ACII in cyclophosphamide-induced immunosuppresion in BALB/c mice an implication in HIV infection. Amala Res. Bull. 24:133

Tharakan ST, Girija K, Ramadasan K (2006) Effect of ACII, an herbal formulation on radiation-induced immunosuppresion in mice. Indian J Exp Biol 44:719–725

Tiwari P, Patel RK (2011) Evaluation of diuretic potential of draksharishta prepared by traditional and modern methods in experimental rats. Pharmacologyonline 3:566–572

Tiwari P, Patel RK (2012) Cardioprotective activity of Ashwagandharishta on isoproterenol induced myocardial infarction. Pharmacologyonline 1:17–24

Verotta L et al (2004) 4-alkyl- and 4- phenylcoumarins from *Mesua ferrea* as promising multidrug resistant antibacterials. Phytochemistry 65:2867–2879

Walia S, Mukerjee SK (1984) Ferrxanthone, a 1,3,5,6-tetraoxygenated xanthone from *Mesua ferrea*. Phytochemistry 23:1816–1817

Yadav SS et al (2010) Anti-inflammatory activity of Shirishavaleha: an Ayurvedic compound formulation. Int J Ayurveda Res 1:205–207

Zakaria R, Choong CY, Faridah-Hanum I (2007) Systematic study on Guttiferae Juss of peninsular Malaysia based on plastid sequences. Tropics 16(2):141–150

Chapter 7
Linum usitatissimum L.: Rich Storehouse of Pharmacologically Active Metabolites in Indian Traditional Medicine

Haram Sarfraz and Iffat Zareen Ahmad

Abstract Indigenous traditional medicines in India use a lot of plant ingredients and their derivatives. *Linum usitatissimum* L. commonly known as 'Alsi' or 'Tisi', in India or 'Flax' in English, is one such pharmacologically rich plants. Phytochemicals (e.g.: phenolic acids, lignans, and flavonoids) are found naturally in *Linum usitatissimum.* While scientific data support flaxseed eating, a sizable portion of the public remains uninformed of the advantages and potential uses. The present chapter summarizes the available information on the traditional uses of *Linum usitatissimum* L. and its pharmacologically active natural constituents.

Keywords *Linum usitatissimum* · Flaxseed · Metabolites · Phenolics · Lignan · Health

7.1 Introduction

Since ancient times, medicinal plants have been employed as a traditional remedy in all societies. Initially, humans relied on plants to supply their dietary needs. Natural flora became an extremely valuable resource for health betterment and the treatment of numerous diseases all over several human populations, and a range of plant species were accessible that are still used in several parts of the globe, including South America, Asia, and Africa, as remedies for a variety of diseases. Even though the World Health Organization estimated that traditional medicines are used by 60% of the world's population, a wide range of plants with potential biological activity remains undiscovered (Li and Vederas 2009). The effectiveness of folk medicines is receiving increasing interest, due to their superior interaction with the human body, as well as fewer adverse effects.

H. Sarfraz · I. Z. Ahmad (✉)
Natural Products Laboratory, Department of Bioengineering, Integral University, Lucknow, Uttar Pradesh, India
e-mail: iffat@iul.ac.in

Á. Máthé, I. A. Khan (eds.), *Medicinal and Aromatic Plants of India Vol. 2*,
Medicinal and Aromatic Plants of the World 9,
https://doi.org/10.1007/978-3-031-44914-7_7

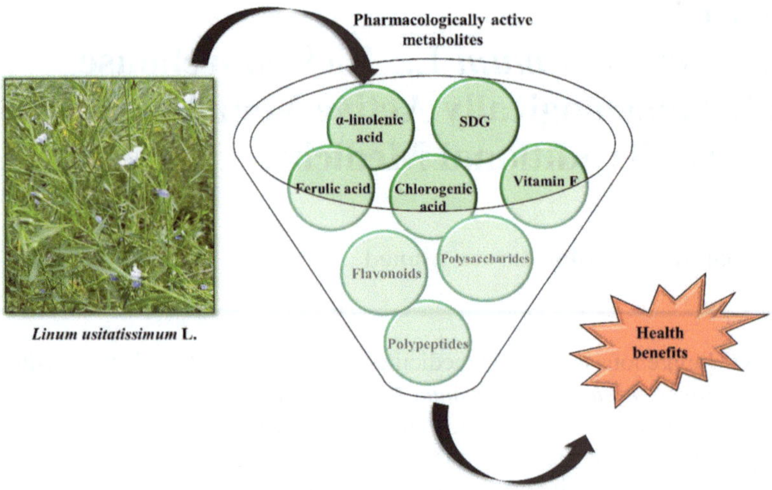

Fig. 7.1 Pharmacologically active metabolites with potential health benefits of *Linum usitatissimum* L. (Image: Sarfraz H)

Linum usitatissimum L. is an annual plant, generally referred to as flaxseed. Due to its possible health advantages, flaxseed has garnered significant attention in the area of food and disease studies (Fig. 7.1). Today, it is cultivated on approximately 2.6 million hectares. Main producers are India, China, the United States, and Ethiopia. In India, it is grown primarily in Uttar Pradesh, Madhya Pradesh, Maharashtra, Rajasthan, and Bihar (Ganorkar and Jain 2013).

7.2 Taxonomic Characteristics

Linum usitatissimum L. belongs to the class Magnoliopsida (dicotyledons), order Linales and family Linaceae. This genus has approximately 230 species in the family Linaceae and is referred by a variety of vernacular names, including common Flax, Flax, Linseed, Flix, Lint Bells (English); Alsi, Tisi (Hindi);Alsi, Katan (Urdu); Bazrul Katan, Bazen, Buzruk, Katan (Arabic);Tukhm-e-Katan, Bazarug, Zaghir, Kuman, Zaghu (Persian); Lifertus, Lisfermoon, (Unani); Atasi, Chanaka, Devi, Haimwati, Atima, Kshauma, Kshaumi, Kshuma, Madotkata, Madagandha, Malina, Masrina, Masruna, Masina, Nilapushpi, (Sanskrit); Alish, Kenu (Kashmir); Alsi, Tisi (Uttar Pradesh); Chikna, Tisi (Behar); Alish, Alsi, Tisi (Punjabi); Alshi, Arasi (Gujarati); Agasebeeja, Agasi, Kain Atish, Semeegara, Agashi (Kannada); Masina, Tisi, Alasi (Bengali); Tisi, Tusi, (Assamese); Agastha, Cheruchana, Vittintevilta (Malayalam); Alashi, Javasa, Javas(Marathi); Bazrag (Sherazi); Alshi, (Tamil); Atushi, Peso (Oriya); Bari Aala (Suryani); Atasi, Madanginjalu, Ullusulu (Telugu) (Khan et al. 2017).

7.3 Crude Drug Used

Flaxseed is mostly eaten as the whole seed. Various flaxseed products, like partially defatted flaxseed oil, completely defatted flaxseed meal, flaxseed mucilage extract, flaxseed hulls, and flaxseed alcohol extracts, are used in traditional medicine systems to treat several illnesses. Additionally, whole flaxseed mill and different solvent-based extracts provide raw material for various formulations, dietary supplements, and decoctions with therapeutic and prophylactic purposes (Kezimana et al. 2018).

7.4 Morphological Description

Linum usitatissimum L., one of the earliest domesticated crops, is now extensively planted for soil, fibre, and food production. Flowers are hermaphroditic and hypogenous, with five petals and filaments, and a complex pistil made of five carpels divided by a false septum. The fruits are a capsule with five carpels and up to ten seeds. The mature flaxseed is rectangular and flattened, measuring around 46 mm in length and 22.5 mm in width, and consists of an embryo having 2 cotyledons covered by a delicate and smooth endosperm, typically lustrous yellow to dark brown seed coat with a mucilaginous and oily flavour. When crushing and moistening seeds, a strong odour and flavour may arise (Evans 2009). Seeds are sown in late March and bloom, in June. Stems are solitary or sparsely corymbose branched, and branches rise to the tip.

7.5 Geographic Distribution

There is widespread agreement that this plant species has its origins in the eastern parts of the Mediterranean Sea, near India (Zeven and Zhukovsky 1975), from where it was carried over to Europe as well as Asia, before reaching the Americas. *Linum usitatissimum* was farmed originally in the Fertile Crescent area, which encompasses the modern-day Egypt, Israel, Iraq, Palestine, Lebanon, Iran, Syria, Jordan, Turkey, and Cyprus (Fu 2011). About 5000 years ago, the use of the crop spread across Europe, ultimately hitting Germany and Switzerland (Barber 1991), and was domesticated in India and China (Cullis 2007). The Phoenicians are said to have brought the crop to Western Europe by their wide commercial link that extended all the way to the British Isles. Charlemagne's (AD 742–814) and later Napoleon's policies favoured the growth of a flax industry in Western Europe.

Fibre flax plants are cultivated in China, the Russian Federation, and Western Europe in cool-temperate climates. Linseed varieties are significantly farmed across a larger area in the United States, India, Canada, China, Argentina, Kazakhstan, Ukraine, and Russia (Kole and Hall 2008). In India, *L. usitatissimum* is cultivated in various states, particularly in Uttar Pradesh, Madhya Pradesh, Bihar, Rajasthan, Himachal Pradesh, Jharkhand, Odisha, Telangana Karnataka, and Andhra Pradesh, which are among the largest producing states. Seed oil production is one of the primary purposes of *L. usitatissimum* farming. In 2014, the world produced 2.65 million metric tonnes of flax, with Canada making up 33% of the total (FAO 2014), while India is among the first, in terms of area under flaxseed cultivation and ranked third in production, in the world. In India, flaxseed is cultivated on around 4.68 lakh ha, and total flaxseed production is about 1.63 lakh metric tonnes (Maurya et al. 2017).

7.6 Ecological Requirements

Flax is a versatile crop that thrives under a variety of ecological conditions and soil types. Flax is also a significant industrial fibre crop, growing mostly on acidic (pH $\leq$ 5.5) non-chernozemic soils (Marchenkov and Rozhmina 2003). The major *L. usitatissimum*-growing regions in India are in the states of Maharashtra, Bihar and Uttar Pradesh and Punjab. Paddy lands of Madhya Pradesh, Chattisgarh, Himachal Pradesh, Rajasthan, Jharkhand, Odisha, Assam, West-Bengal, Karnataka, Andhra Pradesh, and Telangana (Gutte et al. 2015).

T-397, NP (RR) 9, No.55, K-2, S-4, LC 185, M-10, B-67, Jawahar-7 (R-7), Jawahar-17, Mukta, Hira, Neelum, and Mayurbhanj are a few varieties of flax crop used in India (https://krishijagran.com/agripedia/a-complete-guide-on-flaxseed-cultivation/). Due to favourable ecological conditions, the crop is usually seeded between September and October and harvested between March and April. Environment (e.g., drought, waterlogging, and high temperatures during blooming) and ecological (weed pressure) conditions, as well as agronomic management, all have a significant effect on the plants' performance (e.g., weed management, rotations) (Singh and Panda 2005).

Environmental conditions (temperature, soil conditions, cultural practises, and plant diseases, etc.) influence both the quantity and quality of oil produced. The greatest degree of variation was reported in the composition of fatty acids, oleic acid (14–60%), linolenic acid (31–72%) and linoleic acid (3–21%) (Mazza 1998).

7.7 Nutritional Value of Linum Usitatissimum

Due to its high concentration of linolenic acid, flaxseed is a frequently used oilseed in functional diets. Omega-3 fatty acids contains a total of 50% - 55% flaxseed oil. Typically, flaxseed contains 30% diet fibre, 6% hydration, 20% protein, 4% ash, and 40% oil as well as powerful antioxidants, lignans and dietary fibre, which seem to be rich in short-chain omega-3 fatty acids. Additionally, it contains *ca.* 20% monounsaturated fatty acids, 70% Omega-3 fatty acids, and 9–11% saturated fatty acids.

The protein content of flaxseed fluctuates between 20% and 30% (Martinchik et al. 2012).

7.8 Traditional Use and Common Knowledge

Indian Traditional Medicine is among the world's oldest systems of medical practices. India's traditional medicinal systems include Yoga, Ayurveda, Homoeopathy (AYUSH), Siddha, and Unani (Adhikari and Paul 2018). In India, *Linum usitatissimum* L. has been used as medication, as listed in Ayurveda's "Aharavarga" and "Tailvarga". *Linum usitatissimum* is also a well-known Unani drug (Katan) that is used to treat a variety of pathological disorders (Khan et al. 2017). Uncrushed flaxseeds soaked in water are traditionally used for the preparation of flaxseed tea, which is beneficial for asthma, bronchitis, bad coughs, and dysphonia. Flaxseed drink is also being used for constipation and is prepared with flaxseed powder. Further, flaxseed flour (10 g) in a form of paste is taken for the concerned illness along with honey, which is helpful in treating splenomegaly, pulmonary tuberculosis, hemoptysis, stomach ulcers, inflammations of the intestines, and abdominal pain (Goyal et al. 2014).

7.9 Major Chemical Constituents and Bioactive Metabolites

Every portion of the *Linum usitatissimum* L. plant is used for a variety of purposes, and the plant has garnered greater attention in food and disease research because of the possible health advantages related with many of its bioactive compounds (Gutte et al. 2015). Several flaxseed products are eaten, including the whole seed, flaxseed oil (partially defatted), flaxseed meal (completely defatted), flaxseed mucilage extract, flaxseed hulls, and flaxseed alcohol extracts. People also eat flaxseed oil that has been extracted from the seeds (Rubilar et al. 2010).

7.9.1 Lipids

The seed oil is extracted by pressing raw flax seeds. It is good source of alpha linolenic acid (ALA), an omega-3 fatty acid that accounts for approximately 55% of all fatty acids and are involved in a variety of regulatory processes throughout the body, including the blood pressure, pulse, fat breakdown, and immune response (Bloedon and Szapary 2004). The interest in flaxseed started with the discovery that the cold pressed oil contained 50% omega-3 fatty acids. An alternate extraction approach is to use liquid CO_2 at room temperature. The resultant extract contains linolenic (55%) and linolic (20%) acid glycerides, as well as oleic (17.8%), palmitic (6.7%), lauric (0.25%), and stearic (0.25%) acid glycerides (Mazza 1998).

Though India, Europe, and China have used flaxseed oil for millennia, it is considered bad oil, in most western nations, owing to its quick oxidation and polymerization. As a result, improvements to flaxseed oil were made to strengthen the stability of oil (e.g.: in paints). Lowering the linoleic glycerides concentration in flaxseed oil, below the 3% limit, by using the partial catalytic hydrogenation process, is generally done to use the flaxseed oil as a nutrition component. Unfortunately, this process is supported with partial esterification of double bonds in poly unsaturated fatty acid residues, resulting in the creation of trans-isomers. As a consequence, the hydrogenated oleic and linoleic acid include mixes of their respective *cis*, and *trans* isomers (Simopoulos 1996). Flaxseed has 5.5 times the amount of ALA found in other sources (Bloedon and Szapary 2004).

7.9.2 Proteins

L. usitatissimum has an average of 22 g of protein per 100 g of seed. The circumstances under which seeds are dehusked or defatted have an effect on the protein level of the produce. Due to the low protein concentration of the husk, meal synthesis from husks and defatted seeds results in a high protein isolate. In flax, globulins are the predominant protein, accounting for 18.6% of total protein, whereas albumin accounts for 17.7%. It changes according to environmental as well as genetic influences. Cold growing conditions result in low level of proteins, while warm growing conditions lead to a high level of proteins (Chung et al. 2005). While the protein component of *Linum usitatissimum* includes all essential amino acids in adequate amounts, it is deficient in threonine, lysine, and tyrosine. Additionally, it contains methionine and cysteine. *Linum usitatissimum* protein concentrate and isolate are not commercially available and are manufactured in the laboratory (Mazza 1998). Flaxseed proteins are composed of 70–85% of globulins (Oomah et al. 2006). Flax protein has an appropriate emulsion activity and stability as well as water absorption capacity. Linseed protein isolated through alkali extraction and acid precipitation has a high capacity for water and oil absorption. Additionally in research assessing the flaxseed protein and soy protein effects on uric acid and plasma triglycerides levels, it was discovered that flaxseed had a considerably greater hypo-tri-glyceremic impact than soy protein in addition to, flaxseed reduced triglyceride concentrations by twice as much as soy protein. Linseed has a strong influence on blood uric acid levels, but soy protein has the opposite effect, it has a high foaming capacity. Linseed proteins, on the other hand, are more lipophilic and have less trypsin inhibitory action than soy proteins (Borgmeyer et al. 1992).

7.9.3 Soluble Polysaccharides

Fibre, which gives most foods, their volume and shape, is not hydrolyzed in the digestive system. During digestion, fibre retains water and stops cholesterol from being absorbed. Insoluble fibre is made of components such as cellulose, hemicellulose, and lignin. Whole-grain cereals have the most of this kind of fibre. Soluble fibre makes a gel when it comes into contact with water. This gel is made of gums, pectin, and sugars, which together make mucilage (8% of flaxseed's dry weight) (Rubilar et al. 2010). The seed is a rich source of soluble polysaccharides due to its approximately 28 g of dietary fibre per 100 g. Flaxseed gum may be extracted by heating flaxseed and then precipitating it with alcohol and lastly, freeze-drying it (Klotzbach-Shimomura 2001). The fibrous rind constitutes between 30% and 39% of the flaxseed's weight. It contains relatively little fat as well as protein, however, it is high in polysaccharides. The shell's exterior is covered with epidermis containing mucilage, while the inside is filled with endosperm. The seed mucilage is a heterogeneous polysaccharide that accounts for a significant portion of the fibre component that are soluble and similarly possesses hypoglycemic properties in humans (Chung et al. 2005) and the soluble polysaccharides are composed mostly of glucose, rhamnose, galactose, xylose, fructose as well as arabinose. Additionally, mucilage of flaxseed is a stabilising and thickening agent (Wanasundara and Shahidi 1997).

7.9.4 Minerals

Linum usitatissimum is high in phosphorus, potassium, calcium, magnesium, and sodium. The seeds contain comparable potassium characteristics, and studies indicate that flaxseed is a good potassium source. Phosphorus is a critical macronutrient in flax seeds. Vanadium, fluorine, manganese, cobalt, iron, copper, chromium, selenium, zinc, iodine, and molybdenum are elements that are present in trace amount in flaxseed. While trace elements are vital for biological construction, excessive amounts may be hazardous (Fraga 2005).

Zinc is the greatest abundant micronutrient in flaxseed and is necessary for wound healing and is essential for the metabolic function of 300 body enzymes. It is also required for protein and DNA synthesis during cell division. Zinc bioavailability is fully dependent on the presence of phytate in the diet, and meals rich in phytate have a beneficial effect on zinc bioavailability (Hambidge et al. 2008). Copper, another critical element, is covalently linked to other metals in the shape of the enzyme. There are signs of copper deficiency anaemia in the body, as well as mental and skeletal system diseases (Davis 1967). Lithium content in linseed is between 4.20 and 5.50 g/g, whereas lead content, often recognised as heavy metal, is between 4 and 5 g/g. The content of cadmium, which is another heavy element is 0.22–0.55 g/g (Goldhaber 2003) while the chromium content ranges from 0.06 to 0.55 g/g which is needed to keep the body's glucose metabolism in check.

7.9.5 Tocopherols

Tocopherols are the supreme potent naturally occurring antioxidants that are oil soluble. Flaxseed oil consists of tocopherols, with a total tocopherol level of 40–50 mg/100 g (Mazza 1998). Gamma tocopherol (more than 80% of total tocopherol) in oil of flaxseed exhibits just 10–20% of the biological vitamin-E action of tocopherol, though having additional *in vitro* antioxidant activity (Javouhey-Donzel et al. 1993). The tocopherol content in flaxseed varies according to culture, variety, as well as climatic circumstances. Flax seed bark contains 26% of total tocopherol. Tocopherols prevent polyunsaturated fatty acids inside the cell membrane from oxidation and therefore contribute to the antioxidant ability of the cell by maintaining selenium in its reducing state. Vitamin E and nitrosamines, on the other hand, have been shown to have a bad effect on the development of children (Mazza 1998). Vitamin E and Vitamin A in linseed were described as 18.17 μg/g and 5.85 μg/g, respectively (Wiesenfeld et al. 2003).

7.9.6 Phenolic Acids

Phenolics are secondary metabolic products in plants that are involved in not only reproduction, physiology, and the coloration of flowers and fruits, but also serve as a key defence against diseases, parasites, and predators (Yang and Liu 2013). Phenolic acids are categorised into two primary groups: those derived from hydroxybenzoic acid and those derived from hydroxycinnamic acid. Linseed has 810 g of total phenolic acid per kg, 3–5 g of etherified phenolic acid per kg, and 5 g of esterified phenolic acid per kg while the amount of esterified and total phenolic acids in skinless as well as oil-free flaxseed is 81 and 73.9 mg per 100 g, respectively (Oomah et al. 1995). The primary phenolic acids present in *L. usitatissimum* are hydroxycinnamic acid derivatives. Such phenolic acids were found to be mostly linked to cell walls through ester bonds. Trans-ferulic, trans-synapic, p-coumaric, and trans-caffeic are present abundantly in 46%, 36%, 7.5%, and 6.5%, respectively. Seasonal variations in the phenolic acid concentration are to be expected. Chlorogenic acid, gallic acid, as well as 4-hydroxybenzoic acid (at negligible levels) are the phenolic acids found in flaxseed powder (Fiuza et al. 2004). Esters of cinnamic acid, including caffeic acid benzyl and phenethyl esters, protocatechuic acid, vanillic acid, syringic acid, and synaptic acid, have been shown to inhibit the proliferation of some kinds of carcinoma cells. Gallic acid and its esters have hydroxybenzoic derivatives that are being used abundantly in food and medicine as antioxidants (Özcan and Değerli 2019). In one study, the total caffeic acid concentration in flaxseed samples ranged from 5.18 to 8.70 g/g, but ferulic acid was plentiful in all flaxseed cultivars and accounted for a greater proportion of the bound extract, extending from 78.9–93.0% (Wang et al. 2017).

7.9.7 Flavonoids

Flavonoids are a class of naturally occurring benzo-γ-pyran derivatives present in photosynthetic cells (Cook and Samman 1996) that provide biological properties which protect against inflammation, allergies, hepatotoxins, free radicals, viruses, bacteria, platelet grouping, and ulcers. When a significant intake of flavonoid i.e., about 30 mg/day is linked to a low intake of flavonoid i.e., 19 mg/day, it was found that a 50% decrease in coronary heart disease occurs. The mechanism of this impact is presumably due to suppression in LDL oxidation of platelet aggregation (Collins et al. 2003). The primary flavonoids are flavan C- and O-glycosides in *L. usitatissimum* (Mazza 1998). Flax seed contains between 35 and 71 mg of flavonoids per 100 g. This level is influenced by cultural heterogeneity and environmental influences. According to one study, the bound flavonoid content of different cultivars ranges from 131.69 mg to 250.60 mg CE/100 g, with two cultivars, Shuangya12 and Ningya17, having significantly greater bound content of flavonoids than other *L. usitatissimum* cultivars. The levels of free flavonoids varied from 102.64 to 249.49 mg CE/100 g. Shuangya12 had the greatest content of free flavonoid, whereas Zhongya2 had the lowest. Overall, Shuangya12, a fibre flaxseed cultivar, had the greatest total flavonoid content (500.42 mg CE/100 g) of all *L. usitatissimum* cultivars, whereas Zhongya2 had the lowest level of total content of flavonoid (250.06 mg CE/100 g), which is also a fibre *L. usitatissimum* cultivar. The role of bound flavonoid to total flavonoid content ranged from 43.1% to 65.4%, indicating that the bound fraction was still used. The research found that the flavonoid concentration of fibre and oil flaxseeds was about the same, which suggests that fibre flaxseed could be a good source of flavonoids in the future (Wang et al. 2017).

7.9.8 Lignans

Linum usitatissimum contains 100 times the amount of lignans found in its nearest rival (Collins et al. 2003). Lignans have phenolic compounds made by combining two cinnamic acid residues to form a 2, 3-dibenzylbutane nucleus. Gut bacteria turn plant lignans linked to carbohydrates into enterolactone and enterodioli, which are thought to be lignans found in mammals (Bloedon and Szapary 2004). Flax seed has a high concentration of secoisolaricircinol diglucoside (SDG), a mammalian lignan precursor i.e., 7 mg/g or 3.7 mg/g SECO. It contains between 75 and 800 times the amount of SDG found in other meals. The lignan concentrations of flaxseeds may vary according to geographic regions, cultivars and harvest (Westcott and Muir 1996). Morris (2004) discovered that ground flaxseed and whole seed contain 0.7–1.9% SDG, or 56–152 mg SDG and 77–209 mg SDG per tablespoon of ground flaxseed and whole seed, respectively. SDG has an aglycone component, SECO. SDG and SECO both have maximal UV absorption at 280 nm, which is typical of lignans. When SDG in flaxseed is ingested, it is dihydroxylated and de-sterilized in the colon by facultative aerobic bacteria. Following that, enterodiol is oxidised to become enterolactone (Mazza 1998). Intestinal bacteria create these lignans, which are subsequently absorbed and transferred to the liver and gallbladder. Furthermore, flaxseed contains trace amounts of lignans such as isolariciresinol, matairesinol, and pinoresinol (Bloedon and Szapary 2004). Although the precise method through which the lignans in flax and its composition suppress tumour growth is unknown, numerous hypotheses are advanced in this area. Lignans, also known as phytoestrogens, are a class of plant-derived compounds that have a structure similar to the oestrogen hormone found in the body (Kris-Etherton et al. 2002).

7.10 Health Benefits of *Linum usitatissimum*

Linum usitatissimum L. has a great nutritional value in protecting against obesity, cardiovascular diseases, different malignancies, and a variety of other ailments. In recent years, the food industry has developed products that are fortified with dietary fibre, omega-3 fatty acids, vitamins, carbohydrates, protein, amino acids, and minerals that benefit human health. A study has reported that the unsaturated fatty acids that are found in flaxseed are readily digested by the human digestive system, hence they are incorporated into a variety of dietary produce consumed with the aim of enhancing the omega-3 fatty acid content intake (Yuksel et al. 2014). *Linum usitatissimum* has been extensively studied for its anti-cancer as well as heart-protective properties. Flax lignans binding to gut bacteria generate two oestrogen-like substances and inhibits the growth of few oestrogen-dependent cancers. In rats, flaxseed decreased colon, breast, and lung cancers. It is said that by taking 10 g of flaxseed daily, the risk of developing breast cancer may be lowered (Rickard et al. 1999). In prostate cancer, flaxseed has been demonstrated to decrease the overall quantity of testosterone and the free androgen index values (Bloedon and Szapary 2004). Some of the major pharmacologically active compounds present in flaxseed, with their health benefits, are listed in Table 7.1.

7.11 Conclusions

Plants have served as the foundation for the complex systems of traditional medicine. Many of the currently recognised lead compounds for medicinal medicines are natural chemicals or their derivatives. *Linum usitatissimum* L. is a significant medicinal herb that has been used in India for a variety of health and nutritional conditions, since ancient times. The modes of medication are contained in Ayurveda's "Aharavarga" and "Tailvarga". Numerous beneficial compounds have been identified in *L. usitatissimum*, including alpha-linoleic acid, lignans, flavonoids, magnesium, zinc, calcium, salt, vitamin E, vitamin K, etc. Scientific studies

Table 7.1 Pharmacologically active compounds of *Linum usitatissimum* with health benefits

Compound name	Structure	Heath benefits	References
α-linolenic acid	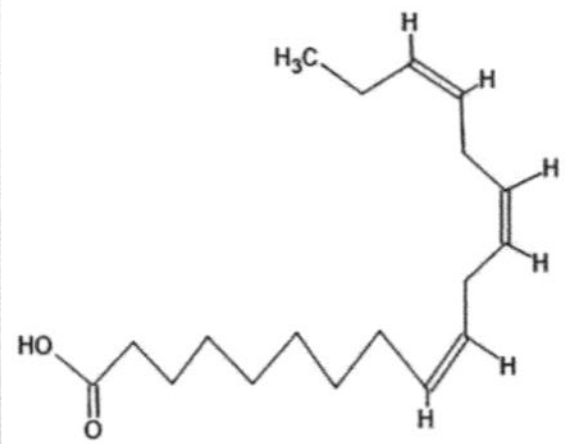	Cardiovascular benefits and atherosclerosis, rheumatoid arthritis and asthma prevention.	Rodriguez-Leyva et al. (2010), Raghuwanshi et al. (2019)
Secoisolariciresinoldiglycoside	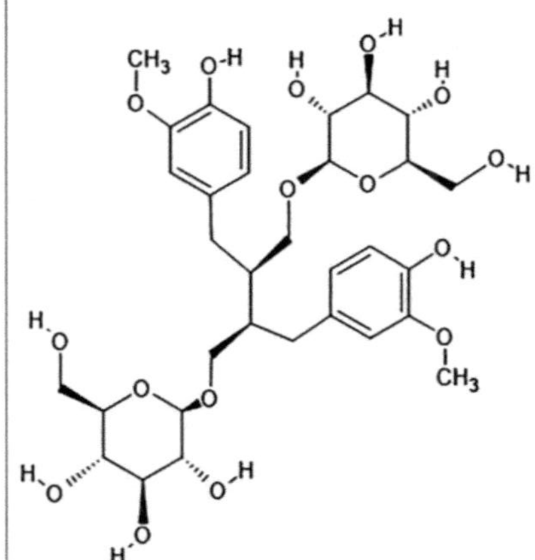	Prevention of cardiovascular diseases, cancer, diabetes, and mental stress.	Kezimana et al. (2018)
Ferulic acid		Anti-inflammatory, anticancer and antidiabetic activity.	Oliveira Silva and Batista (2017)

(continued)

Table 7.1 (continued)

Compound name	Structure	Heath benefits	References
	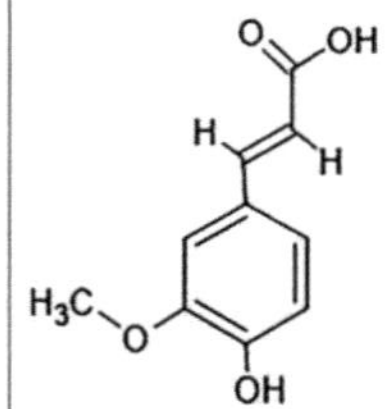		
p-coumaric acid	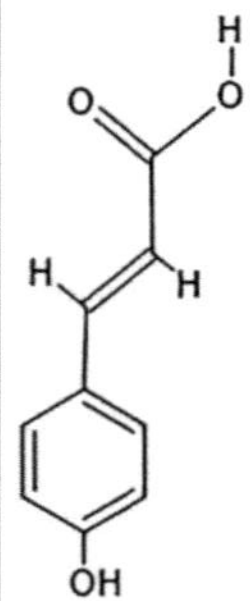	Antioxidant and antimicrobial activities.	Stojković et al. (2013)

Trans-caffeic acid	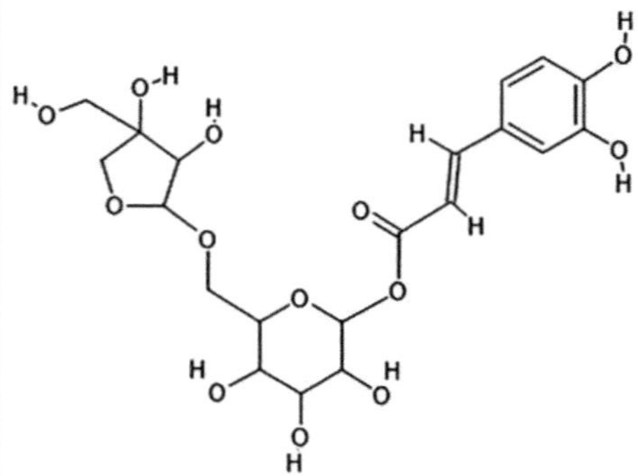	Protective role in osteoporosis and rheumatoid arthritis.	An et al. (2016)
Chlorogenic acid	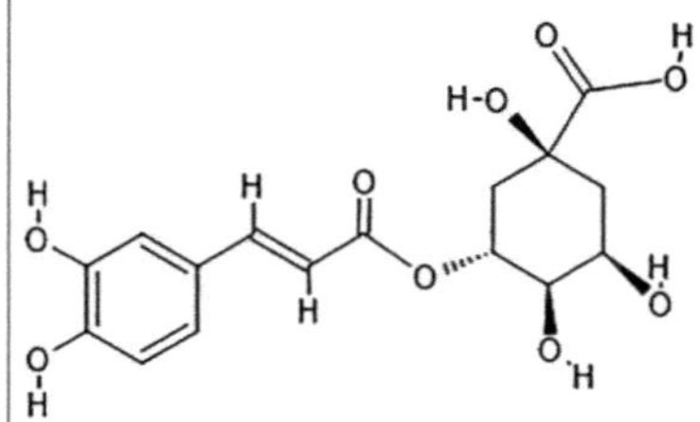	Neuroprotective effects.	Heitman and Ingram (2017)
Gallic acid	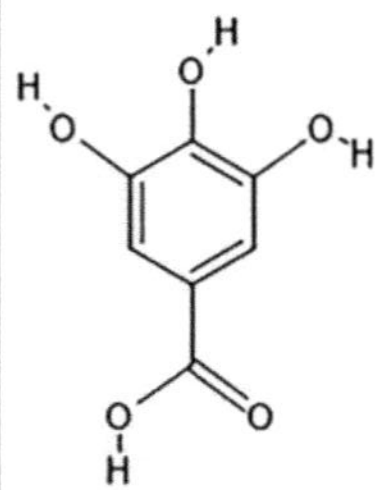	Antidiabetic and wound healing properties.	Yang et al. (2016)

(continued)

Table 7.1 (continued)

Compound name	Structure	Heath benefits	References
Kaempferol	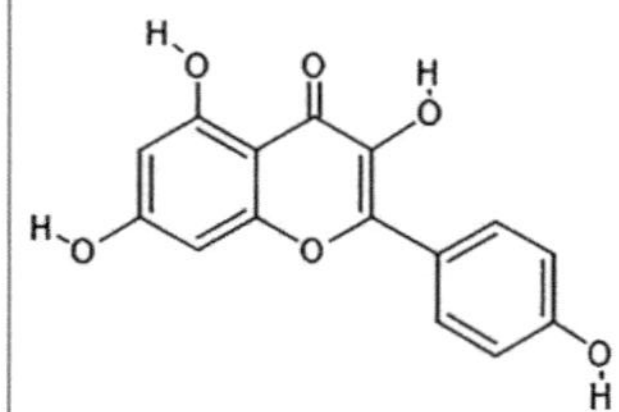	Anti-inflammatory, anticancer, antioxidant, and antidiabetic activity.	Yang et al. (2021)

on *L. usitatissimum* have established numerous health benefits in curing traditional medicinal claims. Health promoting characteristics are linked to anti-inflammatory, anti-oxidant, anti-carcinogenic actions, as well as to cholesterol reduction, cardiovascular disease prevention, and diabetes prevention. Despite of the long history of this popular and widely consumed and used crop, more progress has to be made in order to completely comprehend the way in which the pharmacologically active metabolites of *L. usitatissimum* can be best exploited.

References

Adhikari PP, Paul SB (2018) History of Indian traditional medicine: a medical inheritance. History 11(1):421

An J, Hao D, Zhang Q, Chen B, Zhang R, Wang Y, Yang H (2016) Natural products for treatment of bone erosive diseases: the effects and mechanisms on inhibiting osteoclastogenesis and bone resorption. Int Immunopharmacol 36:118–131

Barber EJW (1991) Prehistoric textiles: the development of cloth in the Neolithic and bronze ages with special reference to the Aegean. Princeton University Press

Bloedon LT, Szapary PO (2004) Flaxseed and cardiovascular risk. Nut Rev 62(1):18–27

Borgmeyer JR, Smith CE, Huynh QK (1992) Isolation, characterization of a 25 kda antifungal protein from flax seeds. Biochem Biophysic Res Comm 187(1):480–487

Chung MWY, Lei B, Li-Chan ECY (2005) Isolation and structural characterization of the major protein fraction from NorMan flaxseed (*Linum usitatissimum* L.). Food Chem 90(1–2):271–279

Collins TFX, Sprando RL, Black TN, Olejnik N, Wiesenfeld PW, Babu US, Bryant M, Flynn TJ, Ruggles DI (2003) Effects of flaxseed and defatted flaxseed meal on reproduction and development in rats. Food Chem Toxic 41(6):819–834

Cook NC, Samman S (1996) Flavonoids chemistry, metabolism, cardioprotective effects, and dietary sources. J Nutr Biochem 7(2):66–76

Cullis CA (2007) Flax. In: Oilseeds. Springer, Berlin, Heidelberg, pp 275–295

Davis PH (1967) Linum L. In: Davis PH (ed) Flora of Turkey and the East Aegean Islands, vol 2. Edinburgh University Press, Edinburgh, pp 425–450

Evans WC (2009) Trease and Evans pharmacognosy. Saunders, Edinburgh; New York

FAO, UN Food and Agriculture Organization (1994) 1993 production yearbook. FAO, Rome

Fiuza MS, Gomes C, Teixeira LJ, Girao da Cruz MT, Cordeiro MNDS, Milhazes NB, Marques MPM (2004) Phenolic acid derivatives with potential anticancer properties-a structure activity relationship study. Bioorg Med Chem 12:3581–3589

Fraga CG (2005) Relevance, essentiality and toxicity of trace elements in human health. Mol Asp Med 26(4–5):235–244

Fu YB (2011) Genetic evidence for early flax domestication with capsular dehiscence. Genet Resour Crop Evol 58(8):1119–1128

Ganorkar PM, Jain RK (2013) Flaxseed – a nutritional punch. Int Food Res J 20(2)

Goldhaber SB (2003) Trace element risk assessment: essentiality vs. toxicity. Regul Toxicol Pharmacol 38(2):232–242

Goyal A, Sharma V, Upadhyay N, Gill S, Sihag M (2014) Flax and flaxseed oil: an ancient medicine & modern functional food. J Food Sci Tech 51:1633–1653

Gutte KB, Sahoo AK, Ranveer RC (2015) Bioactive components of flaxseed and its health benefits. Int J Pharm Sci Rev Res 31(1):42–51

Hambidge KM, Miller LV, Westcott JE, Krebs NF (2008) Dietary reference intakes for zinc may require adjustment for phytate intake based upon model predictions. J Nutr 138(12):2363–2366

Heitman E, Ingram DK (2017) Cognitive and neuroprotective effects of chlorogenic acid. Nutr Neurosci 20(1):32–39

Javouhey-Donzel A, Guenot L, Maupoil V, Rochette L, Rocquelin G (1993) Rat vitamin e status and heart lipid peroxidation: effect of dietary α-linolenic acid and marine ω−3 fatty acids. Lipids 28(7):651–655

Kezimana P, Dmitriev AA, Kudryavtseva AV, Romanova EV, Melnikova NV (2018) Secoisolariciresinol diglucoside of flaxseed and its metabolites: biosynthesis and potential for nutraceuticals. Front Genet 9:641

Khan ZJ, Khan NA, Naseem I, Nami SA (2017) Therapeutics, phytochemistry and pharmacology of Tukhm-e-Katan (*Linum usitatissimum* L.). Int J Adv Pharm Med Bioallied Sci 111

Klotzbach-Shimomura K (2001) Functional foods: the role of physiologically active compounds in relation to disease. Top Clin Nutr 16(2):68–78

Kole C, Hall TC (2008) Compendium of transgenic crop plants. Wiley

Kris-Etherton PM, Hecker KD, Bonanome A, Coval SM, Binkoski AE, Hilpert KF, Griel AE, Etherton TD (2002) Bioactive compounds in foods: their role in the prevention of cardiovascular disease and cancer. Am J Med 113(9):71–88

Li JWH, Vederas JC (2009) Drug discovery and natural products: end of an era or an endless frontier? Science 325(5937):161–165

Marchenkov A, Rozhmina T (2003) Cultivation of flax. In: Flax. CRC Press, pp 86–103

Martinchik A, Baturin V, Zubtsov V, Vlu M (2012) Nutritional value and functional properties of flaxseed. Vopr Pitan 81(3):4–10

Maurya AC, Raghuveer M, Goswami G, Kumar S (2017) Influences of date of sowing on yield attributes and yield of linseed (*Linum usitatissimum* L.) varieties under dryland condition in eastern Uttar Pradesh. Int J Curr Microbiol App Sci 6(7):481–487

Mazza G (1998) Functional foods: biochemical and processing aspects, vol 1. CRC Press

Morris DH (2004) Flax – a health and nutrition primer. https://www.flaxcouncil.ca/. Last Accessed January 2022

Oliveira Silva E, Batista R (2017) Ferulic acid and naturally occurring compounds bearing a feruloyl moiety: a review on their structures, occurrence, and potential health benefits. Comp Rev Food Sci Food Safety 16(4):580–616

Oomah BD, Kenaschuk EO, Mazza G (1995) Phenolic acids in flaxseed. J Agri Food Chem 43(8): 2016–2019

Oomah BD, Der TJ, Godfrey DV (2006) Thermal characteristics of flaxseed (*Linum usitatissimum* L.) proteins. Food Chem 98(4):733–741

Özcan MM, Değerli Z (2019) Effect on human health and bioactive components of linseed. Пищевая промышленность: наука и технологии 12(3):85–92

Raghuwanshi V, Agrawal R, Mane K (2019) Flaxseed as a functional food: a review. J Pharmacogn Phytochem 8:352–354

Rickard SE, Yuan YV, Chen J, Thompson LU (1999) Dose effects of flaxseed and its lignan on N-methyl-N-nitrosourea-induced mammary tumorigenesis in rats. Nutr Cancer 35(1):50–57

Rodriguez-Leyva D, Bassett CM, McCullough R, Pierce GN (2010) The cardiovascular effects of flaxseed and its omega-3 fatty acid, alpha-linolenic acid. Can J Cardiol 26(9):489–496

Rubilar M, Gutiérrez C, Verdugo M, Shene C, Sineiro J (2010) Flaxseed as a source of functional ingredients. J Soil Sci Plant Nutr 10(3):373–377

Simopoulos AP (1996) Flax the next decade. Conf Proceed:5–28

Singh MP, Panda H (2005) Medicinal herbs with their formulations. Daya Books

Stojković D, Petrović J, Soković M, Glamočlija J, Kukić-Marković J, Petrović S (2013) In situ antioxidant and antimicrobial activities of naturally occurring caffeic acid, p-coumaric acid and rutin, using food systems. J Sci Food Agric 93(13):3205–3208

Wanasundara PKJPD, Shahidi F (1997) Removal of flaxseed mucilage by chemical and enzymatic treatments. Food Chem 59(1):47–55

Wang H, Wang J, Qiu C, Ye Y, Guo X, Chen G, Li T, Wang Y, Fu X, Liu RH (2017) Comparison of phytochemical profiles and health benefits in fiber and oil flaxseeds (*Linum usitatissimum* L.). Food Chem 214:227–233

Westcott ND, Muir AD (1996) Variation in the concentration of the flax seed lignan concentration with variety, location and year. In: Proceedings of the 56th Flax Institute of the United States Conference, N.Dak, Fargo, pp 77–80

Wiesenfeld PW, Babu US, Collins TFX, Sprando R, O'Donnell MW, Flynn TJ, Black T, Olejnik N (2003) Flaxseed increased α-linolenic and eicosapentaenoic acid and decreased arachidonic acid in serum and tissues of rat dams and offspring. Food Chem Toxicol 41(6):841–855

Yang J, Liu RH (2013) The phenolic profiles and antioxidant activity in different types of tea. Int J Food Sci Technol 48(1):163–171

Yang DJ, Moh SH, Son DH, You S, Kinyua AW, Ko CM, Miyoung S, Jinhee Y, Yun-Hee C, Ki WK (2016) Gallic acid promotes wound healing in normal and hyperglucidic conditions. Molecules 21(7):899

Yang L, Gao Y, Bajpai VK, El-Kammar HA, Simal-Gandara J, Cao H, Cheng KW, Wang M, Arro RR, Zou L, Farag MA (2021) Advance toward isolation, extraction, metabolism and health benefits of kaempferol, a major dietary flavonoid with future perspectives. Crit Rev Food Sci Nutr:1–17

Yuksel F, Karaman S, Kayacier A (2014) Enrichment of wheat chips with omega-3 fatty acid by flaxseed addition: textural and some physicochemical properties. Food Chem 145:910–917

Zeven AC, Zhukovsky PM (1975) Cannabidaceae. Dictionary of cultivated plants and their centres of diversity. In: *O. guineense*, pp 62–63

Chapter 8
Gymnema sylvestre R. Br.: Phytochemicals and Medicinal Properties

Sharad Vats, Abhijit Dey, Nikkee Bhandari, Krishna Kumari, and Chhavi Kaushal

Abstract *Gymnema sylvestre* R.Br. (Family: Apocynaceae), commonly called 'gurmar', is a commercially important medicinal plant. It is traditionally used as an antidiabetic agent, which is mainly due to the presence of gymnemic acids (triterpenoid glycosides). The plant is also a good source of oleanane-type triterpenoid saponins, flavonoids, steroidal glycosides, sterols, etc. Gurmarin, a protein, is known to suppress sweet taste. Its bioactive compounds have shown diverse therapeutic potentials viz., anticancer, anti-inflammatory, antidiabetic, radio-protective, antimicrobial, hypolipidaemic. Some of these medicinal properties have been validated through clinical trials. The present review focuses on the traditional phytochemical, medicinal properties, pharmacological and clinical trials undertaken on *G. sylvestre*.

Keywords *Gymnema sylvestre* · Gymnemic acid · Phytochemistry · Pharmacology · Clinical trials

8.1 Introduction

Diabetes mellitus (DM) is a common endocrine disorder detected in people of all age groups. In Type I diabetes mellitus, the pancreatic β-cell are not able to synthesize insulin whereas in Type II DM, the body is either not able to produce enough insulin or use it effectively (Steppan et al. 2001). DM has a pleiotropic effect and is associated with diabetic nephropathy, retinopathy, neuropathy and cardiac disorders, which have led to significant mortality globally (Babel and Dandekar 2021). It has been reported that the prevalence of this dreaded disease will increase from

S. Vats (✉) · N. Bhandari · K. Kumari · C. Kaushal
Department of Bioscience and Biotechnology, Banasthali Vidyapith, Banasthali, Rajasthan, India

A. Dey
Department of Life Sciences, Presidency University, Kolkata, West Bengal, India

Á. Máthé, I. A. Khan (eds.), *Medicinal and Aromatic Plants of India Vol. 2*, Medicinal and Aromatic Plants of the World 9,
https://doi.org/10.1007/978-3-031-44914-7_8

38.2 crores in 2013 to 59.2 crores by 2035 (Guariguata et al. 2014). Expected expenses attributed to diabetes may be more than $132 billion in the United States alone (Wild et al. 2004). The available oral antidiabetic drugs (e.g. biguanides, sulfonylureas and glinides) they possess side effects. Thus, the search for safer natural antidiabetic agents remains to be important (Hussain et al. 2021).

Candidate plants with known antidiabetic properties are *Momordica charantia* L., *Trigonella foenum-greacum* L., *Panax quinquefolius* L., *Aloe vera* L., *Allium sativum* L., *Gymnema sylvestre* R.Br. and others. Also, plant products including phenolic compounds, flavonoids, alkaloids, coumarins, terpenoids, etc. are known to reduce blood glucose level (Ramírez-Alarcón et al. 2021).

The present review is focused on an important medicinal plant *Gymnema sylvestre* (GS). The plant is used as a potent antidiabetic agent traditionally, which has been confirmed through scientific studies. The review meticulously explores the ethnobotanical uses, phytoconstituents and their efficacies.

8.2 Taxonomic Description and Traditional Uses

Gymnema sylvestre is a slow growing, perennial, woody climber (Fig. 8.1) species belonging to the Family Apocynaceae.

Morphological Description The leaves of the plant are opposite (elliptic or ovate), with transverse and reticulate venation and 1–2 cm long petiole. Flowers are small, yellow in colour with umbellate cymes, follicles terete, lanceolate up to 3 inches in length. Flowering occurs from August to March. The stem is branched, cylindrical, about 10 mm in diameter (Pramanick 2016). The lamina is ovate, elliptic or

Fig. 8.1 *Gymnema sylvetsre*

ovate-lanceolate with both surfaces pubescent. Seeds (1.3 cm long) are flat with a thin marginal wing and narrowly ovoid-oblong. Plant propagation through seeds is difficult due to low viability. Seeds have a peculiar odour and taste slightly bitter and astringent (Kirtikar and Basu 1998; Zhen et al. 2001).

Geographic Distribution It is a widely distributed plant and is found in the tropical forests of Southern and Central India (~600 m in height). It also grows in Sri Lanka, Malaysia, Southern China and Africa (Grover et al. 2002).

Crude Drug Used
Juice of the leaves is used to treat the opacity of cornea and as eye drops. Paste of leaves is applied on eyelids as a remedy for cataract. A mixture of leaves, pepper, salt and garlic is used against ephemeral fever in animals (Jamadagni et al. 2021).

Ecological Requirements
Gymnema sylvestre prefers 600–1000 mm rainfall, well distributed throughout the year. It grows in secondary forests, woodlands and dry shrub savannahs, usually on sandy or loamy soils.

8.3 Major Chemical Constituents and Bioactive Compounds

Pioneer work on the isolation of Gymnemic acid (GA), the marker metabolite of GS, was done by Hooper. He reported that the compound is a glycoside as it lost its characteristic antisweet activity on boiling with dilute HCl. This was followed by a few futile attempts to purify GA. Pfaffman (1959) confirmed the glycosidic nature of hydrolyzed GA by showing the presence of glucose, arabinose and glucoronolactone. The saponin nature of GA was suggested by Yackzan (1966).

Stocklin in 1967 reported the presence of four compounds $A_1 - A_4$ in the extract of *Gymnema* which were identified as D-glucoronides of hexahydroxyolean-12-ene called gymnemagenin (Stocklin et al. 1967; Stöcklin 1969; Rao and Sinsheimer 1968). These compounds were similar to the isolates of Kurihara (1992). He also identified that GA A_1 could be converted to GA A_2 and A_3 by sequential loss of ester groups. Sinsheimer and Rao (1970) identified GA A-D possessing genins G, K, N and gymnestrogenin, respectively, as aglycones. These aglycones were differentially esterified with acids. The structure of gymnestrogenin (pentahydroxytriterpene) was also proposed earlier by Stocklin (1968).

In 1989 a series of GA (I-IV) were isolated from hot water extract of GS leaves. The difference in the structure of isolates was due to acylating groups to the core structure gymnemagenin with attached glucoronic acid (Yoshikawa et al. 1989). Later GA V and VI were identified. GA VII contained gymnestrogenin as aglycone, which was unlike the previous isolated gymnemic acids having gymnemagenin as aglycone. Liu et al. (1992) isolated and characterized five GA, three of which were similar to GA III, IV and V. The other two GA VIII and IX were similar to GA III and IV respectively but with a replaced glucoronic acid by 3'-*O*-*β*-D-arabino-2-hexulopyranosyl group. Further, Yoshikawa group identified GA X-XVIII (Yoshikawa et al. 1992, 1993).

Sahu et al. (1996) isolated and identified gymnemasins A-D. These compounds possessed new aglycone, gymnemanol, which was characterized as 3*β*,16*β*,22α,23,28-pentahydroxyolean-12-ene. One year later, Yoshikawa group elucidated the structures of six triterpene glyocosides, gymnemosides a-f (Murakami et al. 1996; Yoshikawa et al. 1997a, b). Ye et al. (2000) isolated and identified six oleanane triterpene glycosides. Later three more oleanane-type saponins were identified. Zhu et al. (2008) elucidated the structure of yet another new oleanane-type triterpenoid saponins.

Glycosides of kaempferol and quercetin (flavonol) have also been identified from the aerial part of GS (Liu et al., 2004). Zhang et al. (2012) isolated triterpenoid saponins from the stem of GS.

Zhu et al. (2008) identified oleanane-type triterpenoid saponins. Olean type triterpene and a lupane type triterpene (3 *β*,16 *β*,23,28-tetrahydroxylup-20(29)-ene) were identified by Zarrelli et al. (2013). Five pregnane glycosides (gymsylosides A–E) were reported by Kiem et al. (2020). Further, pregnane glycosides named gymsylvestrosides A–D were identified by Xu et al. (2015). Five Olean-15-ene type gymnemic acids having antihyperglycaemic activity were reported by Alkefai et al. (2019). Arylated gymnemic acids were isolated and identified by Alkefai et al. (2018).

Liu et al. (2021) identified four C_{21} steroidal glycosides, sylvepregosides A-D, together with gymnepregoside H, deacetylkidjoladinin, gymnepregoside G and gymnepregoside I from the plant.

Besides this, other plant constituents, which have been reported are Conduritol A, 1-Heptadecanol, Stigmasterol glucoside, d-Quercitol, 1-Octadecanol, Potassium nitrate, Lupeol, cinnamate, Stigmasterol, campesterol, lanosterol, sitosterol, quercetin and kaempferol (Zhen et al. 2008; Vats and Kamal 2013; Vats and Kamal 2013). The structure of various gymnemic acids and other metabolites has been given in Figs. 8.2 and 8.3, respectively and the IUPAC names of some of the isolated metabolites are provided in Table 8.1.

8.3.1 Gurmarin (Medicinal Properties and Mechanism of Action)

Gurmarin is the marker protein of GS. It consists of 35 amino acids (MW: 4000; Sigoillot et al. 2018). It selectively inhibits the neural responses to sweet taste in rats at a concentration of more than 1 × 10–6 M (Imoto et al. 1991; Kurihra 1992). The response of chorda tympani nerve to sweet to taste substance on the application of gurmarin was evaluated by Ninomiya and Imoto (1995). Miyasaka and Imoto (1995) evaluated the effect of anti-sweet peptide electrophysiologically on taste response

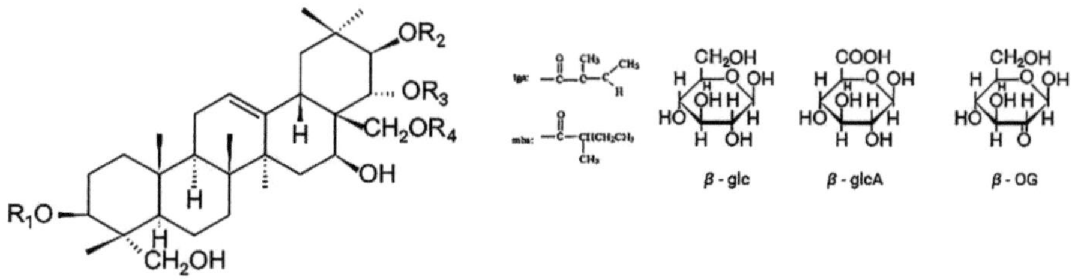

	R_1	R_2	R_3	R_4
Gymnemagenin	H	H	H	H
Gymnemic acid I	β − gla	tga	tga	tga
Gymnemic acid II	β − gla	mba	mba	mba
Gymnemic acid III	β − gla	mba	mba	mba
Gymnemic acid IV	β − gla	tga	tga	tga
Gymnemic acid V	β − gla	tga	tga	tga
Gymnemic acid VI	β − gla^3− β − glc	tga	tga	tga
Gymnemic acid VIII	β − gla^3− β − OG	mba	mba	mba
Gymnemic acid IX	β − gla^3− β − OG	tga	tga	tga
Gymnemic acid X	β − gla	H	H	H
Gymnemic acid XI	β − gla	tga	tga	tga
Gymnemic acid XII	β − gla^3− β − glc	tga	tga	tga
Gymnemic acid XIII	β − gla	H	H	H
Gymnemic acid XIV	β − gla	H	H	H

Fig. 8.2 Chemical structures of various Gymnemic acids and gymnemagenin. (Adapted from Suttisri et al. 1995)

and found that the most effective pH corresponds to the isoelectric point of the peptide (4.5) at which the rat tongue was treated with gurmarin. The maximum effect was shown at a concentration of 5 μM but the activity was still significant at 0.5 μM.

Scientists then focused on the nerve associated with gurmarin sensitivity and insensitivity. Ninomiya et al. (1997) observed that sweet taste receptors are sensitive to gurmarin in mice (C57BL/KsJ strain). The saturation in the effect of the peptide was achieved at 50 μg/mL (approximately 11.9 μM). The inhibitory effect was not witnessed in the glossopharyngeal nerve even at 23.8 μM. Subsequently, Ninomiya et al. (1999) later examined the response of a single fiber of chord tympani nerve to sweeteners in C57BL mice. It was observed that some of the sweet taste receptors were gurmarin-sensitive and others were gurmarin-insentitive. Sensitivity to sweet-tasting compounds (Fructose, Galactose, Glucose, Glycine, Maltose, Sodium saccharin, d-asparagine, d-histidine) was recorded in various neurons in the rat solitary nucleus before and after the application of gurmarin. Almost similar sweet taste suppression was observed across various concentrations of sucrose. However, the differential response was seen in other sweet compounds Lemon et al. (2003).

Gymnema diet induces the presence of serine proteases viz., rat kallikrein 2 (rK2) and rat kallikrein 9 (rK9). The cleavage site in the substrates is quite similar for these two proteins. Significant inhibition of immunoreaction between gurmarin and antigurmarin antiserum was exhibited by rK2 and rK9. This may be due to the cleavage or association of these serine proteases with gurmarin (Yamada et al. 2006).

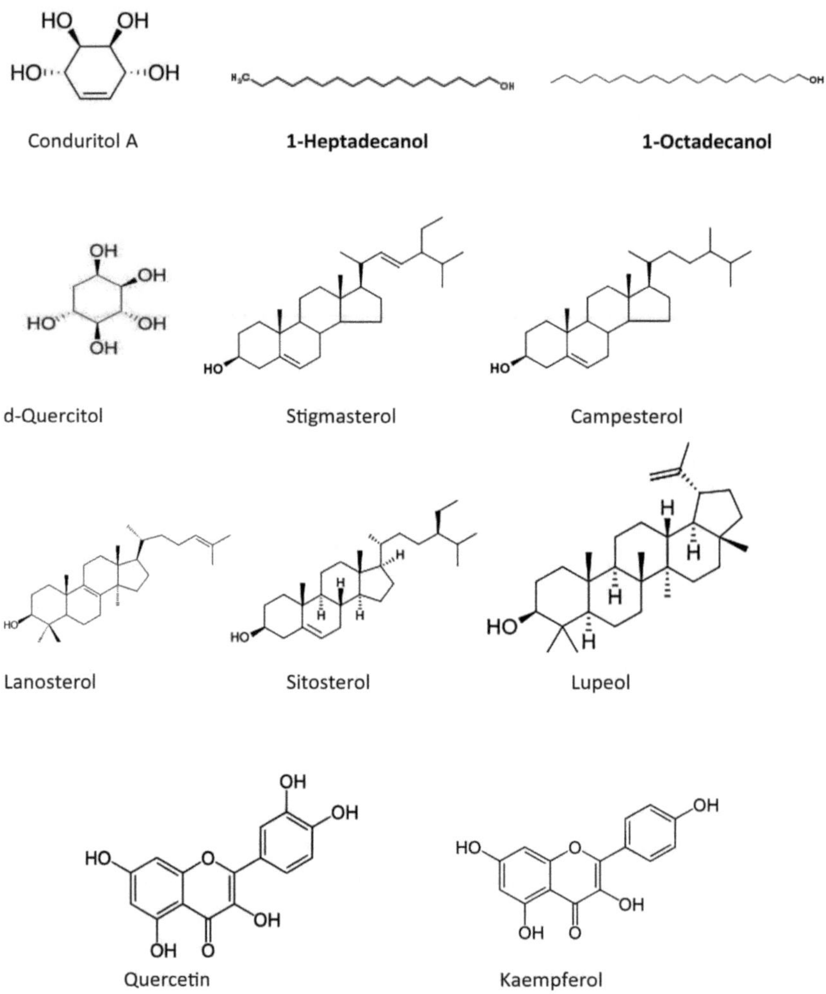

Gurmarin

Glu- Gln- Cys- Val- Lys^{5}- Lys- Asp- Glu- Leu-Cys^{10}- Ile-Pro-Tyr- Tyr-Leu^{15}- Asp- Cys- Cys- Glu-Pro^{20}- Leu- Glu- Cys-Lys-Lys^{25}- Val- Asn- Trp- Trp- Asp^{30}- His- Lys- Cys- Ile- Gly^{35}

Fig. 8.3 Structure of various phyto-constituents of *G. sylvestre*

In another study, the chorda tympani nerve was crushed and its regeneration was examined in rodents. It was revealed that mice possess gurmarin-sensitive and gurmarin-insensitive sweet-reception (Yasumatsu et al. 2007).

A study was done to identify substances that suppress the effect of gurmarin. Effect of gurmarin on behavioral responses with respect to the number of licks for

Table 8.1 Phytocompounds identified in *G. sylvestre*

Chemical group	Phytocompound	References
Triterpenoid saponin (Gymnemasins A-D)	3-*O*-[β -D-glucopyranosyl(1 → 3)- β -D-glucuronopyranosyl]- 22-*O*-tigloyl-gymnemanol; 3-*O*-[β -D-glucopyranosyl (1 → 3)- β-D-glucuronopyranosyl]-gymnemanol; 3-*O*-β-D-glucuronopyranosyl-22-O-tigloyl-gymnemanol and 3-O-β -D-glucuronopyranosyl-gymnemanol, respectively	Sahu et al. (1996)
Triterpene glyocosides (gymnemosides a-f)	21-*O*-tigloyl-22-*O*-acetylgymnemagenin 3-*O*-β-D-glucopyranosiduronic acid; 16-*O*-acetyl-21-*O*-tigloylgymnemagenin 3-*O*-β-D-glucopyranosiduronic acid; 21-*O*-benzoyl-28-*O*-acetylgymnemagenin 3-*O*- β -D-glucopyranosiduronic acid; 23-*O*-[β-D-xylopyranosyl (1 → 6)-β-D-glucopyranosyl (1 → 6)- β -D-glucopyranosyl] gymnestrogenin; 23-*O*-[β-D-xylopyranosyl (1 → 6)-β-D-glucopyranosyl (1 → 6)-β-D-glucopyranosyl]-28-*O*-[β-D-glucopyranosyl (1 → 6)-β-D-glucopyranosyl] 23-hydroxylongispinogenin; 23-*O*-[β-D-xylopyranosyl (1 → 6)- β -D-glucopyranosyl (1 → 6)-β-D-glucopyranosyl]-28-O-[β-O-glucopyranosyl (1 → 6)-β-D-glucopyranosyl] 3 β,16 β,23,28-tetrahydroxyolean-18-ene, respectively	Murakami et al. (1996), Yoshikawa et al. (1997a, b)
Oleanane triterpene glycosides	Longispinogenin 3-*O*-β-D-glucuronopyranoside; 21 β -benzoylsitakisogenin 3-*O*-β -D-glucuronopyranoside; 3-*O*- β -D-glucopyranosyl(1 → 6)- β -D-glucopyranosyl oleanolic acid 28-*O*-β-D-glucopyranosyl ester; oleanolic acid 3-*O*- β -D-xylopyranosyl (1 → 6)- β -D-glucopyranosyl(1 → 6)- β -D-glucopyranoside; 3-*O*- β -D-xylopyranosyl (1 → 6)- β -D-glucopyranosyl (1 → 6)-β-D-glucopyranosyl oleanolic acid 28-*O*-β-D-glucopyranosyl ester and 3-*O*-β-D-glucopyranosyl(1 → 6)-β-D-glucopyranosyl oleanolic acid 28- β -D-glucopyranosyl (1 → 6)-β-D-glucopyranosyl ester	Ye et al. (2000)
Oleanane-type triterpenoid saponins	16 β-hydroxyl olean-12-en-3-*O*-[β-D-glucopyranosyl (1 → 6)-β-D-glucopyranosyl]-28-*O*-β-D-glucopyranoside and 16 β, 21 β, 28-trihydroxyl-olean-12-ene-3-*O*-glucoronopyranoside	Zhu et al. (2008)
Olean type triterpene	3β,16β,21β,23-tetrahydroxyolean-12-ene; 3 β,16 β,23,28-tetrahydroxyolean-13,18-ene; 16 β,23,28-trihydroxyolean-12-en-3-one;	Zhu et al. (2008)

(continued)

Table 8.1 (continued)

Chemical group	Phytocompound	References
	16 *β*,21 *β*,23,28-tetrahydroxyolean-12-en-3-one; 16 *β*,21 *β*,22α,23,28-pentahydroxyolean-12-en-3-one	
Triterpenoid saponins	*β*,16 β,22*α*-trihydroxy-olean-12-ene 3-*O*- *β* -D-xylopyranosyl-(1 → 6)- *β* -D-glucopyranosyl-(1 → 6)- *β* -D-glucopyranoside	Zhang et al. (2012)
Arylated gymnemic acids	23-*O*-*β*-D-glucopyranosyl-21-*O*-tigloyl-28-*O*-benzoyl-16,22-dimethoxygymnemagenin; 3-*O*-*β*-D-glucuronopyranosyl-16-*O*-acetyl-21-*O*-hydrocoumaroyl-16*β*,21*β*,23,29-tetrahydroxyoleanolic acid 28-*O*-*β*-D-glucopyranosyl ester; 3-*β*-*O*-D-glucopyranosyl-21-*O*-hydrocinnamoyl-16*β*,21*β*,23,29-tetrahydroxyoleanolic acid 28-*O*- *β*-D-glucopyranosyl ester and 3-*O*-*β*-D-glucuronopyranosyl- 21-*O*-hydrocinnamoyl-7*β*-hydroxygymnemagenin	Alkefai et al. (2018)
C_{21} steroidal glycosides (sylvepregosides A-D)	2-*O*-cinnamoyl-20-*O*-benzoyl-heptahydroxy-(20*S*)-pregn-6-enyl-3-*O*-*β* cymaropyranoside-(1 → 4)-*β*-cymaropyranoside; 12-*O*-cinnamoyl-20-*O*-(*E*)-2-methyl-2-butenoyl-heptahydroxy-(20*S*)-pregn-6-enyl-3-*O*-*β*-cymaropyranoside-(1 → 4)-*β*-cymaropyranoside; 12-*O*-acetyl-20-*O*-(*E*)-2-methyl-2-butenoyl-sarcostin-(20S)-3-*O*-β-cymaropyranoside-(1 → 4)-β-cymaropyranoside; 12-*O*-acetyl-20-*O*-(*E*)-2-methyl-2-butenoyl-sarcostin-(20*S*)-3-*O*-*β*-cymaropyranoside, respectively	Liu et al. (2021)

quinine hydrochloride (QHCL) mixed sweet substances was evaluated by Murata et al. (2003) in mice (C57BL). The application of gurmarin considerably reduced the number of licks for sucrose mixed with QHCl. Normal behavior was restored after administering β-cyclodextrin orally. Ninomiya et al. (1998) reported that α and γ cyclodextrins have no marked effect on the suppression of gurmarin activity. However, gurmarin mixed with β-cyclodextrin showed decreased suppression of gurmarin on sucrose response. This may be due to the formation of an inclusion complex between gurmarin and β- cyclodextrin, which in turn hinders the binding site of the peptide on sweet taste receptors. On investigating the preference for taste solutions and saliva composition in rats fed with *Gymnema* containing diet, Katsukawa et al. (1999) observed that after an initial decrease in the preference for 0.01 M sucrose and a mixture of 0.03 M sucrose and 0.03 mM QHCl for 1–2 days, it returned close to the control levels in 7 days. This was related to the presence of certain gurmarin binding proteins (15, 16, 45, 60 and 66 kDa), induced by *Gymnema* diet, in submandibular saliva of rats. Later, the interaction of cyclodextrin with gurmarin was investigated. Significant variations in the UV absorption spectrum of gurmarin were induced by β-cyclodextrin probably due to the complex formation of amino acids (tyrosine and tryptophan) with β- cyclodextrin. Also, retention of gurmarin in the gel matrix was observed during gel filtration analysis. Thus, it can be suggested that outwardly projecting amino acids, which form a hydrophobic cluster, might interact with sweet taste receptors sensitive to gurmarin in rodents (Imoto et al. 2001).

Gurmarin (10 μg/mL) was reported to suppress the phasic taste responses to sugars viz., fructose, maltose, sucrose and lactose including saccharin sodium recorded from the greater superficial petrosal nerve innervating palatal taste buds in the rat. Sweet D-amino acids such as Histidine, Aspargine, Phenylalanine, Glutamine were also depressed, whereas such effect was not observed in D-tryptophan and D- Alanine, L-amino acids (Histidine, Aspargine, Phenylalanine, Glutamine, Trptophan and Alanine) and two basic amino acid HCl salts of Arginine and Lysine (Harada and Kasahara 2000). The connection between gurmarin sensitivity and the causative gene has also been explored. Suppression of sweet taste is found to be more in C57BL/6 mice as compared to BALB/c mice. Polymorphism in alleles of sweet response gene, *Sac (Tas1r3)*, maybe one of the reasons for differential sensitivity to Gurmarin. But on the investigation, it was found that another Tas1r3 non-taster strain (129 × 1/Sv) has a similar sensitivity to C57BL/6 mice. Thus, polymorphism of the sweet response gene may not be associated with the inhibition of sweet responses by gurmarin (Sanematsu et al. 2005).

Efforts have also been made to understand the signal transduction pathway of sweet taste response with reference to gurmarin. Gustducin, present in some taste receptor cells, is a G-protein associated with taste and the gustatory system. This protein may be an important molecule related to the pathway for gurmarin-sensitive sweet response (Shigemura et al. 2008). Effect of temperature on Gurmarin inhibition of chorda tympani responses to sweet compounds was examined in knock out mice lacking T1R3, Galphagust, or TRPM5 by Ohkuri et al. (2009). T1R3-knock out mice showed a temperature-dependent increase in Gurmarin sensitivity in responses to sucrose and glucose. A similar effect was seen in TRPM5-knock out mice in response to glucose. On the other hand, in Galphagust-knock out mice temperature-dependent increase in Gurmarin insensitive in responses to sucrose and glucose was observed. All three knock out mice exhibited unmeasurable responses to calorie-free artificial sweetener (SC45647). Also, their responses to saccharin did not show a temperature-dependent increase. Lingual application of pronase (sweet response inhibitor) almost fully inhibited responses to sucrose and glucose, however, the responses to saccharin remained unaffected. The study highlights the presence of multiple transduction pathways for responses to sweet taste.

The binding site of gurmarin was studied in taste buds in rat circumvallate papillae through histochemical technique. Localization of the protein was observed only in a certain, limited number of taste hairs (Yoshie et al. 1994).

8.4 Biological Activity of Plant Extracts and Bioactive Compounds

8.4.1 Antidiabetic Activity

Gymnema sylvestre is known for its antidiabetic activity, which has been validated through several types of research done all over the world. In an experiment conducted by Gupta and Seth (1962), a high carbohydrate diet was given to rats for 15 days and later injected with anterior pituitary extract (100 mg/kg) subcutaneously/day for 10 days, which induced hyperglycemic. These animals were given an oral dose (100 mg/kg.b.wt) of ethanolic extract of GS. The blood sugar level did not reduce significantly in normal rats. On the other hand, the blood sugar level showed a significant reduction in anterior pituitary-treated hyperglycemic rats. Significant influence of the plant extract was observed with respect to the carbohydrate metabolism in hyperglycemic animals. Moreover, the effect of the plant extract was comparable to that of the standard antidiabetic medicine tolbutamide (50 mg/kg) in hyperglycemic rats. Marked reduction in hyperglycemia in response to anterior pituitary extract (100 mg/kg) was observed in GS alcoholic extract (100 mg/kg) fed albino rats as compared to other traditional antidiabetic plants viz., *Coccinia indica, Pterocarpus marsupium* and *Momordia charantia* (Gupta 1963).

Prakash et al. (1986) fed rats with GS leaves powder for 10 days and continued for 15 days post beryllium nitrate treatment (i.v.). The level of blood glucose did not reduce comprehensively as compared to rats treated only with beryllium nitrate. Also, no change in the blood glucose level was observed in normal rats fed with leaf powder for 25 days. It can be concluded that GS has a normalizing effect on blood glucose. Thus, its use can be safer than conventional oral hypoglycemic drugs.

Gymnema sylvestre has been reported to have a positive effect on enzymes viz., phosphorylase, gluconeogenic enzymes and sorbitol dehydrogenase affording the uptake of glucose by insulin-dependent pathways in alloxan diabetic rabbits. A reversal in the pathological alterations in the liver during the high blood sugar phase was also observed. This shows that the plant may correct the metabolic changes in diabetic rabbit kidney, liver and muscle (Shanmugasundaram et al. 1983).

Different doses of GA IV (3.4–13.4 mg/kg) were administered to streptozotocin (STZ)-diabetic mice. A significant reduction in the blood glucose levels by 13.5–60.0% was observed, which was comparable to the potency of glibenclamide. Plasma insulin also increased in tested mice at 13.4 mg/kg dose of GA IV showing the possible insulinotropic effect of GA (Sugihara et al. 2000).

The insulinotropic and regenerative capacity of pancreatic islets were reported by Shanmugasundaram et al. (1990). Two water-soluble extracts (GS_3 and GS_4) were tested on STZ -treated rats. The fasting blood sugar level was restored to normal after 60 and 20 days of oral dose of GS_3 and GS_4, respectively. An increment in the serum insulin level and twice the number of beta cells and islets of Langerhans was also observed. The report suggested the possible insulinotropic, regeneration of pancreatic cells and glucose homeostatic action of *Gymnema*. Few other workers also reported the hypoglycemic effect of GS (Khare et al. 1983; Srivastava et al. 1985; Dogar et al. 1988).

Persaud et al. (1999) showed a profound effect of GS extract (GS_4) on insulin secretion in pancreatic islets and on three different β cell lines viz., HIT-T15, MIN6 and RINm5F. GS_4 extract (2 mg/mL) led to enhanced insulin secretion from islets (>16 ng/islet/h). This high output may not be due to the physiological effect of GS_4. On examining the islets β-cell integrity after application of using trypan blue exclusion assay, it was found that GS4 stimulated insulin secretion together with enhanced uptake of dye. The study suggested that GS extract stimulates the release of insulin by enhancing cell permeability. GS extract restored the reduced area occupied by β-endocrinocytes in the pancreatic islets in streptozotocin-induced diabetic rats. The insulinotropic effect of extract of *Gymnema* leaves is partly due to Ca^{2+} influx. An experiment was conducted on MIN6 β-cell line and isolated human islets of Langerhans. Low concentration (0.06–0.25 mg/mL) of an aqueous extract of *Gymnema* leaves extract termed OSA, showed no adverse effect on the viability of β cells. Higher concentrations ($\geq$ 0.5 mg/mL) caused increased Trypan blue uptake. Increased Ca^{2+} level in β cell through voltage-operated calcium channels and reversible insulin secretion stimulation from isolated human islets was also observed. The presence of extracellular Ca^{2+} ions partially caused insulin secretagogue effects in cell lines (Liu et al. 2009). A triterpene glycoside fraction having a mixture of GA I, IV, and VII and gymnemagenin showed more than one-fold increase in glucose-stimulated insulin secretion in MIN6 cell lines. Also, enhanced expression of GLUT2 was observed (Shenoy et al. 2018). In another study, it was highlighted that GA I protects MIN6 cells from apoptosis under high glucose stress by inducing autophagy. Inhibition of caspase 3 activity and phosphorylation activity of mTOR was also observed (Wu et al. 2019). Sylvepregosides A-D showed significant glucose uptake in L6 cells (Rat skeletal muscle) and promoted GLUT4 fusion with the cell membrane of the test cells (Liu et al. 2021).

One of the reports indicates that antihyperglycaemic activity of GS extract is not mediated through thyroid hormone (Gholap and Kar 2003). Daisy et al. (2009) identified yet another antidiabetic compound dihydroxy gymnemic triacetate. Oral administration of 20 mg/kg body weight dose of the isolated compound for 45 days to STZ-induced diabetic rats showed a significant effect on various parameters studied such as plasma glucose, insulin, tissue glycogen, glycated hemoglobin together with lipid parameters (triglycerides, total cholesterol, low-density lipoprotein (LDL)-cholesterol, high-density lipoprotein (HDL)-cholesterol) and activities of hepatic marker enzymes viz., aspartate aminotransferase (AST), alanine aminotransferase (ALT), alkaline phosphatase and acid phosphatase). The results, after comparison with that of normal rats, suggested possible hypoglycemic as well as hypolipidemic activity of dihydroxy gymnemic triacetate. Antidiabetic activity of *Gymnema* was also reported by Yadav et al. (2010) and Ahmed et al. (2010). Treatment of STZ-induced diabetic rats with *GS* leaves extract had a marked effect in lowering the plasma glucose, ALT, AST, triglycerides, total cholesterol, LDL-cholesterol, malondialdehyde (El Shafey et al. 2013). On the other hand, the extract significantly increased insulin, HDL- cholesterol and erythrocyte superoxide

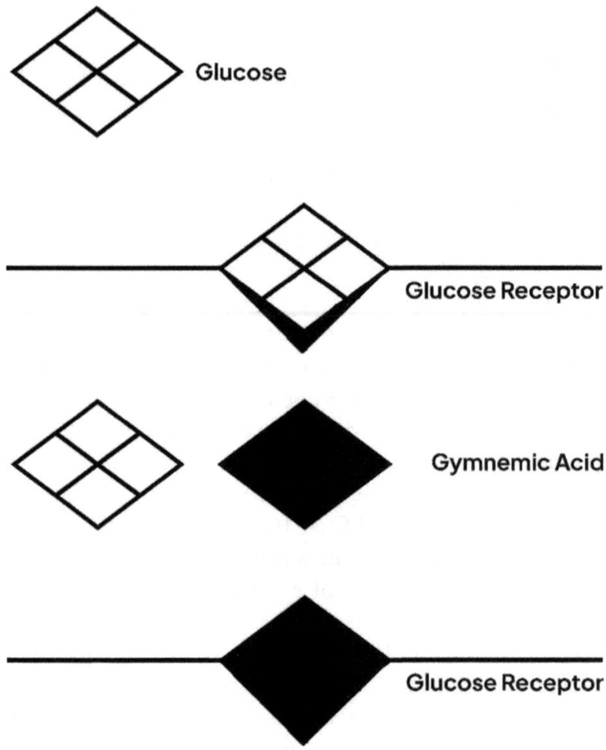

Fig. 8.4 Competitive inhibition of glucose absorption through glucose receptor-A schematic representation

dismutase levels compared to the control (untreated rats). Similar results on different parameters assessed in STZ-induced diabetic rats/mice suggest potential antidiabetic activity of GS (Kumar et al. 2017; Mishal et al. 2020; Mahadeva Rao et al. 2020).

GA is similar in structure to glucose. It competitively binds to glucose receptors of the tongue and intestine preventing the activation of the receptor by glucose, thereby suppressing the glucose uptake (Fig. 8.4). This leads to the subdued ill effect of sugar in the body (Sahu et al. 1996).

8.4.2 Antiobesity and Weight Control

Obese individuals have a body mass index (BMI) greater than 30 kg/m^2. Obesity is increasing tremendously in all parts of the world and the affected individuals are from all age groups. Diabetes is one of the pathological consequences of obesity (Flier 2004). The risk for diabetes increases from two-fold in mildly obese to ten-fold in severely obese persons (United States Report 1975). The risk for the occurrence of diabetes increases with the duration of obesity (Callaghan et al. 2020). According to the National Health and Nutrition Examination Survey (NHANES) II data relative risk for developing diabetes was 2.9 times higher for 20–75 years old obese persons

(Van Itallie 1985). Incidence of diabetes increases with age, diabetic family history, and obesity (Nayak et al. 2014). However, despite the risk of diabetes, many obese individuals do not get diabetic, which suggests the probable role of genetic and/or environmental factors in the incidence of this disease.

Gymnema extract (25–100 mg/kg) was orally administered in experimentally induced hyperlipidaemic rats for 14 days. Reduction in the increased serum triglyceride, total cholesterol, very low-density lipoprotein (VLDL) and LDL-cholesterol was observed in a dose-dependent manner. The potential of the extract (100 mg/kg) as an antiatherosclerotic agent was almost at par with that clofibrate (lipid-lowering drug) (Bishayee and Chatterjee 1994). The effect of gymnemic acids on lipid metabolism has also been reported. Long-term administration of aqueous extract of GS mildly reduced the serum triglyceride and total cholesterol concentration in obese rats (Terasawa et al. 1994). However, Ikemitsu (1990) observed that the intraperitoneal injection of gymnemic acids (25 mg/kg) had no marked effect on serum concentrations of total cholesterol, free cholesterol, triglycerides and free fatty acids. Lowering of cholesterol, LDL and triglyceride and enhanced level of HDL on administering *Gymnema* extract has also been reported in other studies (Rachh et al., 2010; Singh et al., 2017). The lipid-lowering potential of the plant may be attributed to the presence of flavonoids, tannins, saponins, etc. (Rachh et al. 2010; Dholi and Raparla 2014).

The role of *Gymnema* in the treatment of obesity as a hypolipidemic agent has also been reported. The effect of GA on oleic acid absorption was studied by Wang et al. (1998) using intestinal perfusion in rats. GA inhibited the absorption of oleic acid in the intestine in a dose-dependent manner, which was comparable to inhibition of glucose absorption in the intestine.

Effects of GA on fecal steroid excretion in rats and probable role in cholesterol metabolism were investigated for the first time by Nakamura et al. (1999). Three differentially prepared extracts from *G. sylvestre* leaves viz., extract (GSE), acid precipitate (GSA) and column fractionate (GSF) were given orally (0.05–1.0 g/kg) for 22 days. The test rats were put on a cholesterol-free diet. GSA and GSF significantly reduced the body weight gain and intake of food. There were reduced serum and liver lipids concentrations whereas increased fecal neutral steroids and bile acid excretion were observed. This may be attributed to the interaction of saponins with cholesterol and bile acids in the intestine. The results were encouraging but still need further experimentations to elucidate the effect of GA on cholesterol metabolism.

Rats fed with a high-fat diet for 10 weeks showed reduced body weight gain and accumulation of hepatic lipids on the administration of GS extracts. Plasma triglyceride levels decreased in the normal fat diet groups by administration of the extract. The results were at par with chitosan-administered rats. There was no lowering effect on plasma total cholesterol by the extract. The long-term administration of the extract did not influence the hematological and blood chemical parameters (Shigematsu et al. 2001a). Later in the same year, Shigematsu group reported the possible role of *Gymnema* extract on lipid metabolism. Rats fed on a high-fat diet or normal fat diet for 3 weeks were given *Gymnema* extract (GE) orally once/day. A significant reduction in fat digestibility was observed in both the diet groups for the last 14 days of the experimental duration. It was proposed that GE had a marked effect on the lipid metabolism of rats, which was observed by improved serum cholesterol and triglyceride levels (Shigematsu et al. 2001b). High-fat diet-fed male C57BL/6 J mice were administered with methanolic extract of *Gymnema* containing GA. Results showed a decrease in body weight and other parameters of lipid profile. Reduction in epididymal fat weight and adipocyte hypertrophy was also observed (Kim et al. (2016).

8.4.3 Anti-inflammatory and Anti-arthritic Properties

Anti-inflammatory activity of the aqueous extract of *G. sylvestre* leaves was reported by Diwan et al. (1995) on various inflammatory models. The extract showed enhanced production of hepatic enzymes (γ-glutamyl transpeptidase and Superoxide dismutase), which protect the liver against the release of slow-reacting substances and free radicals. The aqueous extract of *G. sylvestre* (300 mg/kg) reduced the volume of paw oedema by 48.5% within 4 h after administration in rats. The result was comparable to Phenylbutazone wherein there was 57.6% decrease in paw oedema as compared to the control. A significant reduction in granuloma weight was also observed (dose: 200 mg/kg and 300 mg/kg) as compared to the control group (Malik et al. 2008). The anti-inflammatory activity of *G. sylveste* mediated TiO_2 nanoparticles was reported by Bharathy et al. (2021).

Arthritis is associated with inflammation, pain and reduced movement. It has been estimated that there will be 60 million patients suffering from arthritis by the year 2020 in America (Ahmed et al. 2005). Anti-inflammatory drugs are the drugs of choice, but risks associated with these drugs advocate herbal intervention in the treatment of this disease. A significant reduction in the carrageenan- and histamine-induced rat paw oedema was shown by the ethanolic extract *G. sylvestre* roots (Shankar and Rao 2008) Malik et al. (2010) reported the anti-arthritic activity of aqueous and petroleum leaf extract of *G. sylvestre* in adjuvant-induced arthritis albino rat model. The activity may be attributed to the plant constituents like triterpenoids, saponin glycosides and steroids.

8.4.4 Antimicrobial Activity

Dental caries (tooth decay) is one of the most prevalent oral health problems affecting individuals of various age groups. It is caused by the action of acids on the surface of the enamel. The reaction of sugars in food products/drinks with bacteria present on the surface of tooth causes loss of calcium and phosphate from the enamel, which is termed demineralization (Akhtar and Bhakuni 2004). Gram-positive bacteria involved in this process are *Streptococcus mutans, S. mitis,*

Staphylococcus aureus together with fungus like *Candida albicans* (Marsh and Martin 1992). GA has been reported to reduce plaque formation by *S. mutants* (Porchezhian and Dobriyal 2003). Different polar and non-polar extracts of GS leaves were tried against the above organisms. Methanolic extract showed maximum antimicrobial activity at 25 mg/ml extract (ParimalaDevi and Ramasubramaniaraja 2010). The results suggest that the plant extract/constituents can be a part of herbal toothpaste as a natural remedy to dental caries.

Antimicrobial activity of ethanolic extract of *GS* has been reported against *Bacillus pumilis, Pseudomonas aeruginosa, B. subtilis* and *Staphylococcus aureus. The extract was* found to be inactive against *Proteus vulgaris* and *Escherichia coli* (Satdive et al. 2003). The fraction of *G. sylvestre* containing pure saponin was effective against *Pseudomonas aeruginosa, E. coli, Salmonella typhi, Klebsiella pneumonia, Proteus mirablis, S. aureus* with MIC in the range of 600–1200 mg/L. The activity was also against *Aspergillus fumigates, A. flavus, A. niger* with MIC 1400 mg/L (Khanna and Kannabiran 2008). In another study, it was found that petroleum ether, chloroform and water:ethanol (1:1) extract was active against gram-positive bacteria (*Bacillus subtilis* and *S. aureus*) but not against *E. coli* (Saumendu et al. 2010). There are other reports on antimicrobial activity of *G. sylvestre* (Bhuvaneswari et al. 2011; Yogisha and Raveesha, 2009). Antimicrobial activity was observed in the chloroform and ethanolic extract of *G. sylvestre* leaves (Tahir et al. 2017) against bacterial poultry pathogens (*S. aureus, Clostridium perfringens type-A, E. coli, S. enterica, Haemophillus paragallinarum*). In another study, it was reported the methanolic extract of the leaves of the plant had significant antimicrobial potential against *E. coli, B. cereus, Candida albicans* and *C. kefyr*. The extract was active against *S. aureus, C. krusei* and *C. kefyr*. However, the chloroform extract did not show antimicrobial activity against the test organisms (David and Sudarsanam 2013). Terpenoid-rich fraction of the leaves of *G. sylvestre* was found to be active against *Shigella flexineri* and *E. coli* followed by *B. licheniformis, S. aureus*, and *S. sonei* (Behuria and Sahu 2020). There are other reports on the antimicrobial potential of the plant (Rajkumar et al. 2020; Behuria et al. 2021; Ghous et al. 2021).

8.4.5 Radioprotective Activity

Radioprotective activity was demonstrated for the first time by the butanol fraction of hydroethanolic extract of *G. sylvestre* leaves. Radiation (8 Gy) in Swiss albino mice induced enhanced lipid peroxidation and decreased glutathione and protein levels in mice brain which was significantly ameliorated by the plant extract. The activity was attributed to the antioxidant potential of components of the extract including phenolic compounds, which was determined by DPPH assay, superoxide radical scavenging assays, inhibition of *in vitro* lipid peroxidation assays and protein carbonyl formation assay with IC50 value 238, 140, 99.46, and 28.03 μg/mL, respectively (Sharma et al. 2009). This study adds to the immense potential of

Gymnema. In similar study protection against radiation-induced biochemical alterations was observed in the liver of Swiss albino mice at the oral treatment dose of 350 mg/Kg.b.wt/day (Bhatia et al. 2008).

Pangasius sutchi (Freshwater fish) was subjected to different doses of gamma radiation. The GE of leaves and its bioactive compound gymnemagenin showed significant radioprotective potential (Sinha et al. 2021).

8.4.6 Anticancer Activity

Saponin (gymnemagenol) from *G. sylvestre* at 50 μg/mL demonstrated cytotoxic activity (63%) against HeLa cells at 48 h with the IC50 value of 37 μg/mL. On the other hand, gymnemagenol was found to be non-toxic to Vero cells. 5-Fluorouracil (positive control) showed 57.5% cell death with the IC50 value of 36 μg/mL (Khanna and Kannabiran 2009). Bio-functionalized gold and silver nanoparticles using aqueous extract of *G. sylvestre* were reported to be cytotoxic to Human HT-29 cells (Arunachalam et al. 2014; Arunachalam et al. 2015). This suggests the preliminary anticancer activity of *Gymnema.* Wu et al. (2012) isolated five water-soluble polysaccharides viz., GSP11, GSP22, GSP33 and GSP44. All the isolated carbohydrates inhibited the proliferation of AGS, SGC and U937 cells except for GSP22 in AGS cells. Reduced incidence of tumor and number of papillomas was observed in 7, 12 – dimethylbenz (a) anthracene (DMBA)-induced papillogenesis in Swiss albino mice on administering the methanolic extract of GS (Agrawal et al. 2016). GS supplemented curd showed decreased growth of liver cancer cell line (HepG2) using MTT assay (Devi et al. 2020).

8.4.7 Wound Healing Activity

Hydro-alcoholic *Gymnema sylvestre* leaves extract using hot maceration method showed antihepatotoxicity in freshly prepared rat hepatocytes at a dose of 200, 400, 600 μg/mL. The hepatotoxicity was induced by D-galactosamine. Significant restoration of changed biochemical parameters as observed ($p < 0.001$) as compared to D-galactosamine treated groups. The treatment dose of 800 μg/ml of the plant extract was found to be cytotoxic (Srividya et al. 2010).

GS has proven wound healing property, which has been demonstrated by Kiranmai et al. (2011). The hydro-alcoholic extract of the plant showed a significant reduction in the duration of epithelization and wound contraction in excision and burn wound albino mice models as compared to the control group. The healing activity may be because of the presence of phenolic compounds (flavonoids) which were confirmed through the chromatographic method. Wound healing properties of *G. sylvestre* in excision, incision and dead space granuloma rat models at a dose of 200 mg/kg has been demonstrated (Malik et al. 2009). Graphene oxide-polyhydroxybutyrate-sodium alginate composite scaffold incorporated with curcumin and GS showed enhanced tissue regeneration in diabetic fibroblast cells making it a good candidate as a diabetic wound healer (Daisy et al. 2020). In a recent study (Subramanian et al. 2021), it was demonstrated that gamma sterilized polycaprolactone (PCL) nanofiber with incorporated *G. sylvestre* leaf extract improved the proliferation of fibroblast cells, which shows its applicability in wound dressing and healing. The activities of different phytoconstituents/extracts of *G. sylvestre* are listed in Table 8.2.

Table 8.2 Bioactivities of different phytoconstituents/extracts of *G. sylvestre*

Gymnema preparation	Activity	References
Gurmarin	Antisweet	Miyasaka and Imoto (1995), Harada and Kasahara (2000)
Gymnemagenin	Radioprotective	Sinha et al. (2021)
Water solube polysaccharides (GSP11, GSP22, GSP33 and GSP44)	Anticancer	Wu et al. (2012)
Gymnemic acid IV	Antidiabetic, insulinotropic	Sugihara et al. (2000)
Dihydroxy gymnemic triacetate	Antidiabetic	Daisy et al. (2009)
Gymnemagenol	Anti-cancer	Khanna and Kannabiran (2009), Arunachalam et al. (2014, 2015)
Sylvepregosides A-D	Antidiabetic	Liu et al. (2021)
GA I	Anti-apoptotic	Wu et al. (2019)
Gymnemic acid	Antiobesity	Wang et al. (1998)
Ethanolic extract of leaves	Antidiabetic	Gupta and Seth (1962)
	Antimicrobial	Satdive et al. (2003)
Aqueous extract of leaves	Anti-arthritic	Malik et al. (2010)
	Anti-oedemic	Malik et al. 2008
	Antihyperglycaemic, insulinotropic, hypolipidemic and antioxidant	El Shafey et al. (2013), Yadav et al. (2010)
Petroleum ether leaf extract	Anti-arthritic	Malik et al. (2010)
	Antimicrobial	Saumendu et al. (2010)
Methanolic extract of *G. sylvestre*	Remedy to dental caries	ParimalaDevi and Ramasubramaniaraja (2010)
	Antidiabetic, regeneration of β-cells	Ahmed et al. (2010), Kumar et al. (2017), Mishal et al. (2020)
	Antiobesity	Kim et al. (2016)
	Antimicrobial	David and Sudarsanam (2013), Behuria et al. (2021)
	Anticancer	Agrawal et al. (2016)

(continued)

Table 8.2 (continued)

Gymnema preparation	Activity	References
Saponin fraction, chloroform and extract of leaves	Antimicrobial	Satdive et al. (2003), Khanna and Kannabiran (2009), Yogisha and Raveesha (2009), Saumendu et al. (2010), Bhuvaneswari et al. (2011)
Butanol fraction of hydroethanolic extract of leaves	Radioprotective	Bhatia et al. (2008), Sharma et al. (2009)
Hydro-alcoholic extract of leaves	Antihepatotoxicity	Srividya et al. (2010)
	Wound healer	Kiranmai et al. (2011), Malik et al. (2009)
	Antimicrobial	Saumendu et al. (2010)
	Insulinotropic, regeneration of pancreatic cells and glucose homeostatic action	Shanmugasundaram et al. (1990), Persaud et al. (1999), Liu et al. (2009)
	Antihyperlipidemic	Rachh et al. (2010)
	Antiobesity/weight control	Shigematsu et al. (2001a, b)
	Antiatheroscelrotic, hypolipidaemic	Bishayee and Chatterjee (1994);
Triterpene fraction	Antidiabetic	Shenoy et al. (2018)
	Antimicrobial	Behuria and Sahu (2020)
Ethanolic extract of roots	Anti-arthritic	Shankar and Rao (2008)
Gymnema based nanoparticles	Anti-inflammatory	Bharathy et al. (2021)
G. sylvestre loaded graphene oxide-polyhydroxybutyrate-sodium alginate	Wound healing	Daisy et al. (2020)
GS leaf extract fused Polycaprolactone nanofiber	Wound dressing	Subramanian et al. (2021)

8.5 Clinical Trials Validating Pharmacological Efficacy of *G. sylvestre*

A handful of reports on clinical trials supports the results observed in the various experimental models. Reduction in fasting Blood Glucose Level (BGL) and post-prandial BGL was observed in diabetic patients after administering *Gymnema* leaf extract (6–10 g) for 15–21 days (Leach 2007; Khare et al. 1983). The effective role of *Gymnema* in diabetes is supported by two significant studies done in the year 1990. In the first trial, effects of *Gymnema* extract (GS_4–400 mg/day) on 22 patients with type II diabetes for 18–20 months was evaluated. Significant reduction in blood glucose ($p < 0.001$), glycosylated hemoglobin ($p < 0.001$), and glycosylated plasma proteins ($p < 0.001$) was observed over the 18–20-month period due to the use of *Gymnema*, however, under conventional treatment, these values increased or remained unaffected (glibenclamide or tolbutamide; n = 25). Moreover, five subjects discontinued taking their conventional antidiabetic drugs and the blood glucose homeostasis was maintained (Baskaran et al. 1990).

GS_4 was also tested on 27 insulin-dependent diabetes mellitus (IDDM) patients using insulin therapy for 6–30 months. Significant reduction in glycosylated plasma protein ($p < 0.001$) in the first 6–8 months and serum amylase ($p < 0.001$) at 16–-18 months was noted. Serum C-peptide levels increased within 16–18 months ($p < 0.001$) as compared to conventional therapy. The conventional treatment group showed no significant improvement in glycemic control during the duration of the treatment. In addition to the above results patients taking GS_4 in both studies reported enhanced well-being, alertness and exhaustion. Also, reduced intake of conventional drugs was seen during the treatment period. No such effect was evident in either of these parameters in the conventional therapy group (Shanmugasundaram et al. 1990). Balasubramaniam et al. (1992) evaluated the hypoglycemic effect of *G. sylvestre* in 16 normal subjects and in 43 mild diabetic patients. Leaf powder at a dose of 10 g/day was given for 7 days. The hypoglycemic effect was at par with tolbutamide. Significant reduction in free fatty acids, serum triacylglycerol and cholesterol levels were observed in diabetic patients, whereas the tested parameters remained unaffected in normal individuals. Increased ascorbic acid and iron content were observed in both groups. Excretion of creatine was reduced in patients suffering from diabetes and no such effect was found in normal healthy volunteers. In another trial, 65 patients suffering from IDDM and NIDDM were each treated with leaf extract of GS (400 mg b.i.d.) for 90 days. Preprandial BGL, postprandial BGL and HbA1c decreased by 11%, 13% and 0.6%, respectively, as compared to the baseline value (Joffe 2001).

GSE (400 mg) in combination with hydroxycitric acid [HCA-SX (4667 mg)] and niacin-bound chromium [NBC (4 mg)] was given to 60 moderately obese subjects in the age group 21–50 (BMI-26 kg/m^2) for 8 weeks. At the end of the trial, BMI and body weight of the subjects decreased by 5–6%. Other parameters viz., food intake, total cholesterol, LDL and triglycerides amount were significantly reduced. On the other hand, increased levels of HDL and excretion of urinary fat metabolites was also observed (Preuss et al. 2004). The study shows that GSE + HCA-SX is a good formulation that can lead to obesity control and reduced blood lipid levels.

The GS extract, Om Santal Adivasi (OSA®), was given to patients with Type II DM at a dose of 1 g/day for 60 days and its effect on plasma insulin, glucose and C-peptide was studied. It was observed that the level of insulin and C-peptide significantly increased with reduction in fasting glucose level (from 162 to 119 mg/dL) and postprandial blood glucose levels (from 291 to 236 mg/dL (Al-Romaiyan et al. 2010).

Eight diabetic human subjects were given capsules containing GS leaf powder (1 g/day dose) for 30 days. A significant reduction in blood glucose, triglyceride, cholesterol and LDL levels was observed in all the patients (Li et al. 2015).

Suppression of sweet taste was done by rinsing the oral cavity with aqueous solution of GS of eight individuals. Thereafter, the subjects were administered with

200 g of 15% glucose solution in total. The effect of sweet taste suppression on postprandial gastrointestinal blood flow and gastric emptying was observed, which may affect carbohydrate metabolism (Kashima et al. 2017). Fifteen patients with impaired glucose tolerance were given *G. sylvestre* capsules (600 mg/day) orally for 12 weeks. Significant reduction in 2-h oral glucose tolerance test, A1Cbody weight and LDL cholesterol was observed (Gaytán Martínez et al. 2021).

8.6 Conclusions

Gymnema sylvestre has been described to contain potent antidiabetic agents and possess several therapeutic activities, like anti-arthritic, anti-inflammatory, antimicrobial, radioprotective, immunomodulatory, etc. (Fig. 8.5) None of the presently available antidiabetic drugs is capable of exerting such a diverse effect. This clearly reveals the ability of this "wonder plant" to manage/alleviate diabetes and its associated disorders. Quite a few commercially available formulations contain *Gymnema* as the only or one of the ingredients e.g. Now foods – *Gymnema sylvestre*, Holistix – *Gymnema sylvestre* Extract, Puritan's Pride – *Gymnema Sylvestre*, Himalaya – Meshashringi, Nutrigold – *Gymnema* gold, etc. Although these findings are strong indications of the antidiabetic potential of *Gymnema*, still further rigorous, specific, and methodological studies are needed to make this *wonder plant* a part of modern pharmacopeia.

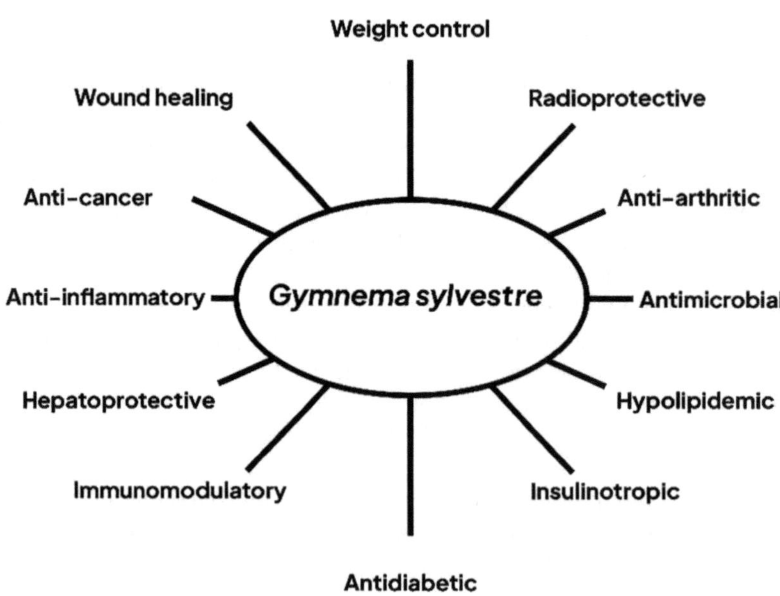

Fig. 8.5 Bioactivities of *Gymnema sylvestre*

References

Agrawal RC, Soni S, Jain N, Rajpoot J, Maheshwari SK (2016) Chemopreventive effect of *Gymnema sylvestre* in Swiss albino mice. Int J Sci Res Publ 6(1):78–83

Ahmed S, Anuntiyo J, Malemud CJ, Haqqi TM (2005) Biological basis for the use of botanicals in osteoarthritis and rheumatoid arthritis: a review. Evid Based Complement Alternat Med 2(3): 301–308

Ahmed AB, Rao AS, Rao MV (2010) *In vitro* callus and *in vivo* leaf extract of *Gymnema sylvestre* stimulate β-cells regeneration and anti-diabetic activity in Wistar rats. Phytomedicine 17(13): 1033–1039

Akhtar MS, Bhakuni V (2004) *Streptococcus pneumoniae* hyaluronate lyase: an overview. Curr Sci 86(2):285–295

Alkefai NH, Ahamad J, Amin S, Sharma M, Mir SR (2018) Arylated gymnemic acids from *Gymnema sylvestre* R. Br. as potential α-glucosidase inhibitors. Phytochem Lett 25:196–202

Alkefai NH, Amin S, Sharma M, Ahamad J, Mir SR (2019) New Olean-15-ene type gymnemic acids from *Gymnema sylvestre* (Retz.) R. Br. and their antihyperglycemic activity through α-glucosidase inhibition. Phytochem Lett 32:83–89

Al-Romaiyan A, Liu B, Asare-Anane H, Maity CR, Chatterjee SK, Koley N, Biswas T, Chatterji AK, Huang G-C, Amiel SA, Persaud SJ, Jones PM (2010) A novel *Gymnema sylvestre* extract stimulates insulin secretion from human islets *in vivo* and *in vitro*. Phytother Res 24:1370–1376

Arunachalam KD, Arun LB, Annamalai SK, Arunachalam AM (2014) Biofunctionalized gold nanoparticles synthesis from *Gymnema sylvestre* and its preliminary anticancer activity. Int J Pharm Pharm Sci 6(4):423–430

Arunachalam KD, Arun LB, Annamalai SK, Arunachalam AM (2015) Potential anticancer properties of bioactive compounds of Gymnema sylvestre and its biofunctionalized silver nanoparticles. Int J Nanomedicine 10:31–41

Babel RA, Dandekar MP (2021) A review on cellular and molecular mechanisms linked to the development of diabetes complications. Curr Diabetes Rev 17(4):457–473

Balasubramaniam KB, Arasaratnam R, Nageswaran A, Anushiyanthan S, Mugunthan N (1992) Studies on the effect of *Gymnema sylvestre* on diabetes. J Nat Sci Coun Srilanka 20(1):81–89

Baskaran K, Kizar Ahamath B, Radha Shanmugasundaram K, Shanmugasundaram E (1990) Antidiabetic effect of a leaf extract from *Gymnema sylvestre* in non-insulin–dependent diabetes mellitus patients. J Ethnopharmacol 30:295–305

Behuria HG, Sahu SK (2020) An anti-microbial Terpenoid fraction from *Gymnema sylvestre* induces flip-flop of fluorescent-phospholipid analogs in model membrane. Appl Biochem Biotechnol 192(4):1331–1345

Behuria HG, Arumugam GS, Pal CK, Jena AK, Sahu SK (2021) Lipid Flip-flop-inducing antimicrobial phytochemicals from *Gymnema sylvestre* are bacterial membrane permeability enhancers. ACS Omega 6:35667–35678

Bharathy MS, Jeyaleela GD, Vimala JDR, Agila A, Sheela SAM (2021) Comparison of anti-inflammatory activities of biogenic gymnema sylvestre-and panicum sumatrense-mediated titanium dioxide nanoparticles. Biomed Biotechnol Res J 5(4):405–411

Bhatia AL, Kamal R, Verma G, Sharma KV, Vats S, Jain M (2008) Radioprotective role of gymnemic acid on mice: study on hepatic biochemical alterations. Asian J Exp Sci 22(3): 427–432

Bhuvaneswari CH, Rao K, Giri A (2011) Evaluation of *Gymnema sylvestre* antimicrobial activity in methanol. Recent Res Sci Technol 3(8):73–75

Bishayee A, Chatterjee M (1994) Hypolipidaemic and antiatherosclerotic effect of oral *Gymnema sylvestre* R.Br. leaf extract in albino rats fed on a high fat diet. Phytother Res 8(2):118–120

Callaghan BC, Reynolds EL, Banerjee M, Chant E, Villegas-Umana E, Gardner TW, Votruba K, Giordani B, Pop-Busui R, Pennathur S, Feldman EL (2020) The prevalence and determinants of cognitive deficits and traditional diabetic complications in the severely obese. Diabetes Care 43(3):683–690

Daisy P, Eliza J, Mohamed Farook KA (2009) A novel dihydroxy gymnemic triacetate isolated from *Gymnema sylvestre* possessing normoglycemic and hypolipidemic activity on STZ-induced diabetic rats. J Ethnopharmacol 126(2):339–344

Daisy EAC, Rajendran NK, Houreld NN, Marraiki N, Elgorban AM, Rajan M (2020) Curcumin and *Gymnema sylvestre* extract loaded graphene oxide-polyhydroxybutyrate-sodium alginate composite for diabetic wound regeneration. React Funct Polym 154:104671

David BC, Sudarsanam G (2013) Antimicrobial activity of *Gymnema sylvestre* (Asclepiadaceae). J Acute Dis 2(3):222–225

Devi J, Arumugam M, Arivarasu A, Dhinakaran AK, Suresh P (2020) Preparation of herbal curd with *Gymnema sylvestre* and its characterization for the treatment of liver cancer abstract J Food Process Eng 43(3). https://doi.org/10.1111/jfpe.v43.3; https://doi.org/10.1111/jfpe.13338

Dholi SK, Raparla R (2014) *In vivo* antidiabetic evaluation of gymnemic acid in streptozotocin induced rats. Pharma Innov 3(7):82–86

Diwan PV, Margaret I, Ramakrishna S (1995) Influence of *Gymnema sylvestre* on inflammation. Inflammopharmacology 3(3):271–277

Dogar IA, Ali M, Yaqub M (1988) Effect of *Grewia asiatica*, *Gossypium herbacium* and *Gymnema sylvestre* on blood glucose, cholesterol and triglycerides levels in normoglycaemic and alloxan diabetic rabbits. J Pak Med Assoc 38:289–295

El Shafey AA, El-Ezabi MM, Seliem MM, Ouda HH, Ibrahim DS (2013) Effect of *Gymnema sylvestre* R. Br. leaves extract on certain physiological parameters of diabetic rats. J King Saud Univ Sci 25(2):135–141

Flier JS (2004) Obesity wars: molecular progress confronts an expanding epidemic. Cell 116: 337–350

Gaytán Martínez LA, Sánchez-Ruiz LA, Zuniga LY, González-Ortiz M, Martínez-Abundis E (2021) Effect of Gymnema sylvestre administration on glycemic control, insulin secretion, and insulin sensitivity in patients with impaired glucose tolerance. J Med Food 24(1):28–32

Gholap S, Kar A (2003) Efficacy of some plant extracts in regulating corticosteroidinduced hyperglycaemia in mice. Pharm Biol 41:315

Ghous T, Akhtar K, Andleeb S, Khizar S, Ali S, Mustafa RG, Shafique I, Naseer A (2021) Evaluation of α-glucosidase inhibition, antioxidant and antibacterial effects of *Gymnema sylvestre* R. Br. Bang J Bot 50(1):61–68

Grover JK, Yadav S, Vats V (2002) Medicinal plants of India with anti-diabetic potential. J Ethnopharmacol 81(1):81–100

Guariguata L, Whiting DR, Hambleton I, Beagley J, Linnenkamp U, Shaw JE (2014) Global estimates of diabetes prevalence for 2013 and projections for 2035. Diabetes Res Clin Pract 103(2):137–149

Gupta SS (1963) Effect of *Gymnema sylvestre* and *Pterocarpus marsupium* on glucose tolerance in albino rats. Ind J Med Res 51:716–721

Gupta SS, Seth CB (1962) Experimental studies on pituitary diabetes, part II. Comparison of blood sugar level in normal and anterior pituitary extract induced hyperglycaemic rats treated with a few Ayurvedic remedies. Ind J Med Res 50:708–714

Harada S, Kasahara Y (2000) Inhibitory effect of gurmarin on palatal taste responses to amino acids in the rat. Am J Physiol Regul Integr Comp Physiol 278(6):R1513–R1517

Hussain H, Nazir M, Saleem M, Al-Harrasi A, Green IR (2021) Fruitful decade of fungal metabolites as anti-diabetic agents from 2010 to 2019: emphasis on α-glucosidase inhibitors. Phytochem Rev 5:1–35

Ikemitsu H (1990) Effects of gymnemic acids in a large dose on the plasma glucose concentrations of rats. J Yonago Med Ass 41:414–431

Imoto T, Miyasaka A, Ishima R, Akasaka K (1991) A novel peptide isolated from the leaves of *Gymnema sylvestre* – I. Characterization and its suppressive effect on the neural responses to sweet taste stimuli in the rat. Comp Biochem Physiol A Comp Physiol 100(2):309–314

Imoto T, Sasamoto K, Ninomiya Y (2001) Beta-cyclodextrin inhibits the sweet taste suppressing activity of gurmarin by the formation of an inclusion complex with aromatic residues in gurmarin. Can J Physiol Pharmacol 79(10):836–840

Jamadagni PS, Pawar SD, Jamadagni SB, Gautam M, Gaidhani SN, Prasad GP, Gurav AM (2021) Recent updates in research on *Gymnema sylvestre*. Pharmacog Rev 15(30):128–133

Joffe D (2001) *Gymnema sylvestre* lowers HbA1c. Diabetes Control Newslett 76

Kashima H, Eguchi K, Miyamoto K, Fujimoto M, Endo MY, Aso-Someya N, Kobayashi T, Hayashi N, Fukuba Y (2017) Suppression of oral sweet taste sensation with *Gymnema sylvestre* affects postprandial gastrointestinal blood flow and gastric emptying in humans. Chem Senses 42(4):295–302

Katsukawa H, Imoto T, Ninomiya Y (1999) Induction of salivary gurmarin-binding proteins in rats fed gymnema-containing diets. Chem Senses 24(4):387–392

Khanna VG, Kannabiran K (2008) Antimicrobial activity of saponin fractions of the leaves of *Gymnema sylvestre* and *Eclipta prostrata*. World J Microbiol Biotechnol 24(11):2737–2740

Khanna VG, Kannabiran K (2009) Anticancer-cytotoxic activity of saponins isolated from the leaves of *Gymnema sylvestre* and *Eclipta prostrata* on HeLa cells. Int J Green Pharm 3:227–229

Khare A, Tondon R, Tewari J (1983) Hypoglycaemic activity of an indigenous drug (*Gymnema sylvestre, "Gurmar"*) in normal and diabetic persons. Indian J Physiol Pharmacol 27:257–258

Kiem PV, Yen DT, Hung NV, Nhiem NX, Tai BH, Trang DT, Yen PH, Ngoc TM, Minh CV, Park S, Lee JH (2020) Five new pregnane glycosides from *Gymnema sylvestre* and their α-glucosidase and α-amylase inhibitory activities. Molecules 25(11):2525

Kim HJ, Hong SH, Chang SH, Kim S, Lee AY, Jang Y, Davaadamdin O, Yu KN, Kim JE, Cho MH (2016) Effects of feeding a diet containing *Gymnema sylvestre* extract: attenuating progression of obesity in C57BL/6J mice. Asian Pac J Trop Med 9(5):437–444

Kiranmai M, Kazim SM, Ibrahim M (2011) Combined wound healing activity of *Gymnema sylvestre* and *Tagetes erecta* Linn. Int J Pharm Appl 2:135–140

Kirtikar K, Basu B (1998) Indian medicinal plants, vol 3. International Book Distributors, Deharadun

Kumar P, Rani S, Arunjyothi B, Chakrapani P, Rojarani A (2017) Evaluation of antidiabetic activity of *Gymnema sylvestre* and *Andrographis paniculata* in Streptozotocin induced diabetic rats. Int J Pharmacogn Phytochem Res 9:22–25

Kurihara Y (1992) Characteristics of antisweet substances, sweet proteins, and sweetness-inducing proteins. Crit Rev Food Sci Nutr 32(3):231–252

Leach MJ (2007) *Gymnema sylvestre* for diabetes mellitus: a systematic review. J Altern Complement Med 13(9):977–983

Lemon CH, Imoto T, Smith DV (2003) Differential gurmarin suppression of sweet taste responses in rat solitary nucleus neurons. J Neurophysiol 90(2):911–923

Li Y, Zheng M, Zhai X, Huang Y, Khalid A, Malik A, Shah P, Karim S, Azhar S, Hou X (2015) Effect of *Gymnema sylvestre*, *Citrullus colocynthis* and *Artemisia absinthium* on blood glucose and lipid profile in diabetic human. Acta Pol Pharm 72(5):981–985

Liu HM, Kiuchi F, Tsuda Y (1992) Isolation and structure elucidation of gymnemic acids, antisweet principles of Gymnema sylvestre. Chem Pharm Bull 40:1366–1375

Liu X, Ye W, Yu B, Zhao S, Wu H, Che C (2004) Two new flavonol glycosides from *Gymnema sylvestre* and *Euphorbia ebracteolata*. Carbohydr Res 339(4):891–895

Liu B, Asare-Anane H, Al-Romaiyan A, Huang G, Amiel SA, Jones PM, Persaud SJ (2009) Characterisation of the insulinotropic activity of an aqueous extract of *Gymnema sylvestre* in mouse β-cells and human islets of Langerhans. Cell Physiol Biochem 23(1–3):125–132

Liu M, Zhou T, Zhang J, Liao G, Lu R, Yang X (2021) Identification of C21 steroidal glycosides from *Gymnema sylvestre* (Retz.) and evaluation of their glucose uptake activities. Molecules 26(21):6549

Mahadeva Rao US, Shanmuga SC, Srinivasan S (2020) Gymnemic acid mitigates hyperglycemia by attenuating the hepatic glucose metabolic enzymes in high fat diet fed-low dose streptozotocin-induced experimental rodents. Res J Pharm Technol 13(2):719–726

Malik JK, Manvi FV, Alagawadi KR, Noolvi M (2008) Evaluation of anti-inflammatory activity of *Gymnema sylvestre* leaves extract in rats. Int J Green Pharm 2(2):114

Malik JK, Manvi FV, Nanjware BR, Singh S (2009) Wound healing properties of alcoholic extract of *Gymnema sylvestre* R. Br. leaves in rats. J Pharm Res 2(6):1029–1030

Malik JK, Manvi FV, Nanjware BR, Dwivedi DK, Purohit P, Chouhan S (2010) Anti-arthritic activity of leaves of *Gymnema sylvestre* R. Br. leaves in rats. Pharm Lett 2:336–341

Marsh P, Martin M (1992) Oral microbiology, vol 3. Chapman and Hall, London

Mishal A, Saravanan R, Atchitha SS, Santhiya K, Rithika M, Menaka SS, Thiruvalluvan T (2020) Effect of *Gymnema sylvestre* leaf extract on Streptozotocin induced diabetic rats. J Pharmacogn Phytochem 9(4):20–23

Miyasaka A, Imoto T (1995) Electrophysiological characterization of the inhibitory effect of a novel peptide gurmarin on the sweet taste response in rats. Brain Res 676(1):63–68

Murakami N, Murakami T, Kadoya M, Matsuda H, Yamahara J, Yoshikawa M (1996) New hypoglycemic constituents in "gymnemic acid" from *Gymnema sylvestre*. Chem Pharm Bull 44(2):469–471

Murata Y, Nakashima K, Yamada A, Shigemura N, Sasamoto K, Ninomiya Y (2003) Gurmarin suppression of licking responses to sweetener-quinine mixtures in C57BL mice. Chem Senses 28(3):237–243

Nakamura Y, Tsumura Y, Tonogai Y, Shibata T (1999) Fecal steroid excretion is increased in rats by oral administration of gymnemic acids contained in *Gymnema sylvestre* leaves. J Nutr 129(6):1214–1222

Nayak BS, Sobrian A, Latiff K, Pope D, Rampersad A, Lourenço K, Samuel N (2014) The association of age, gender, ethnicity, family history, obesity and hypertension with type 2 diabetes mellitus in Trinidad. Diabetes Metab Syndr 8(2):91–95

Ninomiya Y, Imoto T (1995) Gurmarin inhibition of sweet taste responses in mice. Am J Physiol Regul Integr Comp Physiol 268(4):R1019–R1025

Ninomiya Y, Inoue M, Imoto T, Nakashima K (1997) Lack of gurmarin sensitivity of sweet taste receptors innervated by the glossopharyngeal nerve in C57BL mice. Am J Phys 272(3): R1002–R1006

Ninomiya Y, Inoue M, Imoto T (1998) Reduction of the suppressive effects of gurmarin on sweet taste responses by addition of beta-cyclodextrin. Chem Senses 23(3):303–307

Ninomiya Y, Imoto T, Sugimura T (1999) Sweet taste responses of mouse chorda tympani neurons: existence of gurmarin-sensitive and -insensitive receptor components. J Neurophysiol 81(6): 3087–3091

Ohkuri T, Yasumatsu K, Horio N, Jyotaki M, Margolskee RF, Ninomiya Y (2009) Multiple sweet receptors and transduction pathways revealed in knockout mice by temperature dependence and gurmarin sensitivity. Am J Physiol Regul Integr Comp Physiol 296(4):R960–R971

ParimalaDevi B, Ramasubramaniaraja R (2010) Pharmacognostical and antimicrobial screening of *Gymnema sylvestre* R.Br, and evaluation of Gurmar herbal tooth paste and powder, composed of *Gymnema sylvestre* R.Br, extracts in dental caries. Int J Pharm BioSci 1(3):1–16

Persaud SJ, Al-Majed H, Raman A, Jones PM (1999) *Gymnema sylvestre* stimulates insulin releases in vitro by increased membrane permeability. J Endocrinol 163:207–212

Pfaffman C (1959) In: Field J (ed) Handbook of physiology, vol 1. American Physiological Society, Washington, DC

Porchezhian E, Dobriyal RM (2003) An overview on the advances of *Gymnema sylvestre*: chemistry, pharmacology and patents. Pharmazie 58(1):5–12

Prakash AO, Mather S, Mathur R (1986) Effect of feeding Gymnema sylvestre leaves on blood glucose in beryllium nitrate treated rats. J Ethnopharmacol 18:143–146

Pramanick DD (2016) Anatomical studies on the leaf of *Gymnema sylvestre* (Retz.) R. Br. ex Schult. (Apocynaceae): A magical herbal medicine for diabetes. Int J Herb Med 4(1):70–72

Preuss HG, Bagchi D, Bagchi M, Rao CVS, Dey DK, Satyanarayana S (2004) Effects of a natural extract of (−)-hydroxycitric acid (HCA-SX) and a combination of HCA-SX plus niacin-bound chromium and *Gymnema sylvestre* extract on weight loss. Diabetes Obes Metab 6:171–180

Rachh PR, Rachh MR, Ghadiya NR, Modi DC, Modi KP, Patel NM, Rupareliya MT (2010) Antihyperlipidemic activity of *Gymnema sylvestre* R. Br. leaf extract on rats fed with high cholesterol diet. Int J Pharmacol 6(2):138–141

Rajkumar PV, Prakasam A, Rajeshkumar S, Gomathi M, Anbarasan PM, Chandrasekaran R (2020) Green synthesis of silver nanoparticles using *Gymnema sylvestre* leaf extract and evaluation of its antibacterial activity. S Afr J Chem Eng 32(1):1–4

Ramírez-Alarcón K, Victoriano M, Mardones L, Villagran M, Al-Harrasi A, Al-Rawahi A, Cruz-Martins N, Sharifi-Rad J, Martorell M (2021) Phytochemicals as potential Epidrugs in type 2 diabetes mellitus. Front Endocrinol 12:656978

Rao GS, Sinsheimer JE (1968) Structure of gymnemagenin Chem Comm 24:1681–1682

Sahu NP, Mahato SB, Sarkar SK, Poddar G (1996) Triterpenoid saponins from *Gymnema sylvestre*. Phytochemistry 41(4):1181–1185

Sanematsu K, Yasumatsu K, Yoshida R, Shigemura N, Ninomiya Y (2005) Mouse strain differences in Gurmarin-sensitivity of sweet taste responses are not associated with polymorphisms of the sweet receptor gene, Tas1r3. Chem Senses 30(6):491–496

Satdive RK, Abhilash P, Fulzele DP (2003) Antimicrobial activity of *Gymnema sylvestre* leaf extract. Fitoterapia 74(7):699–701

Saumendu DR, Sarkar K, Dipankar S, Singh T, Prabha B (2010) *In vitro* antibiotic activity of various extracts of *Gymnema sylvestre*. Int J Pharm Res Dev 2:1–3

Shankar KR, Rao BG (2008) Anti-arthritic activity of *Gymnema sylvestre* root extract. Biosci Biotechnol Res Asia 5(1):469–471

Shanmugasundaram KR, Panneerselvam C, Samudram P, Shanmugasundaram ER (1983) Enzyme changes and glucose utilisation in diabetic rabbits: the effect of *Gymnema sylvestre*, R. Br. J Ethnopharmacol 7(2):205–234

Shanmugasundaram ERB, Rajeswari G, Baskaran K, Kumar BRR, Shanmugasundaram KR, Ahmath BK (1990) Use of *Gymnema sylvestre* leaf extract in the control of blood glucose in insulin dependent diabetes mellitus. J Ethnopharmacol 30:281–294

Sharma KV, Singh U, Vats S, Priyadarsini KI, Bhatia AL, Kamal R (2009) Evaluation of evidenced-based Radioprotective efficacy of *Gymnema sylvestre* leaves in mice brain. J Environ Pathol Toxicol Oncol 28(4):313–326

Shenoy RS, Prashanth KV, Manonmani HK (2018) *In vitro* antidiabetic effects of isolated triterpene glycoside fraction from *Gymnema sylvestre*. Evid Based Complement Alternat Med. https://doi.org/10.1155/2018/7154702

Shigematsu N, Asano R, Shimosaka M, Okazaki M (2001a) Effect of long term-administration with Gymnema sylvestre R. Br. on plasma and liver lipid in rats. Biol Pharm Bull 4(6):643–649

Shigematsu N, Asano R, Shimosaka M, Okazaki M (2001b) Effect of administration with the extract of *Gymnema sylvestre* R. Br leaves on lipid metabolism in rats. Biol Pharm Bull 24(6): 713–717

Shigemura N, Nakao K, Yasuo T, Murata Y, Yasumatsu K, Nakashima A, Katsukawa H, Sako N, Ninomiya Y (2008) Gurmarin sensitivity of sweet taste responses is associated with co-expression patterns of T1r2, T1r3, and gustducin. Biochem Biophys Res Commun 367(2): 356–363

Sigoillot M, Brockhoff A, Neiers F, Poirier N, Belloir C, Legrand P, Charron C, Roblin P, Meyerhof W, Briand L (2018) The crystal structure of gurmarin, a sweet taste–suppressing protein: identification of the amino acid residues essential for inhibition. Chem Senses 43(8): 635–643

Singh DK, Kumar N, Sachan A, Lakhani P, Tutu S, Nath R et al (2017) Hypolipidaemic effects of *Gymnema sylvestre* on high fat diet induced dyslipidaemia in Wistar rats. J Clin Diagn Res 11(5):FF01–FF05

Sinha P, Arunachalam KD, Nagarajan SK, Madhavan T, Jayakumar AR, Saiyad Musthafa M (2021) Radio-protective efficacy of *Gymnema sylvestre* on *Pangasius sutchi* against gamma (60Co) irradiation. Int J Rad Biol 2021:13. https://doi.org/10.1080/09553002.2022.1998701

Sinsheimer JE, Rao GS (1970) Constituents from *Gymnema sylvestre* leaves VI: acylated genins of the gymnemic acids-isolation and preliminary characterization. J Pharma Sci 59:629–632

Srivastava Y, Bhatt HV, Prem AS (1985) Hypoglycemic and life-prolonging properties of Gymnema sylvestre leaf extract in diabetic rats. Isr J Med Sci 21:540–542

Srividya AR, Varma SK, Dhanapal SP, Vadivelan R, Vijayan P (2010) *In vitro* and *in vivo* evaluation of hepatoprotective activity of *Gymnema sylvestre*. Int J Pharm Sci Nanotech 2:768–773

Steppan CM, Bailey ST, Bhat S, Brown EJ, Banerjee RR, Wright CM, Patel HR, Ahima RS, Lazar MA (2001) The hormone resistin links obesity to diabetes. Nature 409(6818):307–312

Stocklin W (1968) Gymnestrogenin, ein neues Pentahydroxytriterpen aus den Blättern von *Gymnema sylvestre* R. BR. Glykoside und Aglykone, 309. Mitteilung. Helv Chim Acta 51(6): 1235–1242

Stöcklin W (1969) Gymnemagenin, Struktur und O-Isopropylidenderivate Glykoside und Aglykone. 313. Mitteilung Helv Chim Acta 52(2):365–370

Stocklin W, Weiss E, Reichstein T (1967) Gymnemasiure, das antisaccharine Prinzip von Gymnema sylvestre. lsolierungen und Identifizierungen. Helv Chim Acta 50:474–490

Subramanian S, Karuppannan SK, Ramalingam R, Dowlath MJH, Khalith SM, Musthafa SA, Chitra V, Munuswamy-Ramanujam G, Arunachalam KD (2021) Effect of gamma sterilization on *Gymnema sylvestre* leaf extract fused Polycaprolactone nanofiber for effective wound dressing applications. Mater Lett 300:130145

Sugihara Y, Nojima H, Matsuda H, Murakami T, Yoshikawa M, Kimura I (2000) Antihyperglycemic effects of gymnemic acid IV, a compound derived from Gymnema sylvestre leaves in streptozotocin-diabetic mice. J Asian Nat Prod Res 2:321–327

Suttisri R, Lee IS, Kinghorn AD (1995) Plant-derived triterpenoid sweetness inhibitors. J Ethnopharmacol 47:9–26

Tahir M, Rasheed MA, Niaz Q, Ashraf M, Anjum AA, Ahmed MU (2017) Evaluation of antibacterial effect of *Gymnema sylvestre* R. Br. Species cultivated in Pakistan. Pak Vet J 37(3):245–250

Terasawa H, Miyoshi M, Imoto T (1994) Effects of long-term administration of *Gymnema sylvestre* watery-extract on variations of body weight, plasma glucose, serum triglyceride, total cholesterol and insulin in Wistar fatty rats. Yonago Acta Med 37:117–127

United States (1975) National Commission on Diabetes. Report of the National Commission on Diabetes to the Congress of the United States. Bethesda, Maryland: U.S. Department of Health, Education and Welfare. Publication no. 76–1021, Vol. 1

Van Itallie TB (1985) Health implications of overweight and obesity in the United States. Ann Intern Med 103:983–988

Vats S, Kamal R (2013) *In vivo* and *in vitro* evaluation of sterols from *Gymnema sylvestrte* R. Br. Pak J Biol Sci 16:1771–1775

Wang LF, Luo H, Miyoshi M, Imoto T, Hiji Y, Sasaki T (1998) Inhibitory effect of gymnemic acid on intestinal absorption of oleic acid in rats. Can J Physiol Pharmacol 76(10–11):1017–1023

Wild S, Roglic G, Green A, Sicree R, King H (2004) Global prevalence of diabetes: estimates for the year 2000 and projections for 2030. Diabetes Care 27:1047–1053

Wu X, Mao G, Fan Q, Zhao T, Zhao J, Li F, Yang L (2012) Isolation, purification, immunological and anti-tumor activities of polysaccharides from *Gymnema sylvestre*. Food Res Int 48(2): 935–939

Wu Y, Hu Y, Yuan Y, Luo Y, Lai D, Zhou H, Tong Z, Liu D (2019) Gymnemic acid I triggers mechanistic target of rapamycin-mediated β cells cytoprotection through the promotion of autophagy under high glucose stress. J Cell Physiol 234(6):9370–9377

Xu R, Yang Y, Zhang Y, Ren F, Xu J, Yu N, Zhao Y (2015) New pregnane glycosides from *Gymnema sylvestre*. Molecules 20(2):3050–3066

Yackzan KS (1966) Biological effects of Gymnema sylvestre fractions. Alabama Journal of Medical Sciences 3:1–9

Yadav M, Lavania A, Tomar R, Prasad GBKS, Jain S, Yadav H (2010) Complementary and comparative study on hypoglycemic and antihyperglycemic activity of various extracts of Eugenia jambolana seed, Momordica charantia fruits, Gymnema sylvestre, and Trigonella foenum graecum seeds in rats. Appl Biochem Biotechnol 160(8):2388–2400

Yamada A, Nakamura Y, Sugita D, Shirosaki S, Ohkuri T, Katsukawa H, Nonaka K, Imoto T, Ninomiya Y (2006) Induction of salivary kallikreins by the diet containing a sweet-suppressive peptide, gurmarin, in the rat. Biochem Biophys Res Commun 346(2):386–392

Yasumatsu K, Kusuhara Y, Shigemura N, Ninomiya Y (2007) Recovery of two independent sweet taste systems during regeneration of the mouse chorda tympani nerve after nerve crush. Eur J Neurosci 26(6):1521–1529

Ye WC, Zhang QW, Liu X, Che CT, Zhao SX (2000) Oleanane saponins from *Gymnema sylvestre*. Phytochemistry 53(8):893–899

Yogisha S, Raveesha KA (2009) *In-vitro* antibacterial effect of selected medicinal plant extracts. J Nat Prod 2:64–69

Yoshie S, Miyasaka A, Imoto T (1994) Histological localization of the sweet taste receptor in rat taste buds by the use of gurmarin, a sweet taste-suppressing peptide. Arch Histol Cytol 57(5): 531–534

Yoshikawa K, Amimoto K, Arihara S, Matsuura K (1989) Structure studies of new anti-sweet constituents from *Gymnema sylvestre*. Tetrahedron Lett 30:1103–1106

Yoshikawa K, Nakagawa M, Yamamoto R, Arihara S, Matsuura K (1992) Antisweet natural products V. Structures of gymnemic acids VIII-XII from Gymnema sylvestre. Chem Pharm Bull 40:1779–1782

Yoshikawa K, Kondo Y, Arihara S, Matsura K (1993) Antisweet natural products. IX. Structures of gymnemic acids XV–XVIII from Gymnema sylvestre. Chem Pharm Bull 41:1730–1732

Yoshikawa M, Murakami T, Kadoya M, Li Y, Murakami N, Yamahara J, Matsuda H (1997a) Medicinal foodstuffs. IX. The inhibitors of glucose absorption from the leaves of *Gymnema sylvestre* R. BR. (Asclepiadaceae): structures of gymnemosides a and b. Chem Pharm Bull 45(10):1671–1676

Yoshikawa M, Murakami T, Matsuda H (1997b) Medicinal foodstuffs. X. Structures of new triterpene glycosides, gymnemosides-c, -d, -e, and -f, from the leaves of *Gymnema sylvestre* R. Br.: influence of gymnema glycosides on glucose uptake in rat small intestinal fragments. Chem Pharm Bull 45(12):2034–2038

Zarrelli A, DellaGreca M, Ladhari A, Haouala R, Previtera L (2013) New triterpenes from *Gymnema sylvestre*. Helv Chim Acta 96(6):1036–1045

Zhang MQ, Liu Y, Xie SX, Xu TH, Liu TH, Xu YJ, Xu DM (2012) A new triterpenoid saponin from *Gymnema sylvestre*. J Asian Nat Prod Res 14(12):1186–1190

Zhen H, Xu S, Pan X (2001) The pharmacognostical identification of peel of *Gymnema sylvestre*. Zhong Yao Cai 24(2):95–97

Zhen HS, Zhu XY, Lu RM, Liang J, Qiu Q, Meng QM (2008) Research on chemical constituents from stem of *Gymnema sylvestre*. Zhong Yao Cai 31(8):1154–1156

Zhu XM, Xie P, Di YT, Peng SL, Ding LS, Wang MK (2008) Two new triterpenoid saponins from *Gymnema sylvestre*. J Integr Plant Biol 50(5):589–592

Chapter 9
Medicinal and Aromatic Plants of India Used in the Treatment of Skin Disorders

Mohammed Abdul Rasheed Naikodi

Abstract Skin is the largest protective organ of human body. Several studies have been carried out on medicinal plants species, where their phytochemicals played an important role in treating and preventing chronic diseases, in particular skin disorders, like vitiligo, eczema, psoriasis etc. Major groups of phytochemicals are known to help prevent or treat the skin ailments due to their potential antioxidant, anticancer, antifungal, antibacterial, antiviral and immunomodulators properties. Phytochemicals can modify the inflammation processes of skin and intensify the possible anti-aging, anti-cancer effects. They can also be used against atopic dermatitis, psoriasis, and vitiligo. Therefore, as of now, there exists a large scope for further systematic research in screening Indian medicinal plants for potent phytochemicals and assessing their potential against different types of diseases.

Keywords Medicinal plants · Phytochemistry · Active principle · Isolation · Standardization

9.1 Introduction

Nature provides everything for human being needed to survive and thrive. India has a vast biodiversity and rich ancient heritage of traditional medicinal knowledge. Traditional medicine has a long documented history and has been used by people since time immemorial.

In pharmaceutical landscape, plants with a long history of use provide a rich source of ethnomedicinal information on bioactive substances to treat ailments and diseases. Medicinal plants of India are considered to be a rich repository of bioactive compounds possessing varied therapeutic properties including antitumor, antimicrobial, antifungal, anti-inflammatory, antiviral, and analgesic properties etc.

M. A. R. Naikodi (✉)
Drug Standardization Research Unit, National Research Institute of Unani Medicine for Skin Disorders, Under CCRUM, Ministry of Ayush, Hyderabad, Telangana, India

Á. Máthé, I. A. Khan (eds.), *Medicinal and Aromatic Plants of India Vol. 2*,
Medicinal and Aromatic Plants of the World 9,
https://doi.org/10.1007/978-3-031-44914-7_9

The World Health Organization's Traditional Medicine Strategy 2014–2023 aims to strengthen the role of traditional medicine by keeping populations healthy and prioritizing health services and systems, including traditional and complementary medicine products, practices and practitioners (Anonymous 2017; Burton et al. 2015). Since centuries the medicinal plants knowledge has been accumulated in India through Indigenous Systems of Medicines such as Ayurveda, Unani and Siddha (ASU). Traditional healers reported to have used 2500 plant species as medicine in India (Pei 2001). In traditional systems of medicine several potential medicinal plants have been recommended for the treatment of various diseases particularly in skin diseases in day to day practice.

This chapter focuses on medicinal plants used for the treatment of skin disorders, such as *Psoralea corylifolia* L.; *Ammi majus* L.; *Cassia tora* L.; *Curcuma longa* L.; *Ruta graveolens* L.; *Aloe barbadensis Mill* etc. Plant based drugs prepared from them are popular and known or earlier reported to be useful for their therapeutic potentials in treating leucoderma or vitiligo, eczema, leprosy, psoriasis, scabies, ulcers, pigment formation, inflammatory diseases of the skin (Chopra et al. 1956; Kirtikar and Basu 1982; Anonymous 1976; Nadkarni 1976a).

The skin is a largest organ of human body and an ever changing organ which contains several specialized cells. The skin functions as a protective barrier and act as interfaces and sometimes hostile environment. Skin, in spite of serving as a protective shield against heat, light, injury and infections also regulates human body temperature, stores water, fat, vitamin D. It can also feel and sense, painful and pleasant stimuli etc. Skin is one of the most vulnerable organs of the body. It is visible that skin disorders can lead to physiological stress. Though seldom life threatening, skin disorders become uncomfortable and may cause chronic disabilities. Deficiency of essential nutrients in the human body, such as beta-carotene, vitamin B complex and vitamins C and E and result the drying of the skin.

Skin disorders occur due to improper functioning of internal processes and metabolism in the human body. When certain chemicals present *in vivo* do not react properly, or in certain cases hormones are not properly balanced, the result will manifest itself in the skin with redness, blistering, itchiness etc. Potent medicinal plants are used to treat these symptoms and to prevent the skin conditions from manifesting themselves again. Skin disorders is a most common ailment, and it affects all ages from the neonate to the elderly and produces harm in number of ways such as skin rashes, fungal infections, bacterial infections, viral infections, parasitic infections, pigmentation disorders, tumour or cancer etc. Conventional treatment for such skin diseases are through topical use of antibacterial and antifungal agents etc., and for oral treatment use antibiotics, antifungal and antiviral agents, corticosteroids (autoimmune diseases), immunosuppressants (ex. azathioprine and methotrexate) to treat different conditions including acute to chronic cases of psoriasis and eczema as in Fig. 9.1.

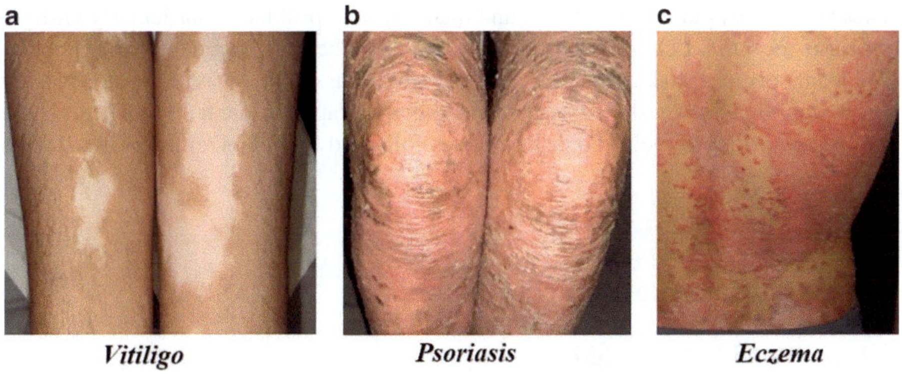

Fig. 9.1 Types of skin diseases (**a**) Vitiligo (**b**) Psoriasis (**c**) Eczema

9.2 Common Indian Home Remedies for Skin Care

Home remedies to look after the skin has been recorded by several authors. According to Frankova (2016) and Anonymous (1996) the major treatments are as follows

1. *Azadirachta indica* aqueous leaf extract is used topical externally on boils and blisters shows quick response in healing.
2. Sesame seeds and turmeric to be grind in equal ratio or quantity along with sufficient quantity of water. Apply to the face or discolored skin to nourish.
3. Regularly before going to bed, the face should be washed thoroughly and then apply a paste made up of one teaspoon of coriander juice mixed with a pinch of turmeric powder which keeps the skin good and smooth.
4. Coconut oil has anti-inflammatory, antioxidant, and healing properties and help to keep the skin moisturize and soothe your skin barrier.
5. A paste of fresh fenugreek leaves can be applied over the skin and then washed off with warm water before going to bed helps to prevents pimples formation over the skin.
6. Apply externally papaya juice or take the mashed papaya to the affected areas over the skin to nourish.
7. Aloe vera may stimulate towards the new cell growth showed healing properties. It is effective in soothes the skin and moisturizes without clogging pores.
8. Turmeric acts as an antioxidant that helps skin in achieving wonderful glow and rejuvenated. Also curcumin from turmeric acts an anti-inflammatory agent and helps to remove puffiness.

9.3 Plants in India Used for the Treatment of Skin Diseases

The most common plant species that are easily available in the surroundings of Indian households include *Curcuma longa, Aloe vera, Ocimum sanctum, Azadiracta indica, Cassia tora, Ruta graveolens* etc. which are widely used in dermatology.

These medicinal plants help to cure and maintain skin problems, particularly vitiligo, psoriasis, eczema, itching, roughness of skin etc. The most frequently used medicinal plant species for skin disorders in India are discussed below with regards to their known application in skin diseases and other therapeutic areas. The species, are ranked according to their vernacular names followed by botanical name in brackets are as follows.

*9.3.1 Babchi (***Psoralea corylifolia** *L.)*

9.3.1.1 Botanical Description

The *Psoralea* genus comprises around 130 species. It belongs to the Leguminosae family. *Psoralea corylifolia* L. is available in semi-arid regions. It is a small and erect annual herb found throughout the loamy, sandy plains of East and Central India growing up to 60–120 cm high. Seeds are flat, oblong and blackish brown in colour. Leaf is round, simple and at apex appears as mucronate. Flowers are yellowish purple in colour, fruits are about 5–6 mm long and compressed. It is the one of the most potential widely used since ages for the treatment of skin disorders in traditional system of medicine (Yadava and Verma 2005) and widely exploited for its enchanted effect against skin diseases such as vitiligo, psoriasis, and leprosy (Sah et al. 2006; Sharma et al. 2001; Khushboo et al. 2010). The plant has been used in the Indian traditional systems of medicine such as Ayurveda, Unani, Siddha etc. and very popular in Ayurveda known as *Bakuchi* and in Unani called as *Babchi* which is widely used in Ayurveda and Unani to treat vitiligo, infective and inflammatory diseases of the skin.

9.3.1.2 Crude Drug Used

Psoralea corylifolia seed possesses various biological and pharmacological properties like chemoprotective, antioxidant, stimulant, anti-bacterial, anti-inflammatory, aphrodisiac, cytotoxic, cardiac, astringent, diuretic in nature and blood purifying properties. The seed powder and seed oil used for leprosy and leukoderma and applied externally in the form of paste or ointment (Panda 2000; Nadkarni 1976b) in case of vitiligo treatment. It is also helpful in the mucomembranous disorders, dermatitis, inflammatory diseases, and oedematous skin conditions (Sharma et al. 2001; Rajpal 2005) and possesses blood purifying properties. Seeds are found beneficial in the treatment of various conditions of skin such as itching eruptions, itching red papules, extensive eczema, dermatosis and scabies (Wang and Wang 2007).

Seeds of *P. corylifolia* L. mostly used as one of the ingredient in preparation of polyherbal formulations – relieve various problems such as Vitiligo or Bars or Leucoderma, skin rashes and other infections associated with skin. Skin disorders affect nearly 2% of global population and suffer with other conditions like cardiovascular, nephritis, osteoporosis, and cancer. Seeds are reported to very effective in the treatment and management of vitiligo thus named as Kushtanashini -leprosy destroyer (Cardoso et al. 2002; Bishnoi and Parsad 2018; Kovacs et al. 2018; Anonymous 1999, 2007; Wilson et al. 1999).

9.3.1.3 Major Chemical Constituents and Bioactive Compounds

Chemical investigations led to the isolation of the main active principle psoralen, in 1933 by Jois for the first time: later Späth synthesized this compound (Späth and Holzen 1933; Manga and Mutasim 2007). The class of furanocoumarins compounds in *Psoralea corylifolia* species contain psoralens which promote to the formation of pigmentation (Sebastian 2006). Psoralen reported to stimulate the formation of melanin on exposure of skin to UV or under sun light and used in the treatment of skin diseases particularly vitiligo (Manga and Mutasim 2007).

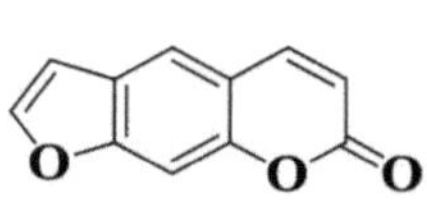

Psoralen

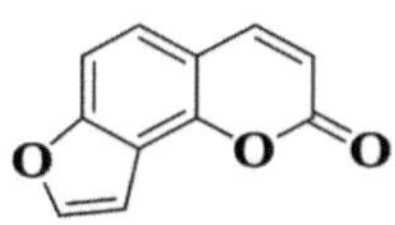

Isopsoralen (or angelicin)

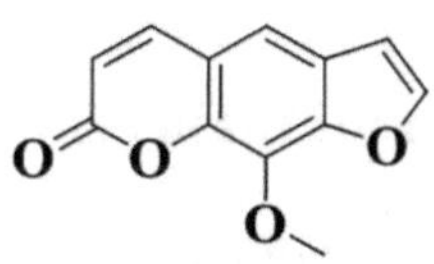

8-methoxypsoralen

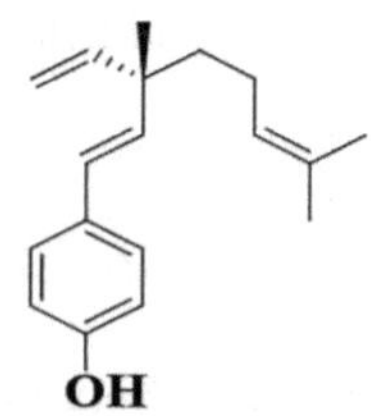

Bakuchiol

The isolated compounds of total 155 bioactive principles belonging to different classes of flavones, flavonoids, chalcones, coumarins, meroterpenes, stigmasterols, etc. from *P. corylifolia* crude extracts. The prominent phytochemicals which are psoralen, Bakuchiol and isopsoralen present in the plants are mainly responsible for the vitiligo and anti-psoriasis properties.

Psoralen is a furocoumarin (IUPAC: *7H*-Furo[3,2-*g*]chromen-7-one) it intercalates with DNA which lastly inhibiting DNA synthesis and cell division. Psoralen majorly used under photochemotherapy in the presence of high-intensity long-wavelength region of UV-A irradiation. The psoralen based herbal products can be use among children topically or orally for well-defined patches or lesions for repigmentation, whereas oral administered psoralens are found effective when treating multiple lesions (Manga and Mutasim 2007).

Isopsoralen (or angelicin) was known to be a photosensitive compound isolated from *P. corylifolia* (Ji and Xu 1995). Psoralen and isopsoralen are photosensitive compounds, and photosensitivity of psoralen was much better when compared to that of isopsoralen. It plays a prominent role in treating vitiligo and in addition other skin diseases which shows good effect on psoriasis and alopecia areata (Wang and Wang 2007). The efficacy of topical psoralen plus UV-A (PUVA) with 8-methoxypsoralen (8-MOP) gel showed more effective in treating the localized forms of eczema, psoriasis and in some cases of palmoplantar pustulosis (a chronic skin condition where blisters and fluid-filled bumps appear over palms, hands and feet soles), which avoids undesirable side-effects of systemic psoralens (Engin and Oguz 2005).

Bakuchiol was also isolated earlier from this plant. It is known to be a potent antimicrobial agent against a range of oral microorganisms. It is used in food additives and mouthwash preparations, and also to treat and prevent dental caries (Katsura et al. 2001). Bakuchiol *(IUPAC: 4-[(1E,3S)-3-ethenyl-3,7-dimethylocta-1,6-dienyl]phenol*) a meroterpene compound and a derivative of resveratrol having antifungal and antitumour effects. Bakuchiol showed potent cytotoxic agent against leukemia cancer cells by exhibiting concentration dependent growth inhibition (Majeed et al. 2012).

Another furocoumarin class of compound is 5-methoxypsoralen, occurs natural in the plant, it is used in combination with UV-A light irradiation (PUVA - as psoralen with UV) to treat vitiligo and psoriasis. In vitiligo patients, PUVA therapy 5-methoxypsoralen produces a dose-related increment in cutaneous photosensitivity (Li et al. 2016; Ruan et al. 2007; Beier et al. 1983; Shailajan et al. 2012; Ali et al. 2008b) and helps to initiate repigmentation of lesions.

9.3.2 *Atrilal* (Ammi majus *L.)*

9.3.2.1 Botanical Description

Ammi majus L. belongs to Apiaceae family commonly known as Bishop's weed and also as Atrilal. It has tap-roots and an erect stem, experimentally cultivated in many parts of India mainly Jammu and Kashmir (Cherian and Bhambri 2010). It is native to Egypt and widely distributed in the regions of Europe, The Mediterranean and West Asia. *A. majus* is an erect branching annual herb of 1.5–2.0 metres height. Macroscopic characters of fruits are glabrous, cylindrical to oblong-obovoid, having dimensions of 2.5–3.0 mm long × 1 mm wide, yellow to brown in colour, pedicel attached, entire cremocarps (schizocarps) and possess separate mericarps, the stylopods on the verge or tip of the fruit are bifid and free ends are curved along the dorsal region. The dorsal region of mericarp appeared convex, having five prominent and longitudinal ridges starting from base to apex and 4 brown colour furrows. In TS endocarpal cells are in rectangular and elongated form. Endosperm is polygonal, thick walled 16–60 μ × 9–30 μ size cells, having fixed oils and round aleurone grains of 5–12 μ in diameter size (Usmani et al. 2021).

Thesowing time of fruits significant rainfall in region, temperature, humidity and crop requires mild cool climate and warm, sometimes dry weather for ideal growth (Bhambri et al. 2012). Fruits are sown in well fertilized soil and raised beds; germination 10 - 12 days after sowing and seedlings are transplanted, fruiting age is attained in 50–60 days. Fruits are harvested when umbels turn to light brown and dried under shades (Bhattacharjee 2004; Kokate et al. 2012).

9.3.2.2 Crude Drug Used

Fruits have emmenagogic, as well as a blood purifying and diuretic effect. It is highly used in the treatment of leprosy or vitiligo, urinary and digestive disorders, skin disorders. Fruits are mainly recommended in the treatment of skin diseases particularly vitiligo which can be used either (i) by oral administration, (ii) by local topical application at the affected sites or lesions followed by sunlight or ultra-violet lamp exposure, or (iii) by a combination of (i) and (ii).

9.3.2.3 Major Chemical Constituents and Bioactive Compounds

Ammi majus contains furanocoumarins (8-methoxypsoralen or xanthotoxin, bergapten, imperatorin, isopimpinellin). Schönberg and Sina have reported xanthotoxin present in the fruits (Schönberg and Sina 1948). The phytochemical examination of *A. majus* revealed the presence of coumarins, furocoumarins, flavonoids that are mainly responsible for various pharmacological activities, in particular to leucoderma and vitiligo. The major phytoconstituents ammoidin, ammajin, furocoumarins, flavonoids, isoimperatorin, iso-pimpinallin are present in the seed and fruit. Some chemical compounds are found in high concentrations: these include nonhydroxylic coumarins, ammajin, marmesin, ammirin, khellin alloimperatorin, and acetylated flavonoids, visnagin, essential oils etc. (Akhtar et al. 2010; Elgamal et al. 1993; Ossenkoppele et al. 1991). Methoxsalen a naturally occurring photosenstitive bioactive molecules isolated with other compounds *viz.* proteins contain 13.83%, fixed oil as 12.92%, oleoresin as 4.76%, acrid oil as 3.2%, glucoside as 1% and tannin as 0.45% (Anonymous 1987; Al-Snafi 2013; Usmani et al. 2021). The photosensitizing activity of furanocoumarins aids to cure certain skin disorders such as leucoderma, psoriasis, and hypopigmentation in the disease called tinea versicolor (Staniszewska et al. 2003; Bartnik and Mazurek 2016). These furanocoumarins stimulate the production of pigmentation on the skin when exposed to ultra-violet light, and also generate melanin pigment thus effective in the treatment of leucoderma or vitiligo. Furanocoumarins is also used to some extent in the herbal treatment in the case of psoriasis (autoimmune skin disease).

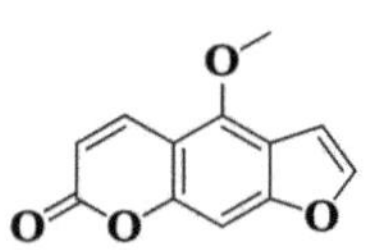

Bergapten

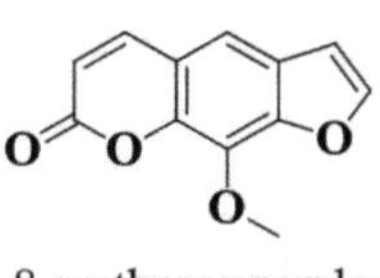

8-methoxypsoralen

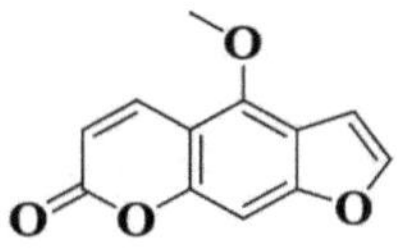

5-methoxypsoralen

Clinical trial studies showed good therapeutic potential in vitiligo and other skin disorders. 5-Methoxypsoralen reported for psoriasis treatment in a comparative clinical trials protocol of parallel design where more than 90% psoriasis clearance rates were observed in the equivalent patients' numbers (as 60–77 percent) who received oral PUVA 5-methoxypsoralen or oral PUVA 8-methoxypsoralen treatments (McNeely and Goa 1998; Rio et al. 2014). El-Mofty studies reported on 8-methoxypsoralen in the treatment of vitiligo for its efficacy (El-Mofty 1952). Some furocoumarins like bergapten, imperatorin, isopimpinellin which possess photosensitizing action and psoralens are still widely applied in the PUVA and photochemotherapy for the treatment of vitiligo, psoriasis, and other skin disorders (Honigsmann 2013).

A. majus fruit showed antibacterial effect against *S. aureus, E. coli* and *Proteus vulgaris* organism extracted in sodium phosphate citrate buffer at pH (6.8) with standard control as chloramphenicol having zone of inhibition 12–14 mm is mainly due to prominent class of compounds such as phenolic acid, terpenes, tannin, and flavonoids (Al-Akeel et al. 2014; Al-Hadhrami and Hossain 2016). *Ammi majus* L. seed extract proved effective against psoriasis-like inflammation induced in male mice by Imiquimod in comparison to clobetasol propionate, changes occur in modulation of skin thickness and inflammation parameters (Ramadhan and Mutlag 2021).

Studies conducted on the crude extract of fruit and its isolated active principles prove the potential action of *A. majus* showed its efficacy in vitiligo and other dermatological disorders. Many potentially bioactive compounds are present in the essential oil; detailed studies for their biological activities are still need to evaluate. There is much more scope to conduct the experimental *in vitro* and *in vivo* and clinical studies to ascertain the efficacy and safety of the plant in the treatment of vitiligo.

*9.3.3 Aloe vera (*Aloe barbadensis *Mill)*

Aloe barbadensis Mill. commonly known as *Aloe vera* is the oldest and extensively used medicinal plant worldwide known for many applications and most effective response towards treating skin disorders. It is popularly used, since ages, for medicinal purposes in various ailments such as gastrointestinal disorders, wounds, burns, acne, diabetes, immune modulation and various skin diseases.

9.3.3.1 Botanical Description

Aloe barbadensis is a perennial, succulent and drought-resisting plant which grows upto an height of 1–2.5 feet belongs to Asphodelaceae family. *Aloe barbadensis* or Aloe vera is a semi tropical plant and one among the 250 number of different species of Aloe. The species is found as wild along the coast regions of south India and grows in many parts of India in hot and dry climates. It is called *Ghee kunwar* in

Sanskrit and Gheekawar in Arabic (Manvitha and Bidya 2014). Its leaves are lance-shaped, sharp and pointed, jagged and edged, long and thick, juicy. The leaves edges appear thorny like structure having a thorny tip. It contains a jelly like substance in the leaf which is in the yellow-coloured part of the sap just beneath the rind, having bad odour and bitter in taste. Its leaves mature fully in 3 years, their length ranges from 25 to 30 cm and breadth ranges 3–5 cm. The prolongued storage of harvested leaves without proper refrigeration may cause or generate enzymatic changes and bacterial degradation of its polysaccharides. Due to the potential carcinogenicity of aloins A and B content, the pulp requires filtration or decolorization with activated charcoal or diatomaceous earth: this is a usual, suitable practice that greatly helps in the manufacturing of *Aloe vera* pulp juices. Decolorization eliminates a variety of small-molecular-weight organic compounds, including bitter aloins A and B, from *Aloe vera* leaf pulp (Manvitha and Bidya 2014).

9.3.3.2 Crude Drug Used

Aloe vera leaf pulp is a natural product frequently used in the field of cosmetology. Its crude extracts, juice, gel are employed to treat skin problems (acne, wounds, burns, and anti-inflammatory processes) and for medicinal purposes in several cultures since millennia. *Aloe vera* possesses properties such as anticancer, antioxidant, antidiabetic, and antihyperlipidemic. Aloe vera juice (30 mL/twice a day daily) and Aloe vera gel (5 mg/thrice a day daily) for 3 months reduces the burning sensation. Crude extracts decrease cell viability and induce apoptosis (Sánchez et al. 2020; Hussain et al. 2015).

9.3.3.3 Major Chemical Constituents and Bioactive Compounds

There are about 200 compounds present in *Aloe barbadensis* and the leaf contain different classes of compounds which include anthraquinones (like aloe-emodin), anthrones and their glycosides (such as 10-(1, 5′ anhydroglucosyl)-aloeemodin-9-anthrone, also known as aloin A and B), chromones, lipids, sugars, carbohydrates, proteins, amino acids, glycoproteins, organic acids, minerals and vitamins (Upton et al. 2012; Saeed et al. 2004; Avijgan et al. 2016).

Aloe vera keeps the skin moisturized and mostly found effective for pigmentations, wrinkles, stretch marks. Its extract heals the burns and also owns rejuvenating, analgesic and anti-inflammatory, antioxidant, antiallergic, antibacterial properties etc. (Anonymous 1985a; Reynolds and Dweck 1999; Cárdenas et al. 2006; Ro et al. 2000; Wu et al. 2013). The assay of active constituents of anthraquinones like aloe-emodin and aloin are reported in Aloe vera (Kumar et al. 2017).

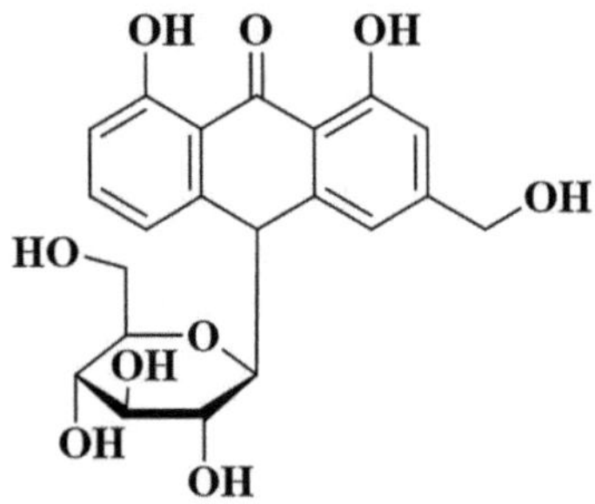

Aloin

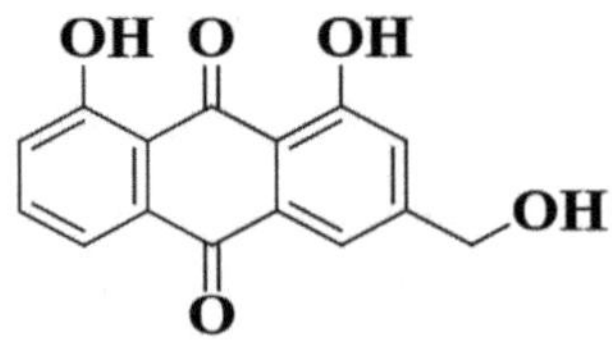

Aloe-emodin

In the indigenous system of medicine, *Aloe vera* was reported a vast role in Ayurveda, Siddha, Unani and Homoeopathy. Traditionally, it is used to treat skin disorders such as burns, wounds, and anti-inflammatory processes. Earlier, *in-vitro* studies reported its ability in skin protection and its active compounds for wound healing. Aloesin shows enhanced wound healing by causing increasing cell migration through phosphorylation of cytokines, Cdc42 and Rak1 and the growth factors (Wahedi et al. 2017). *Aloe vera* gel extract exhibited the improved organization of skin and collagen in Wistar rat experiments (Brandão et al. 2016). Polysaccharide in Aloe at different concentrations of 20, 40, and 80 μg/mL for about 24 h can act as a good agent to treat psoriasis as evidenced by the inhibition of TNF-α levels and IL-8 and IL-12 protein expression when carried in human keratinocyte HaCaT cell line. Another randomized, double-blind, placebo-controlled study with Aloe vera reported to maintain healthy skin benefits with daily oral intake of 40 μg of *Aloe* sterol (cycloartenol and lophenol). The study was done for 12 weeks had showed the improved skin elasticity in men under 46 years upon exposure to sunlight without using sunscreen to protect themselves (Tanaka et al. 2016) and reduced facial wrinkles in Japanese women over 40 years old by stimulating hyaluronic acid and collagen production (Tanaka et al. 2015). The latex contains anthraquinones which are potent laxative and increases intestinal water content, stimulates mucus secretion and enhances the intestinal peristalsis (Ishii et al. 1994).

*9.3.4 Sarpagandha (***Rauvolfia serpentina** *Benth ex Kurz)*

Rauwolfia serpentina known earlier around 1950 and possesses international attention for its extensive pharmacological properties. It is widely used in Ayurveda and Unani, in particular, for the treatment of hypertension, insomnia, insanity and also useful for intractable skin disorder as in a condition of psoriasis.

9.3.4.1 Botanical Description

Rauvolfia serpentina is an evergreen shrub belongs to Apocynaceae family; *Rauwolfia* genus has more than 100 species and is native to moist, deciduous forest of tropical and subtropical regions of the world including Asia, Australia, Africa, Europe and America (Vakil 1955; Lobay 2015). The plant has been mentioned in Indian manuscripts as long ago as 1000 BC and also known as *sarpagandha* and *chandrika* (Yarnell and Abascal 2001). It grows upto a height of 60–90 cm, leaves are elliptical or lanceolate shaped and occur in whorls of 3–5 ranging from 7 to 10 cm long and 3.5 to 5.0 cm wide. It has black or purple colour round fruits of approximately 0.5 cm diameter, contains pink or white small flowers and tuberous, soft taproot length ranges from 30 and 50 cm and 1.2 and 2.5 cm diameter (Brijesh 2011). Its roots are subcylindrical or tapering and rarely branched, yellowish brown outer surface and possess faint longitudinal ridges. Roots have a short fracture, transversely cut surface. Its bark is 0.5–2.0 mm wide and composed of a stratified cork having alternating bands of suberised cork cells which are isodiametric. The parenchyma cell contains 2–3 simple and rounded starch grains, prismatic crystals of calcium oxalate.

9.3.4.2 Crude Drug Used

The roots, leaves and juice are of medicinal importance in Indigenous system of medicine. Reserpine the major indole alkaloid content was found highest in the root and lower in the stems and leaves (Lobay 2015; Ruyter et al. 1991). The plant parts acts as hypnotic, sedative, antihypertensive, and antihypercholesteremic and also used to treat neurological disorders such as vertigo, schizophrenia, insomnia. Its leaves and root extract used to relieve stomachache, pain in the liver, dysentery, and to expel the intestinal worms (Kumari et al. 2013; Ezeigbo et al. 2012). Its root powder significantly reduces the serum lipids including triglycerides (TG), LDL-C, TC and increased in the HDL-C levels when given orally in rabbits for about 12 days (Shah et al. 2020).

9.3.4.3 Major Chemical Constituents and Bioactive Compounds

Rauvolfia serpentina contain phytochemical compounds which mainly include flavonoids, alkaloids, tannins, and phenols. It contains 1.2–1.4% of total indole alkaloids, in which most important are reserpine, deserpidine and rescinnamine. The other major alkaloids are ajmaline, ajmalicine, ajmalimine, deserpidine, indobine, indobinine, reserpiline, rescinnamidine, serpentine, serpentinine and yohimbine. The pharmacological, chemical properties and clinical studies carried out on this plant led to the isolation of a sedative principle an alkaloid termed reserpine (Vakil 1949) which made revolutionary treatment of hypertension, as it showed twin effect of lowering high blood pressure and also acting as a tranquillizer (Woodson et al. 1957). The reserpine found to be high in the root and low content in the stems and leaves (Ruyter et al. 1991). On the other hand, there are other alkaloids including canescine, deserpidine, recanescine, and rescinnamine which also possess biochemical medicinal actions. Reserpine has been reported to be lipid soluble and can penetrate blood-brain barrier and distributed throughout the body to the brain spleen, kidney, liver and adipose tissue. It binds to red blood cells (RBCs) and in the peripheral neuron at its site of action.

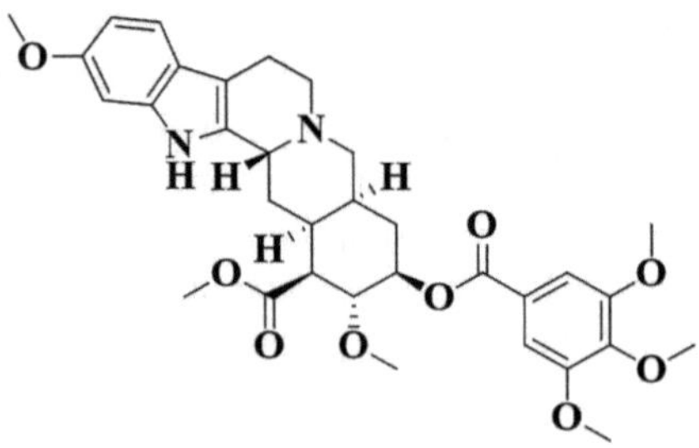

Reserpine

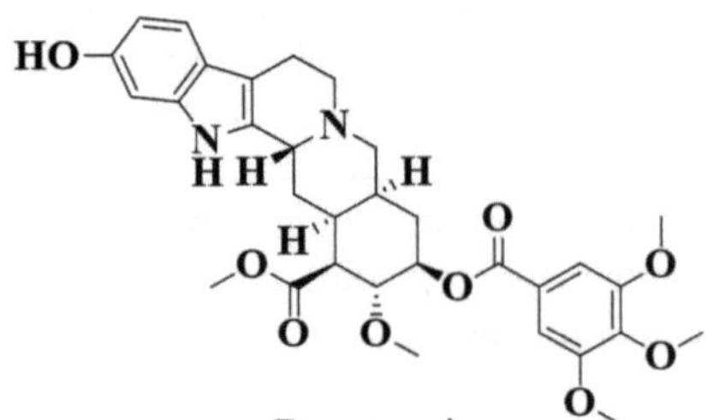

Desreserpine

The antihypertensive properties of *R. serpentina* roots are attributed to reserpine (3,4,5-trimethyl benzoic acid ester of reserpic acid, an indole derivative of 18-hydroxy yohimbine type). It is the most prominent of all alkaloids and used mainly as a natural tranquillizer. (Pullaiah 2002) Later many alkaloid constituents like ajmaline, ajmalicine and deserpidine from this plant shown to possess antihypertensive action. Ajmaline a major alkaloid of *R. serpentina* has significant antiarrhythmic activity and is being used clinically for the treatment of haemodynamically stable ventricular tachycardia. (Gonska 2000).

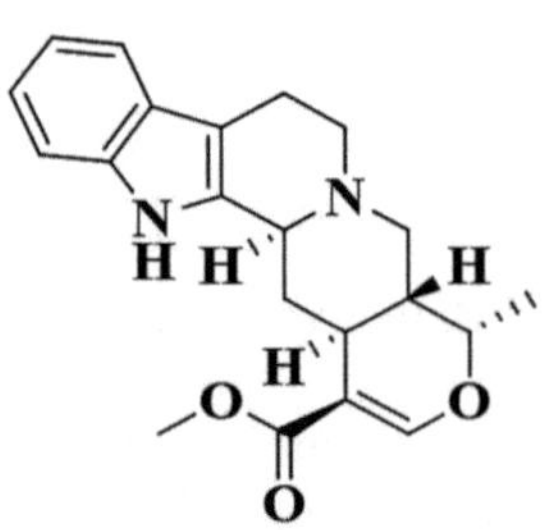

Ajmalicine

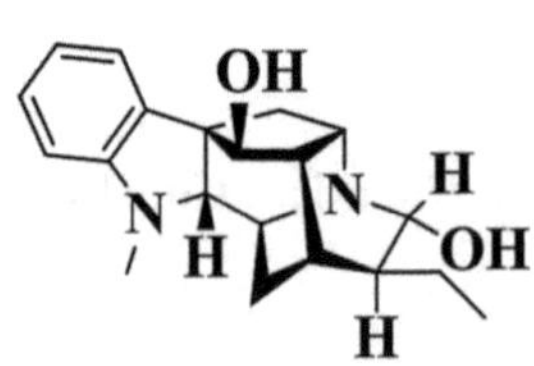

Ajmaline

9.3.4.4 Medicinal Properties

Different regional and ethnic groups use this plant to treat in case of snake bite, poisonous insects or other animal bite, mental illness, hypertension, skin diseases, scabies, blood pressure, pneumonia, fever, malaria, gastrointestinal diseases, asthma, eye diseases, spleen diseases, circulatory disorders, AIDS, rheumatism, body pain, veterinary diseases etc. *R. serpentina* is popular for its antimicrobial, antifungal, anti-inflammatory, antiproliferative, antidiuretic and anticholinergic activities (Kumari et al. 2013). The root extract used to cure liver and abdomen pain, and other gastrointestinal disorders and to expell out the worms of intestine in children. Juice or leaves extract of *A. paniculata* and *Nyctanthes arbortristis* is mixed along with *R. serpentina* juice extracted from root to treat in scabies (Mohanta et al. 2006). The root also used to treat urticaria patients through Unani system of medicine (USM) (Shamsi et al. 2006). Root paste along with *A. paniculata* used topical on itches, boils and eczema by the Kandhas, Kandhamal district in Orissa (Behera et al. 2006). It is also used as a uterine stimulant, febrifuge, and to cure insanity. Its extracts reported to have a broad range of therapeutic effects such as antioxidant, antiaging, antihypertensive, anticancerous, antimalarial, antiinflammatory, antifibrillar, anthelmintic, antiarrhythmic, anticholinergic, antidysentry, antidiarrhoeal, antihypotensive, anticontractile, antipyretic, antidiuretic, sympathomimetic, and antipsychotic (Dey and De 2011; Dutta et al. 2011).

*9.3.5 Guggul (*Commiphora wightii *(Arnott))*

Guggul, the oleo-gum resin from the tree *Commiphora wightii* was very popularly in traditional system of medicine to treatment several disorders such as rheumatoid arthritis, obesity, acne and skin diseases. Forty-four ayurvedic compound preparations contain this gum-resin as an important component (Sastry 1976).

9.3.5.1 Botanical Description

Commiphora wightii is an endangered species due to over exploitation for its oleogum resin, poor seed set declared by World Conservation Union in Red Data List. It is a stunted bush with spinescent branches, found wild in South-West and North-Western regions of India. Guggul the oleo-gum resin exudes from the bark of *Commiphora wightii* (Arnott) its synonyms are *Commiphora mukul* (Hook. Ex Stocks) Engl; *Balsamodendron mukul* (Hook. Ex Stocks) belongs to Burseraceae family. It is a shrub or small tree, grows upto a maximum height of 4 m, contain thin papery bark, and thorny branches. The leaves are simple or trifoliate, the leaflets are ovate with 1–5 cm long and are 0.5–2.5 cm broad, irregularly toothed. The pale-yellow colour aromatic fluid as oleo gum resin was collected, which flows out from bark turn in a stalactic pieces and around weigh 700–900 g resin from each four- to six-foot-tall tree. Oleo gum resin have complicated mixture of minerals, gum, terpenes, sterols (Guggulsterol -I,-II,-III,-IV,-V), essential oils, sterones (Z-, E-, M-guggulsterone, and dehydroguggulsterone-M), ferrulates, lignans, and flavanones found in the balsam canals in the phloem of the large veins of leaf and base of the stem.

9.3.5.2 Crude Drug Used

When injuring the bark of *Commiphora wightii* (Arn.) an oleo-gum-resin exudate is observed. The oleo resin was used to treat inflammation, arthritis, obesity, wound, pain, tumor, microbial infection, fractures and gastrointestinal diseases. Guggulsterones *E* and *Z* are the chief bioactive constituents present in the oleo resin and are accomplished with immense pharmacological value. In traditional system of medicine it has been mentioned that while guggul consumed, one should evade foods such as sour or bitter in taste. It is also recommended to avoid alcohol, excessive exercise, physical and mental strain, anger, and direct exposure of sunlight (Sarup et al. 2015).

9.3.5.3 Major Chemical Constituents and Bioactive Compounds

Commiphora wightii (Arn.) contain different groups of phytoconstituent such as volatile oil containing terpenoidal constituents viz. monoterpenoids, diterpenoids, triterpenoids and sesquiterpenoids, steroids, flavonoids, guggultetrols, lignans, sugars, and amino acids. The amount of phytoconstituents vary as volatile oil (0.6%), resin (61%), gum (29.6%) and insoluble substances (3.2%) (Sarup et al. 2015; Chaudhary 2012). Bioassay-guided fractionation of gum-resin eventually led in 1971 led to the isolation and characterization of two anti-hyperlipoproteinemic compounds, *Z*-guggulsterone and *E*-guggulsterone which are active constitute responsible for lipid lowering properties in human blood by Satyavati et al. (1969) and Kapoor et al. (1979); both *E*- and *Z*- compounds have similar activity in comparison to clofibrate, a synthetic hypolipaemic drug which had been then launched in the market (US). Guggul gum is a mixture of about 61% of resins and 29.3% of gum, 6.1% of water, 0.6% of volatile oil and 3.2% of foreign matter.

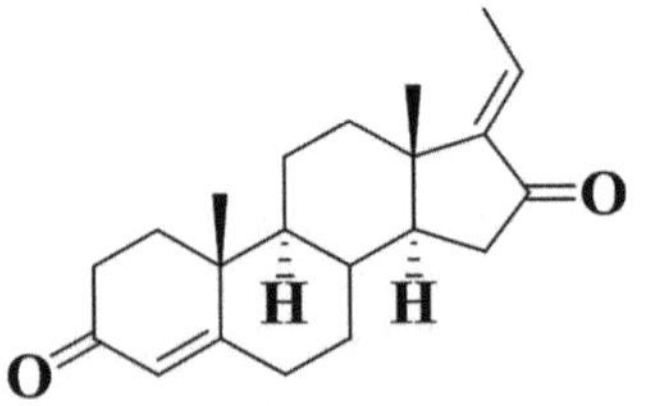

E-Guggulsterone

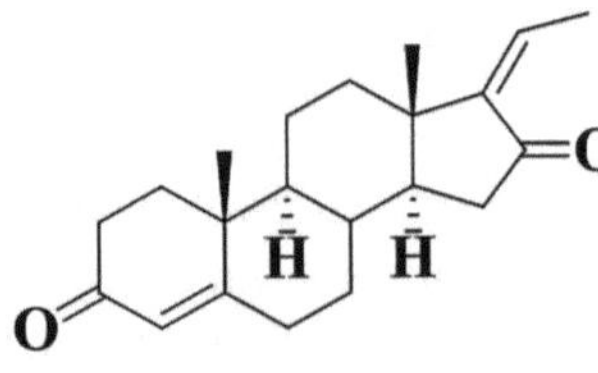

Z-Guggulsterone

Further studies based on pharmacological, biochemical, toxicological, teratogenic and mutagenic, and clinical data have led to a standardized product guggulipid from ethyl acetate extract as a hypolipidemic agent, comprising at least 4% guggulsterones. Guggulipid exhibit dose reliant lowering of serum cholesterol and triglycerides in normal and hyperlipidemic amoung rats, rabbits and monkeys. It is marketed in India since 1987 (Nityanand and Kapoor 1978; Agarwal et al. 1986) and helpful as an antiseptic. *In-vitro* cytotoxicity,y guggul exhibit decreased cell viability towardes MCF-7 (breast) tumor cells, PC-3 (prostate) tumor cells, and in parental and transfected P388 cells (Mesrob et al. 1998). Further the hypolipidemic effect, guggul exhibited beneficial effects on inflammation (Kimura et al. 2001). Guggul aids in the treatment of obesity, inflammation, and skin diseases (Dev 1987; Anurekha and Gupta 2006). A study conducted in 21 patients found that administration of gugulipid was found effective in the treatment of nodulocystic acne as tetracycline in the treatment. The patients with oily faces responded better to the gugulipid treatment (Thappa and Dogra 1994). The therapy derived from the multiple pharmacological activities for cardiovascular aids are linked with the guggul or guggulsterone, remarkably its hypolipidemic nature, antioxidant, and anti-inflammatory activities. The inflammatory responses mainly responsible with the mechanism involving the guggulsterone had found to potently inhibit the activation of nuclear factor-kappaB (NF-kappaB). Gum resin showed diverse pharmacological properties mainly hypolipidemic, antiobesity, anti-tumor effects, anti-inflammatory, neuroprotective, hepatoprotective, cardioprotective, thyroid stimulatory effects etc.

9.3.6 *Sudab (*Ruta graveolens *L.)*

Ruta graveolens L. is an important medicinal plant and possesses a repository of secondary metabolites having morphological resemblance. Its extracts are mostly used to treat dermatitis, rheumatism, pain, and other inflammatory diseases. Its essential oil exhibited good antibacterial and antifungal properties.

9.3.6.1 Botanical Description

Ruta graveolens L is an odoriferous herb of the Rutaceae family: it is cultivated as a medicinal herb and an ornamental herb with non-fringed petals and blunted apices of fruit lobes as main identification characters. Petiole has an arc of separate vascular bundles; stem contains dictyostele with a ring of sclerenchyma and acicular calcium oxalates located in-between the palisade and spongy layers (Kannan and Babu 2012).

9.3.6.2 Crude Drug Used

The plant stated as panacea has high therapeutic values in treating anticular, ear, pharynx, pulmonary, digestive, neurological, renal, gynecological diseases, parasites excretion, pain relief, and also in the treatment of spasm and inflammation (Asgarpanah and Khoshkam 2012; Ratheesh et al. 2013; Baharvand-Ahmadi et al. 2015; Javadi and Emami 2015). Its leaves, roots are of medicinal importance. *Ruta graveolens* has shown neurobiological activity, and contains important metabolite rutin, which is not only confined to one specific action, but also exhibits a wide spectrum of activities. It owns anti-inflammatory properties and also exerts positive effects on the CNS.

9.3.6.3 Major Chemical Constituents and Bioactive Compounds

Ruta graveolens has more than 120 natural compounds which include acridone alkaloids, coumarins, essential oils (2-nonanone, 2-nonylacetate, 2-undecyl acetate), flavonoides, and fluoroquinolones found in the roots and aerial parts of plant. The plant contains a number of compounds belonging to different class of compounds such as rutin, a neuroactive compounds potentially able to promote neuroprotection. Among all the components present in the plant extracts, rutin is a highly abundant important metabolite that positively interacts with the neurophysiologcal effect of CNS. It is particularly efficient against neurotoxicity. Rutin, (3,3′,4′,5,7-pentahydroxyflavone-3-rhamnoglucoside) is the main components present in the water extract of *Ruta graveolens* which displayed monoamine oxidase-B inhibiting property, with adjuvant antioxidant and scavenging properties (Russo et al. 2013). Rutin exerts beneficial effects when given to a number of dopaminergic cellular models of Parkinson's disease. A new compound isolated of furanocoumarin class named as Rutagrarin from *R. graveolens* (Ainiwaer et al. 2022). Extracted the light yellow color volatile oil (1.29% v/w) by hydrodistillation from the aerial parts of *R. graveolens* L in which 13 chemical constituents are characterized by GC-MS consisting of 2-ketones as major such as 2-decanone (yield: 2.58%), 2-tridecanone (yield: 2.59%), 2-undecanone (yield: 43.66%), 2-dodecanone (yield: 2.23%) and 2-nonanone (yield: 16.09%), and in essential oil other compounds present are 2-Undecanol (yield: 2.19%), 2-acetoxy tetradecane (yield: 14.49%), 1-methyl-5, 6-divinyl-1-cyclohexene (yield: 3.38%), aliphatic ester nonyl cyclopropanecarboxylate (yield: 9.22%).

R. graveolens contains psoralens and acts as photoactive in nature and when applied topically or rubbed onto the skin it then reacts with sunlight to produce erythema, hyperpigmentation and blistering (Heskel et al. 1983). The essential oil present in it expresses as a source for antibacterial compounds and its possible applications in the pharmaceutics and industry are due to its antibacterial activity with inhibition zones of 8.30–25.60 mm towards minimum inhibition of concentration ranges from 0.75 to 1.40 μg/mL against Gram-positive and Gram-negative bacteria, and *Bacillus cereus* and *Staphylococcus* aureus are the most susceptible bacterium. (França Orlanda and Nascimento 2015; Reddy and Al-Rajab 2016). *R. graveolens* extract and essential oil (containing 2-nonanone, 2-nonylacetate, 2-undecyl acetate) possess various pharmacological activities such as contraceptive, antimicrobial, antipyretic, antiviral, anti-inflammatory, antioxidant, free radical scavenging, analgesic, antihyperglycemic, hypotensive, and antiplasmodial effects (Malik et al. 2017).

*9.3.7 Panwad (*Cassia tora *L.)*

9.3.7.1 Botanical Description

Cassia tora L. belongs to Caesalpiniaceae family an annual herbaceous foetid herb grows wild under shrub and occurs as wasteland rainy season weed, grows in dry soil in tropical region, hilly areas and plains in India grows up to a height of 30–90 cm high, leaves are pinnate. Flowers are in pair in axils of leaves having five petals and pale yellow in colour. Pods are flattened or four angled, 10–15 cm long. Seeds are 30–50 in a pod, rhombohedral in shape (Jain and Patil 2010).

9.3.7.2 Crude Drug Used

Different parts of the plant like leaves, seed and root are reputed for medicinal values. The seed and leaves constitute a worthful remedy in skin disorders, chiefly for ringworm and itch cases. The paste made from leaves when mixed with lime-juice helps in the treatment of vitiligo. The seeds of *C. tora* are also use to treat vision problem, lowering blood pressure, cholesterol, antiasthenic, and xerophthalmia (Jain and Patil 2010; Gulia and Choudhary 2012). However, the whole plant parts such as leaves, roots, and seeds are used in traditional medicine to treat various diseases such as psoriasis, skin diseases, and cardiovascular diseases (Jain and Patil 2010). In Indigenous medicine it is used as tea by mashing the original or toasted seeds, trailed by boiling in water for diuretic action (Gow-Chin et al. 1998; Ko et al. 2020).

9.3.7.3 Major Chemical Constituents and Bioactive Compounds

C. tora seed have chrysophanol as a marker constituent and also contain other amino acid, fatty acids, aloe-emodin, emodin, rhein and sitosterol. The other compounds such as myricyl alcohol, chrysophanic acid and its 9-anthrone derivative were also isolated from the seeds along with three naphthopyrone glucosides, cassiaside, rubrofusarin, and 6-O-β-D-glucoside, 6-O-β-D-gentiobioside. The norrubrofusarin and its gentiobioside, cassia-lactone, isotoralactone, teraglucoside, apioglucoside, obtusin, quercetin compounds are also isolated from the seeds (Chatterjii and Pakrashi 2006).

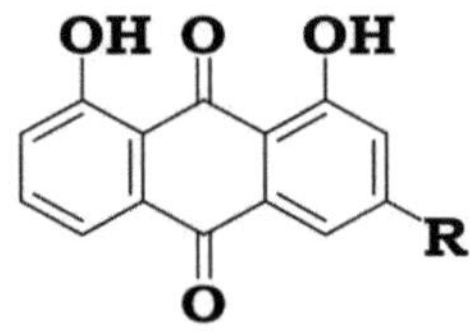

Chrysophanol: R=H

Rhein: R=COOH

C. tora is traditionally recognized as laxative and also useful in the treatment of leprosy, ringworm infection, ophthalmic, liver disorders, in the treatment of Psoriasis and other diseases (Khare 2007; Duke 2002). The topical creams and crude leaves extract have shown potent antipsoriatic activity in UV-B-induced psoriasis in the rat model (Maity et al. 2001; Singhal and Kansara 2012). The acitve chemical compounds such as anthraquinone, naphthopyrone glycosides and phenolic compounds have been isolated from this plant viz., Chrysophanic acid-9-anthrone extracted from seed found to be active against ringworm fungi. Seed acts as blood purifier and therefore used in skin diseases such as leprosy, ringworm, pityriasis. It is also used internally as well as externally in vitiligo and melasma (Ali 1979). Its seeds extract at 500 mg/kg, b.w. administered simultaneously for 15 days showed lipid levels were reduced in plasma and liver of animals following reactivation of plasma post heparin lipolytic activity and hepatic lipoprotein lipase activity in animals. The hypolipidemic activity of *C. tora* seeds was compared with a standard drug guggulipid (200 mg/kg, b.w.) in acute and chronic models showed lipid-lowering effect (Awasthi et al. 2015).

9.3.8 *Haldi* (Curcuma longa *L.*)

9.3.8.1 Botanical Description

The *Curcuma* genus comprises more than 70 species and about 40 species widely cultivated thorough out India in particularly *Curcuma longa* L. commonly called as turmeric. It is a perennial herb erect belongs to Zingiberaceae family, stemless herb with root stock measures up to a height of 1 meter high, its leaves are broadly lanceolate to oblong. Petiole and sheath as long as the blade; flowers are funnel-shaped yellow in colour and reddish at outer boarder (Chanda and Ramachandra 2019). It is spread throughout the tropical and subtropical regions of the world and cultivated in Asiatic countries, mainly in India and China. In India a well-known medicinal plant and largest producer of turmeric (rhizome) which is popularly known as haldi and extensively used as a drug in traditional systems particularly in Ayurvedic and Unani system of medicine (Chopra et al. 1956; Kapoor 2001) for broad medicinal properties.

9.3.8.2 Crude Drug Used

Rhizome of *Curcuma longa* known as turmeric, traditionally used as a spice in Indian food. In turmeric, the active constituents are Curcumin a flavonoid and contain volatile oils, including turmerone, atlantone, and zingiberone etc., are present. Turmeric has been shown remarkable capacity in rhizome powder, extract or an isolated compound, to heal wounds, involvement in correcting biliary hyperplasia, fatty alterations, necrosis, and biliary necrosis. *C. longa* extracts effectively treats diabetes, cancer, gastrointestinal problems, and neurological ailments. Turmeric or rhizome extract may also be applied topical over the skin to alleviate the symptoms of inflammatory skin disorders and allergies by reducing inflammation and irritation. Turmeric is suitable and very much beneficial to the preparation of nanofibers (NF) which include pharmaceuticals, active biological species, food industry components, and other sites, also aid from the potent antioxidant activity of turmeric-based TNF (Turmeric Nanofibers). It is also a useful in the form of ointment to treat several skin disorders like acne, restoration of skin texture, rejuvenating skin property.

9.3.8.3 Major Chemical Constituents and Bioactive Compounds

C. longa rhizome contains a yellow colour phenolic pigment which is composed of three phenolic compounds or active principles Curcumin, Desmethoxycurcumin and Bisdesmethoxycurcumin (Anubala et al. 2014; Rasheed et al. 2017).

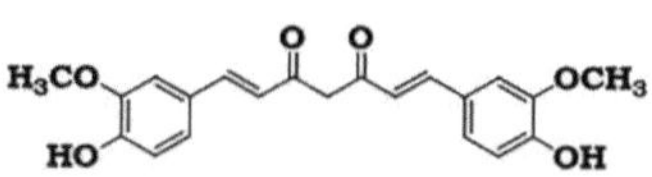

Curcumin

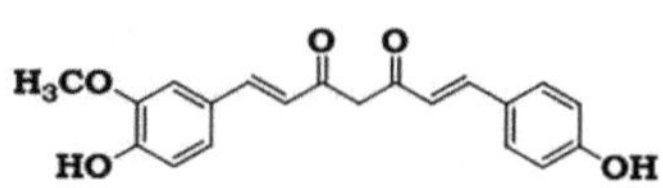

Desmethoxycurcumin

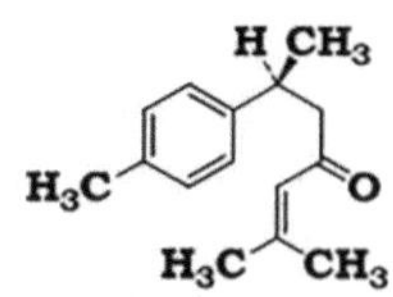

ar-Turmerone

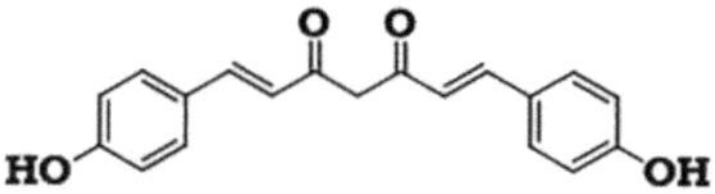

Bisdesmethoxycurcumin

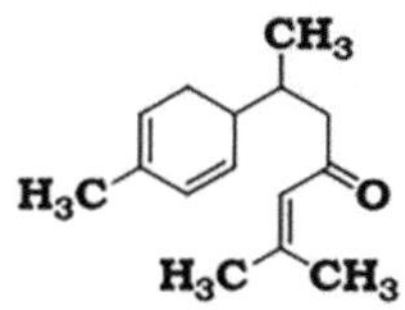

α-Turmerone

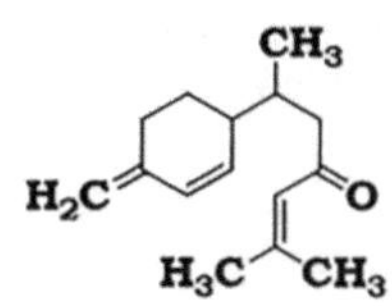

β-Turmerone

Curcumin (7-bis-(4-hydroxy-3-methoxyphenyl)-1, 6- heptadiene-2, 5-dione) is the main active principle content present in the rhizome which possesses antioxidant activity and other biological activities. Curcumin possesses wider range of pharmacological activities such as antioxidant, anti-inflammatory, hypercholesterolemia, ant carcinogenic, wound healing property, anticoagulant, antispasmodic, antitumor, hepatoprotective activities (Maiti et al. 2007; Anand et al. 2008; Goel et al. 2008; Kunnumakkara et al. 2007).

Curcumin antioxidant property prevent rancidity of foods and provide foodstuffs containing less oxidized fat or free radicals and keep curry for a longer time without turning rancid. Turmerones, a component of turmeric oleoresin used to augment the absorption of curcumin. These turmerones from present in turmeric have strong anti-inflammatory and anticancer properties. Turmerones have anti-cancer activity with the whole extract of turmeric. Volatile oils and turmerones in turmeric are reported as a principle flavouring compound; α-turmerone, and β-turmerone, ar-turmerone are the primary components of oleoresin of turmeric rhizome and furnish characteristic pungent smell. Aromatic (ar-) turmerone a natural antimicrobial and antifungal compound possesses antivenom properties. α- and β- turmerones act as antibacterial agents against disease-causing bacterium which are held responsible for serious infections and also to prevent cavities.

It also contains essential oil (about 2–4%); turmerones (about 2–3%), atlantone, and zingiberone. Antony in 2003 reported that it contains sugars, proteins and Oleo resins (25–55%) (Antony 2003). It is popularly used to treat eczema, obstinate itching, infections like ringworm and other parasitic skin diseases (Zanwar et al. 2013). Contact urticaria and contact dermatitis after topical application of curcumin-based creams are reported and curcumin as a non-competitive inhibitor considered suitable for the topical treatment of psoriasis by inhibit Phosphorylase kinase activity. The potential topical and systemic use of curcumin helps in the treatment and prevention of skin aging because of its inflammatory and antioxidant effects (Vollono et al. 2019).

The topical application of turmeric or curcumin were examined for various skin disorders such as atopic dermatitis, oral lichen planus, facial photoaging, pruritus, acne, alopecia, psoriasis and vitiligo radiodermatitis. There has been noted a statistically significant improvement in skin disease with respect to turmeric/curcumin treatment groups when compared with control groups. Overall, there is evidence that turmeric or curcumin products and supplements, administered both oral and topical, have provided therapeutical benefits for skin health (Vaughn et al. 2016). Turmeric ethanol extract showed significant antioxidant effect when compared to that of the aqueous turmeric extract that protects free radicals against damage (Tanvir et al. 2017).

*9.3.9 Neem (***Azadiracta indica** *A. Juss)*

9.3.9.1 Botanical Description

Azadirachta indica commonly called as Neem belongs to Meliaceae family, it grows in tropical and semi tropical regions and widely distributed in Burma, India and - Pakistan. A fast growing, ever green tree reaches up to a height of 15–20 m.

It is widely attributed to health-promoting effect due to rich source of antioxidant and use widely in Unani and Ayurvedic medicines especially in Indian Subcontinent and also worldwide in the treatment and prevention of various diseases. It shed leaves in severe drought its branches are wide and spreading. The fairly dense crown is roundish and may reach a diameter of 20–25 m.

9.3.9.2 Crude Drug Used

A. indica is a well-known tree with immense medicinal properties and most versatile medicinal plant and the active ingredients are present in all parts of the tree but in general, seed, bark, leaves and roots are mostly used for extraction purpose. Leaf, seed and bark extracts of *A. indica* showed significant antioxidant potential. Its root bark extract exhibited higher free radical scavenging effect with 50% scavenging activity and total antioxidant activity against standard ascorbic acid (Kiranmai et al. 2011). Its leaves are used for leprosy, eye disorders, bloody nose, intestinal worms. Bark is used for malaria, stomach and intestinal ulcers, skin diseases, pain, and fever. Its flower and fruits are used for reducing bile, controlling phlegm, and treating intestinal worms and hemorrhoids. Its seed and seed oil are used for leprosy and intestinal worms whereas stem, root bark, and fruit are used as a tonic and astringent. Neem leaf extract is applied to skin to treat head lice, skin diseases, wounds, and skin ulcers; it also acts as a mosquito repellent; and as a skin softener.

9.3.9.3 Major Chemical Constituents and Bioactive Compounds

Azadirachta indica has complex diverse constituents the most important active constituent is azadirachtin and other compounds include are nimbin, nimbidin, nimbidol, nimbolinin, salannin (Ali 1993; Hossain et al. 2011; Kokate et al. 2010).

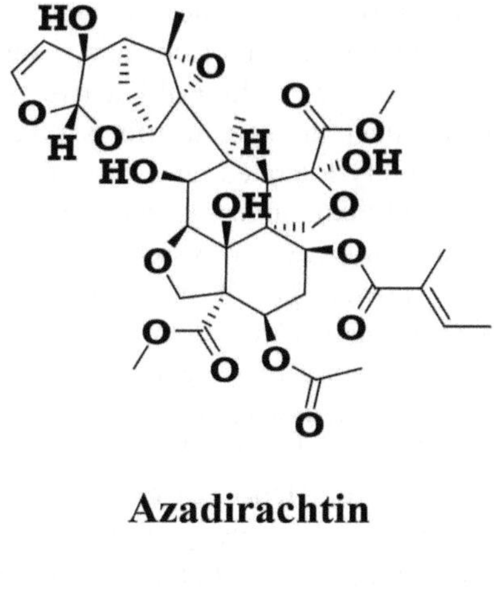

Azadirachtin

Leaves contain quercetin and ß-sitosterol, polyphenolic flavonoids were derived from neem fresh leaves and purified. These compounds are found to have antibacterial and antifungal properties (Govindachari et al. 1998) and seeds contain valuable constituent's gedunin and azadirachtin. A number of diverse biological and pharmacological activities have been reported mainly antibacterial (Singh and Sastry 1997), antifungal (Kher and Chaurasia 1997), including anti-inflammatory, antigastric ulcer, antipyretic, hypoglycemic, antiarthritic and antitumour activities (Bandyopadhyay et al. 2004; Sultana et al. 2007; Ebong et al. 2008; Paul et al. 2011).

Neem extract has shown oxidative stress induced by *A. indica* is evidenced by the depletion of catalase, SOD, and GSH levels in human lymphocytes. It showed a significant increase in DNA damage when compared to control in human lymphocytes ($P < 0.05$). The *A. indica* extract at lower concentration ranges at 0.7–1 mg/mL is found as nontoxic while toxic at higher concentrations ranges from 1.2–2 mg/mL. The anti-inflammatory effect of aqueous extract of neem leaf (400 mg/kg b.wt.) was compared with that of dexamethasone (0.75 mg, intraperitoneally) by administering 1 h before the formalin injection and administered once daily for 7 days in rats. The reduction was statistically significant in each case ($p < 0.001$) (Uzzaman 2020).

Some medicinal plants have been described in Table 9.1 which is effective in treating various skin disorders and other ailments along with their family, common name, extract yield, active chemical constituent, therapeutic uses.

9.3.10 Conclusion

Skin is the largest protective organ of the human body. The pivotal role in aesthetic appearance, skin health has got a significant impact on life quality, and it protects against various environmental assaults and noxious agents. In the present scenario, several people are suffering with the skin disorders viz., vitiligo, psoriasis and eczema particularly in lower economic regions where adequate hygiene is not maintained, ultimately leading to the occurence of acute to chronic skin disorders. There is a vital need for development of safer and economical drugs from the medicinal plants. Phytochemicals are biologically active compounds derived from medicinal plants perform beneficial therapeutic effects towards the several ailments. They are widely available, highly tolerated, and cost-effective. The crude drug or plant parts should be identified taxonomically and authenticated for ultimate utilization. The plant parts are used in several ways viz., powder form, exudates, juices, pulps and extracts which showed potent pharmacological effects. The active phytoconstituents are mainly responsible for the beneficial effects shown by the potent medicinal plants. In literature there had been several reports available in which phytochemicals have been demonstrated to mitigate skin *in-vitro* and *in-vivo* studies conducted and exhibited the diverse pharmacological and biological activities like antioxidant, anticancer etc. Hence, there is a strong need for the suitable standards methods for determining the good quality through identification of active major principle or biomarker compounds with the aid of recent analytical methods to

Table 9.1 Some medicinal plants, constituent, and therapeutic uses or activities

S. no.	Plant name (family)/ traditional/or common name/ Ext. yield	Active chemical constituents	Therapeutic uses/activities	References
1.	*Andrographis paniculata* (Burm. f.) Wall ex Nees. / (Acanthaceae)/ Kalmegh, /Yield:Methanol extract: ~16%	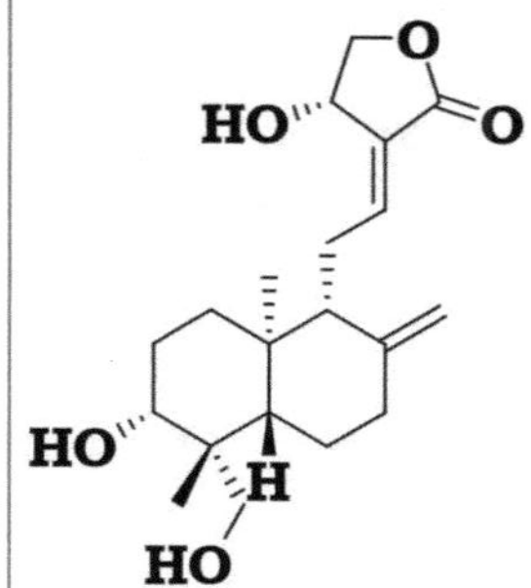 Andrographolide	To treat bronchitis, diabetes, effective in skin diseases *viz.* itching, vitiligo and antidote to snake bite, antioxidant, anticancer activities.	Shen et al. (2006), Pholphana et al. (2004), Islam et al. (2018)
2.	*Azadirachta indica* A. Juss./ (Meliaceae)/ ***Neem*:** Nimba/**Yield** Methanol extract: 20–25%	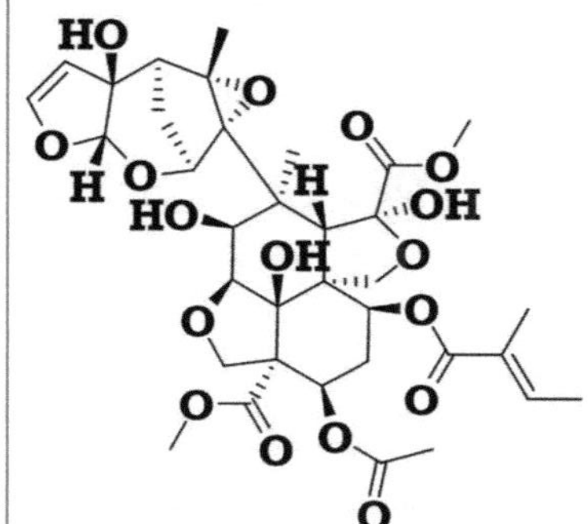 Azadirachtin	Variety of skin afflictions, ulcers, urinary disorders, detoxicant, blood purifier, wound healing properties, alcoholic extract is effective in skin disorders, antiulcer.	Anonymous (1985b), Schaaf et al. (2000), Kawami et al. (2018)

(continued)

Table 9.1 (continued)

S. no.	Plant name (family)/ traditional/or common name/ Ext. yield	Active chemical constituents	Therapeutic uses/activities	References
3.	*Berberis aristata* DC. / (Berberidaceae)/ Daaru haldi/Yield: Methanol extract (bark): ~20%	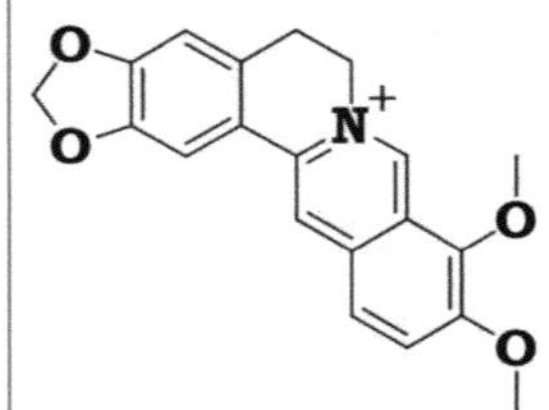 Berberine	Cure piles and haemorrhoids, used in wound healing, dysentery, indigestion, uterine and vaginal disorders.	Janbaz and Gilani (2000); Nimisha et al. (2017)
4.	*Cassia tora* L./ (Caesalpiniaceae)/ Panwad, Chakramarda/Yield: Seeds, leaves, roots are used.	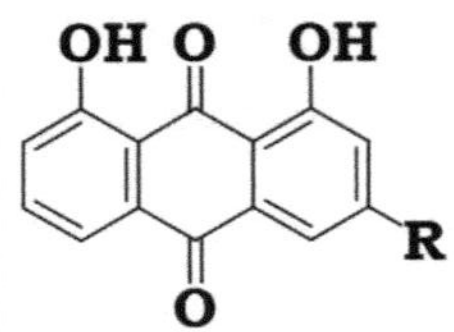 Chrysophanol: R=H Rhein: R=COOH	Seeds are used in liver cirrhosis, gout, piles, laxative property, antibacterial and useful in skin diseases.	Yen and Chuang (2000); Vijayalakshmi and Madhira (2014)
5.	*Emblica ribes* Burm f. / (Myrsinaceae)/ Baobarang; Vidanga;/ Yield: Ethanolic extract (fruit): 4–5%	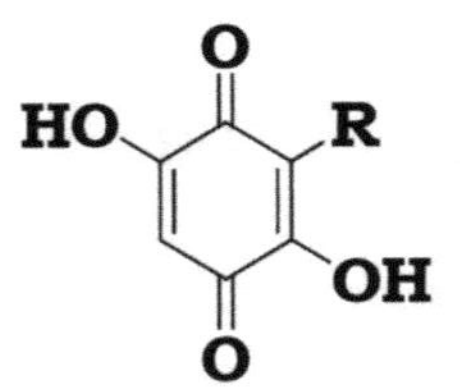 Embelin: R=n-$C_{11}H_{23}$	Recommended for treatment of dyspepsia, colic pain, epilepsy, antispetic and emetic agent. Antibacterial activity, antioxidant activity.	Chitra et al. (2003), Madhavan et al. (2011)

6.	*Gymnema sylvestre* R. Br./ (Asclepiadaceae) Gurmar booti/ Yield: 95% ethanolic extract (leaves): ~8%	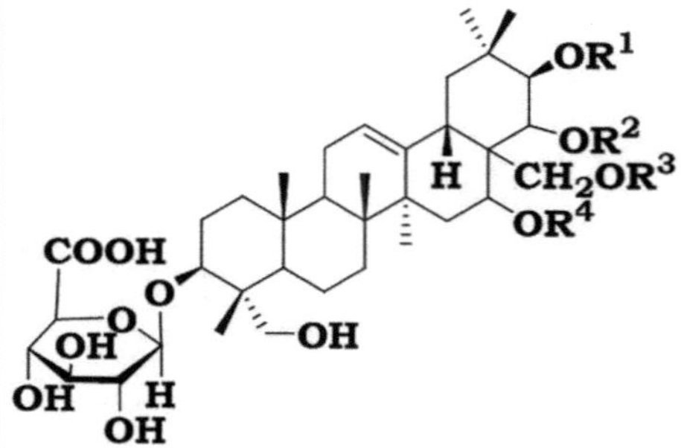 Gymnemic acid I R^1:Tiglol R^2:H R^3:Ac R^4:H	Used in diabetes, chronic skin diseases, obesity, migrane, urinary discharges. Stimulates heart and circulatory system.	Kar et al. (2003), Singh et al. (2017)
7.	*Holarrhena antidysenterica* Wall./ (Apocynaceae)/ Yield: Methanol extract (stem bark): 7–8%	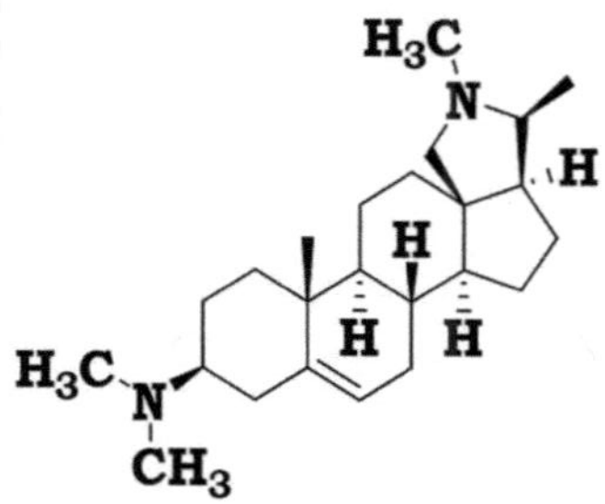 Conessine	Well known for diarrhoea and dysentry. Curative of urinary problems, ulcers, leprosy and other skin diseases.	Anonymous (1959), Patel and Prajapati (2008), Siriyong et al. (2017)

(continued)

Table 9.1 (continued)

S. no.	Plant name (family)/ traditional/or common name/ Ext. yield	Active chemical constituents	Therapeutic uses/activities	References
8.	*Lawsonia inermis* L./ (Lythraceae)/ Mehndi/ Yield: Ethanol extract (dried leaves): 17.8%	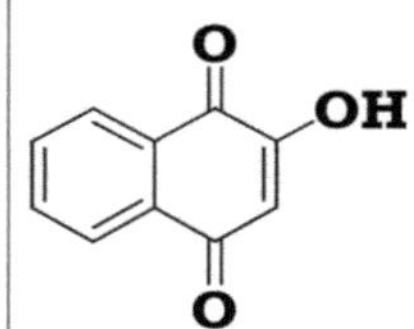 Lawsone	Used for treatment of skin diseases, epilepsy, jaundice, dyeing hair. Possess analgesic activity.	Dasgupta et al. (2003), Charoensup et al. (2017)
9.	*Mallotus philippensis* (Lam.) Muell.-Arg./ (Euphorbiaceae)/ Kamela/ Yield: Ether extract: 68.7%	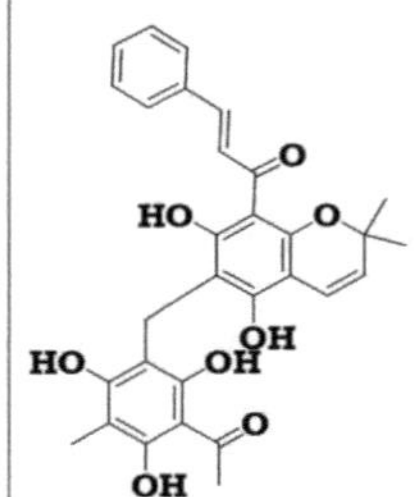 Rottlerin	Used to treat skin diseases, purgative. Antifertility, antioxidant, scabies, ringworm etc., possess antibacterial properties.	Arfan et al. (2007), Oyedemi et al. (2016)
10.	*Ocimum sanctum* L. / (Lamiaceae)/ Tulsi/ Yield: Methanol extract (leaves): 18–21%).	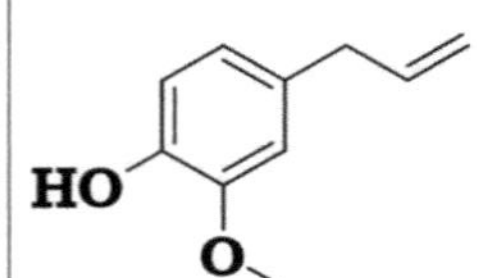 Eugenol	Treatment for catarrh, asthma, skin eruptions, disorders of blood and heart problems. Possess analgesic, antiinflammatory, antibacterial, antioxidant	Anandjiwala et al. (2006), Shokeen (2008)

11.	*Rubia cordifolia* L./ (Rubiaceae)/ Majeeth/ Yield:$CHCl_3$ extracts (root): 3–4%	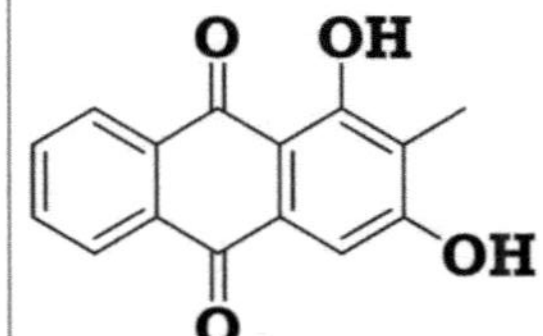 Rubiadin	Powder dried roots taken internally for the treatment of skin diseases and disorders of spleen.	Cai (2004), Antarkar et al. (1983), Rao et al. (2006)
12.	*Zingiber officinale* Rosc./ (Zingiberaceae)/ Sonth/ Yield: Ethanol extract rhizome): 10–13%	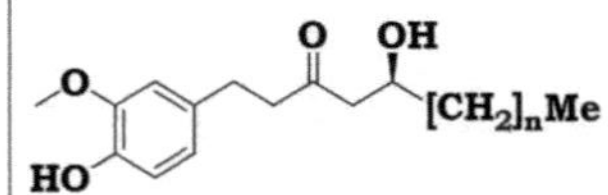 [6]-Gingerol, n=4 [8]- Gingerol, n=6 [10]- Gingerol, n=8 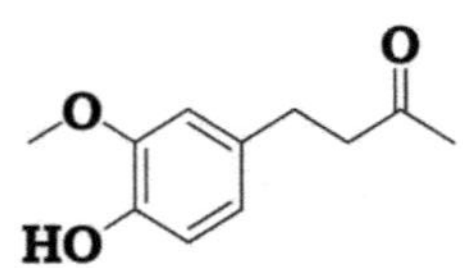 Zingerone	Remedy for gastrointestinal disorders, treatment of chronic skin diseases. Possess antioxidant, antibacterial, Antifungal activities.	Khan et al. (2010), Ali et al. (2008a), Li et al. (2012)
13.	*Santalum album* L./ (Santalaceae)/Sandal Yield: 41–54% of α-santalol and 16–24% of β-santalol	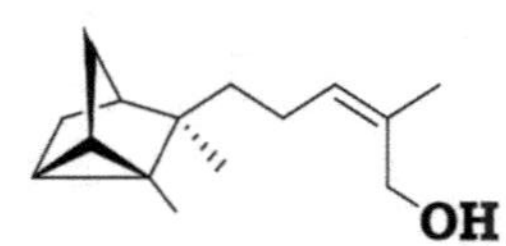 (+)-*cis*-*α*-santalol 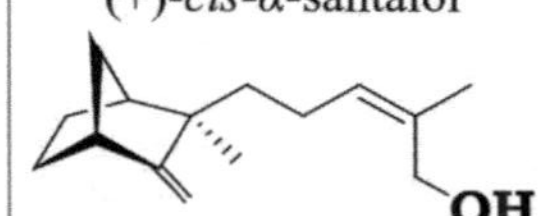 (-)-*cis*-*β*-santalol	Used in skin ailments	Mukherjee (2002), Dev (2012)

generate evidence-based data. The presence of active phytochemical substances with therapeutic activities may provide substantial basis for the use of these plants in ethnomedicine and the development of drug delivery systems to improve efficacy of promising bioactive compounds as effective agents. They can also be used against dermatitis, psoriasis, and vitiligo. Therefore, a huge scope appears for further systematic research in screening Indian medicinal plants for potent phytochemicals and assessing their potential against different types of diseases.

Acknowledgement The author expresses special thanks to Dr. Ahmed Minhajuddin, Director in charge, National Research Institute of Unani Medicine for Skin Disorders, Hyderabad for providing the necessary facility for collection of literature and also for his constant encouragement, support and ardent interest shown. The author is extremely grateful to Prof. Asim Ali Khan, Director General, Central Council for Research in Unani Medicine (CCRUM), Ministry of Ayush, Govt. of India for providing the infrastructure and necessary facilities, and generous inspiration throughout.

Conflict of Interest None.

References

Agarwal RC, Singh SP, Saran RK, Das SK, Sinha N, Asthana OP, Gupta PP, Nityanand S, Dhawan BN, Agarwal SS (1986) Clinical trial of gugulipid – a new hypolipidemic agent of plant origin in primary hyperlipidemia. Indian J Med Res 84:626–634

Ainiwaer P, Nueraihemaiti M, Li Z, Zang D, Jiang L, Li Y, Aisa HA (2022) Chemical constituents of *Ruta graveolens* L. and their melanogenic effects and action mechanism. Fitoterapia 156: 105094. https://doi.org/10.1016/j.fitote.2021.105094

Akhtar P, Ali M, Sharma MP, Waris M, Hasan H, Ali B, Chaudhary N, Khan M, Ali A, Najib S, Farooqi H, Khan HN (2010) Development of quality standards of *Ammi majus* L. Fruit. J Exp Sci 1(11):20–24

Al-Akeel R, Al-Sheikh Y, Mateen A, Syed R, Janardhan K, Gupta VC (2014) Evaluation of antibacterial activity of crude protein extracts from seeds of six different medical plants against standard bacterial strains. Saudi J Biol Sci 21:147–151. https://doi.org/10.1016/j.sjbs.2013.09.003

Al-Hadhrami RMS, Hossain MA (2016) Evaluation of antioxidant, antimicrobial and cytotoxic activities of fruit crude extracts of *Ammi majus* grown in Oman. Egypt J Basic Appl Sci 3:329–334. https://doi.org/10.1016/j.ejbas.2016.08.001

Ali S (1979) Unani advia mufrida. Qaumi Council Brae frogh Urdu Zuban, Government of India, New Delhi, p 103

Ali A (1993) Textbook of Pharmacognosy. Publication and Information Directorate, New Delhi

Ali BH, Blunden G, Tanira MO, Nemmar A (2008a) Some phytochemical, pharmacological and toxicological properties of ginger (*Zingiber officinale* Roscoe): a review of recent research. Food Chem Toxicol 46(2):409–420. https://doi.org/10.1016/j.fct.2007.09.085

Ali J, Akhtar N, Sultana Y, Baboota S, Ahmad S (2008b) Thin-layer chromatographic analysis of psoralen in babchi (*Psoralea corylifolia*) oil. Acta Chromatogr 20:277–282

Al-Snafi AE (2013) Chemical constituents and pharmacological activities of *Ammi majus* and *Ammi visnaga*. A review. Int J Pharm Ind Res 3(3):257–265

Anand P, Thomas SG, Kunnumakkara AB, Sundaram C, Harikumar KB, Sung B, Tharakan ST, Misra K, Priyadarsini IK, Rajasekharan KN, Aggarwal BB (2008) Biological activities of curcumin and its analogues (congeners) made by man and mother nature. Biochem Pharmacol 76(11):1590–1611. https://doi.org/10.1016/j.bcp.2008.08.008

Anandjiwala S, Kalola J, Rajani M (2006) Quantification of eugenol, luteolin, ursolic acid, and oleanolic acid in black (Krishna Tulasi) and green (Sri Tulasi) varieties of *Ocimum sanctum* Linn. Using high-performance thin-layer chromatography. J AOAC Int 89(6):1467–1474

Anonymous (1959) The wealth of India, vol 5. Publication Information Directorate, CSIR, New Delhi, p 103

Anonymous (1976) The wealth of India, vol I–X. CSIR, New Delhi

Anonymous (1985a) The wealth of India, vol 1A. Publication Information Directorate, CSIR, New Delhi, p 192

Anonymous (1985b) The wealth of India, vol 1A. Publication Information Directorate, CSIR, New Delhi, p 504

Anonymous (1987) Standardisation of Single Drugs of Unani Medicine – Part I. Central Council of Research in Unani Medicine (CCRUM) Publication, New Delhi

Anonymous (1996) British herbal pharmacopoeia. Exeter, British Herbal Medicine Association

Anonymous (1999) The Ayurvedic pharmacopoeia of India, 1st ed., Ministry of health and family welfare, Govt. of Inida, Part I, Vol. I, pp 31

Anonymous (2007) The Unani Pharmacopoeia of India, Ministry of health and family welfare, Govt. of India, New Delhi, Part-I, Vol. I, pp 13

Anonymous (2017) Traditional Medicine, Definition, WHO, https://www.who.int/health-topics/traditional-complementary-and-integrative-medicine#tab=tab_1. Accessed on 19 Dec 2021

Antarkar DS, Chinwalla T, Bhatt N (1983) Anti-inflammatory activity of *Rubia cordifolia* Linn. In rats. Indian J Pharmacol 15(3):185–188

Antony MB (2003) Indigenous medicinal plants: their extracts and isolates as a value added export product. Agrobios 1:39–41

Anubala S, Sekar R, Nagaiah K (2014) Development and validation of an analytical method for the separation and determination of major bioactive curcuminoids in *Curcuma longa* rhizomes and herbal products using non-aqueous capillary electrophoresis. Talanta 123:10–17. https://doi.org/10.1016/j.talanta.2014.01.017

Anurekha J, Gupta VB (2006) Chemistry and pharmacological profile of guggulu – a review. Indian J Tradit Knowl 5:478–483

Arfan M, Amin H, Karamać M, Kosinska A, Shahidi F, Wiczkowski W, Amarowicz R (2007) Antioxidant activity of extracts of *Mallotus philippinensis* fruit and bark. J Food Lipids 14:280–297. https://doi.org/10.1111/j.1745-4522.2007.00086.x

Asgarpanah J, Khoshkam R (2012) Phytochemistry and pharmacological properties of *Ruta graveolens* L. J Med Plants Res 6:3942–3949

Avijgan M, Kamran A, Abedini A (2016) Effectiveness of *Aloe vera* gel in chronic ulcers in comparison with conventional treatments. Iran J Med Sci 41:S30

Awasthi VK, Mahdi F, Chander R, Khanna AK, Saxena JK, Singh R, Mahdi AA, Singh RK (2015) Hypolipidemic activity of Cassia tora seeds in Hyperlipidemic rats. Indian J Clin Biochem 30(1):78–83. https://doi.org/10.1007/s12291-013-0412-2

Baharvand-Ahmadi B, Bahmani M, Zargaran A, Eftekhari Z, Saki K, Baharvand-Ahmadi S, Saki K, Kopaei MR (2015) *Ruta graveolens* plant: a plant with a range of high therapeutic effect called cardiac plant. Pharm Lett 7:172–173

Bandyopadhyay U, Biswas K, Sengupta A et al (2004) Clinical studies on the effect of neem (*Azadirachta indica*) bark extract on gastric secretion and gastroduodenal ulcer. Life Sci 75(24):2867–2878. https://doi.org/10.1016/j.lfs.2004.04.050

Bartnik M, Mazurek AK (2016) Isolation of Methoxyfuranocoumarins from *Ammi majus* by centrifugal partition chromatography. J Chromatogr Sci 54(1):10–16. https://doi.org/10.1093/chromsci/bmv098

Behera SK, Panda A, Behera SK, Misra MK (2006) Medicinal plants used by the Kandhas of Kandhamal district of Orissa. Ind J Trad Knowl 5(4):519–528

Beier RC, Ivie GW, Oertli EH, Holt DL (1983) HPLC analysis of linear furocoumarins (psoralens) in healthy celery (*Apium graveolens*). Food Chem Toxicol 21(2):163–165. https://doi.org/10.1016/0278-6915(83)90231-4

Bhambri M, Bajaj A, Cherian KJ (2012) Effect of different sowing periods on the growth and yield of A*mmi majus* L. in the Vidarbha region. Bionano Front 5(2):3

Bhattacharjee SK (2004) Hand book of medicinal plants, 4th revised ed. Pointer Publishers, Jaipur

Bishnoi A, Parsad D (2018) Clinical and molecular aspects of vitiligo treatments. Int J Mol Sci 19(5):1509. https://doi.org/10.3390/ijms19051509

Brandão ML, Reis PRM, Araújo LAD, Araújo ACV, Santos MHDAS, Miguel MP (2016) Evaluation of wound healing treated with latex derived from rubber trees and *Aloe vera* extract in rats. Acta Cir Bras 31:570–577. https://doi.org/10.1590/S0102-865020160090000001

Brijesh KS (2011) Rauwolfia: cultivation and collection. Biotech artic web site https://www.biotecharticles.com/Agriculture-Article/Rauwolfia-Cultivation-and-Collection-892.html. Accessed on 31 Dec 2021

Burton A, Smith M, Falkenberg T (2015) Building WHO's global strategy for traditional medicine. Eur J Integr Med 7(1):13–15. https://linkinghub.elsevier.com/retrieve/pii/S1876382015000037

Cai Y, Sun M, Xing J, Corke H (2004) Antioxidant phenolic constituents in roots of *Rheum officinale* and *Rubia cordifolia*: structure-radical scavenging activity relationships. J Agric Food Chem 52(26):7884–7890. https://doi.org/10.1021/jf0489116

Cárdenas C, Quesada AR, Medina MA (2006) Evaluation of the anti-angiogenic effect of aloe-emodin. Cell Mol Life Sci 63(24):3083–3089. https://doi.org/10.1007/s00018-006-6399-6

Cardoso CA, Honda NK, Barison A (2002) Simple and rapid determination of psoralens in topic solutions using liquid chromatography. J Pharm Biomed Anal 27(1–2):217–224. https://doi.org/10.1016/s0731-7085(01)00537-4

Chanda S, Ramachandra TV (2019) Phytochemical and pharmacological importance of turmeric (*Curcuma longa*): a review. Res Rev A J Pharmacol 9(1):16–23

Charoensup R, Duangyod T, Palanuvej C, Ruangrungsi N (2017) Pharmacognostic specifications and lawsone content of *Lawsonia inermis* leaves. Pharm Res 9(1):60–64. https://doi.org/10.4103/0974-8490.199775

Chatterji A, Pakrashi SC (2006) The treatise of Indian medicinal plants, vol II. CSIR, New Delhi, pp 44–46

Chaudhary G (2012) Pharmacological properties of *Commiphora wightii* arn. Bhandari – an overview. Int J Pharm Pharm Sci 4(3):73–75

Cherian KJ, Bhambri MR (2010) Role of organic nutrients on the yield of *Ammi majus* L. Int J Environ Rehabil Conserv 1(2):16–22

Chitra M, Devi CS, Sukumar E (2003) Antibacterial activity of Embelin. Fitoterapia 74(4):401–403. https://doi.org/10.1016/s0367-326x(03)00066-2

Chopra RN, Nayar SL, Chopra IC (1956) Glossary of Indian medicinal plants. CSIR, New Delhi, p 7

Dasgupta T, Rao AR, Yadava PK (2003) Modulatory effect of henna leaf (*Lawsonia inermis*) on drug metabolising phase I and phase II enzymes, antioxidant enzymes, lipid peroxidation and chemically induced skin and forestomach papillomagenesis in mice. Mol Cell Biochem 245(1–2):11–22. https://doi.org/10.1023/a:1022853007710

Dev S (1987) A modern look at an age old ayurvedic drug guggulu. Sci Age 5:13–18

Dev S (2012) Prime Ayurvedic plant drugs: a modern scientific appraisal, 2nd edn. Ane Books, New Delhi, pp 150–250

Dey A, De J (2011) Ethnobotanical aspects of Rauvolfia serpentina (L). Benth. Ex Kurz. In India, Nepal and Bangladesh. J Med Plants Res 5:144–150

Duke JA (2002) Handbook of medicinal herbs, 2nd edn. CRC Press, Florida

Dutta S, Chowdhury AR, Srivastava SK, Ghosh I, Datta K (2011) Evidence for serpentine as a novel antioxidant by a redox sensitive HABP1 overexpressing cell line by inhibiting its nuclear translocation of NF-κB. Free Radic Res 45:1279–1288

Ebong PE, Atangwho IJ, Eyong EU, Egbung GE (2008) The antidiabetic efficacy of combined extracts from two continental plants: *Azadirachta indica* (A. Juss) (neem) and *Vernonia amygdalina* (Del.) (African bitter leaf). Am J Biochem Biotechnol 4(3):239–244. https://doi.org/10.3844/ajbbsp.2008.239.244

Elgamal M, Shalaby N, Duddeck H, Hiegemann M (1993) Coumarins and coumarin glucosides from the fruits of *Ammi majus*. Phytochemistry 34:819–823

El-Mofty AM (1952) Further study on treatment of leucodermia with *Ammi majus* L. J R Egypt Med Assoc 35:1–19

Engin B, Oguz O (2005) Evaluation of time-dependent response to psoralen plus UVA (PUVA) treatment with topical 8-methoxypsoralen (8-MOP) gel in palmoplantar dermatoses. Int J Dermatol 44(4):337–339. https://doi.org/10.1111/j.1365-4632.2004.02153.x

Ezeigbo I, Ezeja M, Madubuike K et al (2012) Antidiarrhoeal activity of leaf methanolic extract of *Rauwolfia serpentina*. Asian Pac J Trop Biomed 2(6):430–432

França Orlanda JF, Nascimento AR (2015) Chemical composition and antibacterial activity of *Ruta graveolens* L. (Rutaceae) volatile oils, from São Luís, Maranhão, Brazil. S Afr J Bot 99:103–106

Frankova A, Janatova A, Tauchen J, Kokoska L (2016) In vitro antimicrobial and antioxidant activity of extracts from six chemotypes of medicinal cannabis. Planta Med 82(S 01):137

Goel A, Kunnumakkara AB, Aggarwal BB (2008) Curcumin as "Curecumin": from kitchen to clinic. Biochem Pharmacol 75(4):787–809. https://doi.org/10.1016/j.bcp.2007.08.016

Gonska BD (2000) Medikamentöse Therapie ventrikulärer Tachykardien. Drug therapy of ventricular tachycardia. Z Kardiol 89(10):51–57. https://doi.org/10.1007/s003920070008

Govindachari TR, Suresh G, Gopalakrishnan G, Banumathy B, Masilamani S (1998) Identification of antifungal compounds from the seed oil of *Azadirachta indica*. Phytoparasitica 26(2): 109–116. https://doi.org/10.1007/bf02980677

Gow-Chin Y, Horn-Wen C, Pin-Der D (1998) Extraction and identification of an antioxidative component from jue ming zi (*Cassia tora* L.). J Agric Food Chem 46:820–824. https://doi.org/10.1021/jf970690z

Gulia Y, Choudhary M (2012) Antiulcer activity of hydroalcoholic extract of *Cassia tora* Linn using ethanol induced ulcer. Int J Pharm Pharm Sci 4:160–163

Heskel NS, Amon RB, Storrs FJ, White CR Jr (1983) Phytophotodermatitis due to *Ruta graveolens*. Contact Dermatitis 9(4):278–280. https://doi.org/10.1111/j.1600-0536.1983.tb04390.x

Honigsmann H (2013) History of phototherapy in dermatology. Photochem Photobiol Sci 12:16–21. https://doi.org/10.1039/c2pp25120e

Hossain MA, Shah MD, Sakari M (2011) Gas chromatography–mass spectrometry analysis of various organic extracts of *Merremia borneensis* from Sabah. Asian Pac J Trop Med 4(8): 637–641. https://doi.org/10.1016/s1995-7645(11)60162-4

Hussain A, Sharma C, Saniyah K, Kruti S, Shafiul H (2015) *Aloe vera* inhibits proliferation of human breast and cervical cancer cells and acts synergistically with cisplatin. Asian Pac J Cancer Prev 16:2939–2946. https://doi.org/10.7314/APJCP.2015.16.7.2939

Ishii Y, Tanizawa H, Takino Y (1994) Studies of aloe. V: mechanism of cathartic effect. Biol Pharm Bull 17:651–653

Islam MT, Ali ES, Uddin SJ, Islam MA, Shaw S, Khan IN, Saravi SSS, Ahmad S, Rehman S, Gupta VK, Găman MA, Găman AM, Yele S, Das AK, de Castro E, Sousa JM, de Moura Dantas SMM, Rolim HML, de Carvalho Melo-Cavalcante AA, Mubarak MS, Yarla NS, Shilpi JA, Mishra SK, Atanasov AG, Kamal MA (2018) Andrographolide, a diterpene lactone from *Andrographis paniculata* and its therapeutic promises in cancer. Cancer Lett 420:129–145. https://doi.org/10.1016/j.canlet.2018.01.074

Jain S, Patil UK (2010) Phytochemical and pharmacological profile of *Cassia tora* Linn. –an overview. Ind J Nat Prod Res 1(4):430–437

Janbaz KH, Gilani AH (2000) Studies on preventive and curative effects of berberine on chemical-induced hepatotoxicity in rodents. Fitoterapia 71(1):25–33. https://doi.org/10.1016/s0367-326x(99)00098-2

Javadi B, Emami SA (2015) Avicenna's contribution to mechanisms of cardiovascular drugs. Iran J Basic Med Sci 18:721–722

Ji L, Xu Z (1995) Review of constituents in fruits of *Psoralea corylifolia* L. Zhongguo Zhong Yao Za Zhi 20(2):120–128

Kannan R, Babu UV (2012) Identity and pharmacognosy of *Ruta graveolens* Linn. Anc Sci Life 32(1):16–19. https://doi.org/10.4103/0257-7941.113792

Kapoor LD (2001) Handbook of ayurvedic medicinal plants. CRC Press, Boca Raton, p 216

Kapoor NK, Dev S, Nityanand S (1979) Process for obtaining hypolipidemic and antiplatelet fractions from guggul resin. Indian patent no.148265 dt 6.4.1979

Kar A, Choudhary BK, Bandyopadhyay NG (2003) Comparative evaluation of hypoglycaemic activity of some Indian medicinal plants in alloxan diabetic rats. J Ethnopharmacol 84(1): 105–108. https://doi.org/10.1016/s0378-8741(02)00144-7

Katsura H, Tsukiyama RI, Suzuki A, Kobayashi M (2001) *In vitro* antimicrobial activities of bakuchiol against oral microorganisms. Antimicrob Agents Chemother 45(11):3009–3013. https://doi.org/10.1128/AAC.45.11.3009-3013.2001

Kawami M, Yamada Y, Issarachot O, Junyaprasert VB, Yumoto R, Takano M (2018) P-gp modulating effect of *Azadirachta indica* extract in multidrug-resistant cancer cell lines. Pharmazie 73(2):104–109. https://doi.org/10.1691/ph.2018.7116

Khan I, Pandotra P, Gupta AP, Sharma R, Das Gupta B, Dhar JK, Ram G, Bedi YS, Gupta S (2010) RP-thin layer chromatographic method for the quantification of three gingerol homologs of ultrasonic-assisted fresh rhizome extracts of *Zingiber officinale* collected from North Western Himalayas. J Sep Sci 33(4–5):558–563. https://doi.org/10.1002/jssc.200900629

Khare CP (2007) Indian medicinal plants: an illustrated dictionary. Springer Science & Business Media

Kher A, Chaurasia SC (1997) Antifungal activity of essential oils of three medical plants. Indian Drugs 15:41–42

Khushboo PS, Jadhav VM, Kadam VJ, Sathe NS (2010) *Psoralea corylifolia* Linn.- "Kushtanashini". Pharmacogn Rev 4(7):69–76. https://doi.org/10.4103/0973-7847.65331

Kimura I, Yoshikawa M, Kobayashi S, Sugihara Y, Suzuki M, Oominami H, Murakami T, Matsuda H, Doiphode VV (2001) New triterpenes, myrrhanol A and myrrhanone A, from guggul-gum resins, and their potent anti-inflammatory effect on adjuvant-induced air-pouch granuloma of mice. Bioorg Med Chem Lett 11(8):985–989. https://doi.org/10.1016/s0960-894x(01)00111-1

Kiranmai M, Kumar M, Ibrahim M (2011) Free radical scavenging activity of neem tree (*Azadirachta indica* A. Juss Var., Meliaceae) root barks extract. Asian J Pharm Clin Res 4: 134–136

Kirtikar KR, Basu BD (1982) Indian medicinal plants, vol I–VI. Bishan Singh Mahendar Pal Singh, Dehradun

Ko E et al (2020) Cassia tora seed improves pancreatic mitochondrial function leading to recovery of glucose metabolism. Am J Chin Med 48:615–629. https://doi.org/10.1142/s0192415x20500317

Kokate C, Purohit AP, Gokhale SB (2010) Pharmacognosy. Nirali Prakashan, Maharashtra

Kokate CK, Purohit AP, Gokhle SB (2012) Pharmacognosy, vol I & II, 47th edn. Nirali Prakashan, Pune

Kovacs D, Bastonini E, Ottaviani M, Cota C, Migliano E, Dell'Anna ML, Picardo M (2018) Vitiligo skin: exploring the dermal compartment. J Invest Dermatol 138(2):394–404. https://doi.org/10.1016/j.jid.2017.06.033

Kumar S, Yadav M, Yadav A, Rohilla P, Yadav JP (2017) Antiplasmodial potential and quantification of aloin and aloe-emodin in *Aloe vera* collected from different climatic regions of India. BMC Complement Altern Med 17:369

Kumari R, Rathi B, Rani A, Bhatnagar S (2013) *Rauvolfia serpentina* L. Benth. ex Kurz.: phytochemical, pharmacological and therapeutic aspects. Int J Pharm Sci Rev Res 23(2): 348–355

Kunnumakkara AB, Guha S, Krishnan S, Diagaradjane P, Gelovani J, Aggarwal BB (2007) Curcumin potentiates antitumor activity of gemcitabine in an orthotopic model of pancreatic cancer through suppression of proliferation, angiogenesis, and inhibition of nuclear factor-kappaB-regulated gene products. Cancer Res 67(8):3853–3861. https://doi.org/10.1158/0008-5472.CAN-06-4257

Li Y, Tran VH, Duke CC, Roufogalis BD (2012) Preventive and protective properties of *Zingiber officinale* (ginger) in diabetes mellitus, diabetic complications, and associated lipid and other metabolic disorders: a brief review. Evid Based Complement Alternat Med 2012:516870. https://doi.org/10.1155/2012/516870

Li CC, Wang TL, Zhang ZQ, Yang WQ, Wang YF, Chai X, Wang CH, Li Z (2016) Phytochemical and pharmacological studies on the genus *Psoralea*: a mini review. Evid Based Complement Alternat Med 2016:8108643. https://doi.org/10.1155/2016/8108643

Lobay D (2015) *Rauwolfia* in the treatment of hypertension. Integr Med (Encinitas) 14(3):40–46

Madhavan SN, Arimboor R, Arumughan C (2011) RP-HPLC-DAD method for the estimation of embelin as marker in *Embelia ribes* and its polyherbal formulations. Biomed Chromatogr 25(5): 600–605. https://doi.org/10.1002/bmc.1489

Maiti K, Mukherjee K, Gantait A, Saha BP, Mukherjee PK (2007) Curcumin-phospholipid complex: preparation, therapeutic evaluation and pharmacokinetic study in rats. Int J Pharm 330(1–2):155–163. https://doi.org/10.1016/j.ijpharm.2006.09.025

Maity TK, Mandal SC, Bhakta T, Pal M, Saha BP (2001) Metabolism of 1,8-dihydroxy 3-hydroxy methyl anthraquinone (aloe-emodin) isolated from the leaves of *Cassia tora* in albino rats. Phytother Res 15(5):459–460

Majeed R, Reddy MV, Chinthakindi PK, Sangwan PL, Hamid A, Chashoo G, Saxena AK, Koul S (2012) Bakuchiol derivatives as novel and potent cytotoxic agents: a report. Eur J Med Chem 49:55–67. https://doi.org/10.1016/j.ejmech.2011.12.018

Malik S, Moraes DFC, do Amaral FMM, Ribeiro MNS (2017) *Ruta graveolens*: phytochemistry, pharmacology, and biotechnology. In: Jha S (ed) Transgenesis and secondary metabolism. Reference series in Phytochemistry. Springer, Cham. https://doi.org/10.1007/978-3-319-28669-3_4

Manga P, Mutasim D (2007) Vitiligo. In xPharm: The comprehensive pharmacology reference, pp 1–15. https://doi.org/10.1016/B978-008055232-3.60769-X

Manvitha K, Bidya B (2014) *Aloe vera*: a wonder plant its history, cultivation and medicinal uses. J Pharmacogn Phytochem 2(5):85–88

McNeely W, Goa KL (1998) 5-Methoxypsoralen. A review of its effects in psoriasis and vitiligo. Drugs 56(4):667–690. https://doi.org/10.2165/00003495-199856040-00015

Mesrob B, Nesbitt C, Mishra R, Pandey RC (1998) High-performance liquid chromatographic method for fingerprinting and quantitative determination of E- and Z-guggulsterones in *Commiphora mukul* resin and its products. J Chromatography B 720:189–196

Mohanta RK, Rout SD, Sahu HK (2006) Ethnomedicinal plant resources of Simlipal biosphere reserve, Orissa, India. Zoo's Print J 21(8):2372–2374

Mukherjee PK (2002) Quality control of herbal drugs, an approach to evaluation of botanicals, 1st edn. Business Horizons, New Delhi, pp 246–377

Nadkarni KM (1976a) Indian Materia Medica. Popular Prakashan, Bombay

Nadkarni KM (1976b) Mumbai: Popular Prakashan Pvt. Ltd; Indian Materia Medica, vol 1, pp 1019–22

Nimisha, Rizvi DA, Fatima Z, Neema, Kaur CD (2017) Antipsoriatic and anti-inflammatory studies of *Berberis aristata* extract loaded Nanovesicular gels. Pharmacogn Mag 13(Suppl 3):S587–S594. https://doi.org/10.4103/pm.pm_210_17

Nityanand S, Kapoor NK (1978) Effect of *guggul* steroids on cholesterol biosynthesis in rats. Indian J Biochem Biophys 15:77

Ossenkoppele P, Van der Sluis W, Van Vloten W (1991) Phototoxic dermatitis following the use of *Ammi majus* fruit for vitiligo. Ned Tijdschur Geneeskd 135(11):478–480

Oyedemi BO, Shinde V, Shinde K, Kakalou D, Stapleton PD, Gibbons S (2016) Novel R-plasmid conjugal transfer inhibitory and antibacterial activities of phenolic compounds from *Mallotus philippensis* (Lam.) Mull. Arg J Glob Antimicrob Resist 5:15–21. https://doi.org/10.1016/j.jgar.2016.01.011

Panda H (2000) National Institute of Industrial Research; Herbs, cultivation and medicinal uses, New Delhi, pp 479–81

Patel RK, Prajapati AM (2008) Development and validation of a visible absorption densitometry method for quantitation of conessine in *Holarrhena antidysenterica* (Kurchi). J AOAC Int 91(2):339–343

Paul R, Prasad M, Sah NK (2011) Anticancer biology of *Azadirachta indica* L (neem): a mini review. Cancer Biol Ther 12(6):467–476. https://doi.org/10.4161/cbt.12.6.16850

Pei SJ (2001) Ethnobotanical approaches of traditional medicine studies: some experiences from Asia. Pharm Biol 39:74–79

Pholphana N, Rangkadilok N, Thongnest S, Ruchirawat S, Ruchirawat M, Satayavivad J (2004) Determination and variation of three active diterpenoids in *Andrographis paniculata* (Burm.f.) Nees. Phytochem Anal 15(6):365–371. https://doi.org/10.1002/pca.789

Pullaiah J (2002) Medicinal plants in India, vol 2. Regency Publisher, New Delhi, pp 441–443

Rajpal V (2005) Standardization of botanicals, vol 2. Eastern Publishers, New Delhi, pp 284–295

Ramadhan AA, Mutlag SH (2021) Modulated effect of *Ammi majus* L. on psoriasis-like skin inflammation caused by Imiquimod *in vivo*. J Res Med Dent Sci 9(12S):9–13

Rao GM, Rao CV, Pushpangadan P, Shirwaikar A (2006) Hepatoprotective effects of rubiadin, a major constituent of *Rubia cordifolia* Linn. J Ethnopharmacol 103(3):484–490. https://doi.org/10.1016/j.jep.2005.08.073

Rasheed NMA, Srividya GS, Nagaiah K (2017) HPTLC method development and quantification of curcumin content in different extracts rhizome of *Curcuma longa* L. Ann Phytomed 6(2):74–81. https://doi.org/10.21276/ap.2017.6.2.6

Ratheesh M, Sindhu G, Helen A (2013) Anti-inflammatory effect of quinoline alkaloid skimmianine isolated from *Ruta graveolens* L. Inflamm Res 62(4):367–376

Reddy DN, Al-Rajab AJ (2016) Chemical composition, antibacterial and antifungal activities of *Ruta graveolens L.* volatile oils. Cogent Chem 2:1220055. https://doi.org/10.1080/23312009.2016.1220055

Reynolds T, Dweck AC (1999) Aloe vera leaf gel: A review update. J Ethnopharmacol 68(1–3):3–37. https://doi.org/10.1016/s0378-8741(99)00085-9

Rio JA, Del-Diaz L, Garcia-Bernal D, Blanquer M, Ortuno A, Correal E, Moraleda JM (2014) Furanocoumarins: biomolecules of therapeutic interest. Stud Nat Prod Chem 43:145–195. https://doi.org/10.1016/B978-0-444-63430-6.00005-9

Ro JY, Lee BC, Kim JY, Chung YJ, Chung MH, Lee SK, Jo TH, Kim KH, Park YI (2000) Inhibitory mechanism of aloe single component (alprogen) on mediator release in Guinea pig lung mast cells activated with specific antigen-antibody reactions. J Pharmacol Exp Ther 292(1):114–121

Ruan B, Kong LY, Takaya Y, Niwa M (2007) Studies on the chemical constituents of *Psoralea corylifolia* L. J Asian Nat Prod Res 9(1):41–44. https://doi.org/10.1080/10286020500289618

Russo P, Frustaci A, Del Bufalo A, Fini M, Cesario A (2013) From traditional European medicine to discovery of new drug candidates for the treatment of dementia and Alzheimer's disease: acetylcholinesterase inhibitors. Curr Med Chem 20(8):976–983. https://doi.org/10.2174/0929867311320080002

Ruyter CM, Akram M, Illahi I, Stöckigt J (1991) Investigation of the alkaloid content of *Rauwolfia serpentina* roots from regenerated plants. Planta Med 57(4):328–330

Saeed MA, Ahmad I, Yaqub U, Akbar S, Waheed A, Saleem M, Din N (2004) *Aloe Vera*: a plant of vital significance. Sci Vis 9:1–13

Sah P, Agrawal D, Garg SP (2006) Isolation and identification of furocoumarins from the seeds of *Psoralea corylifolia* L. Indian J Pharma Sci 68:768–771. https://doi.org/10.4103/0250-474X.31012

Sánchez M, González-Burgos E, Iglesias I, Gómez-Serranillos MP (2020) Pharmacological update properties of *Aloe Vera* and its major active constituents. Molecules 25(6):1324. https://doi.org/10.3390/molecules25061324

Sarup P, Bala S, Kamboj S (2015) Pharmacology and Phytochemistry of oleo-gum resin of *Commiphora wightii* (Guggulu). Scientifica (Cairo) 2015:138039. https://doi.org/10.1155/2015/138039

Sastry VV (1976) History of Guggulu based on Ayurvedic literature. Bull Indian Inst Hist Med Hyderabad 6(2):102–116

Satyavati GV, Dwarakanath C, Tripathi SN (1969) Experimental studies on the hypocholesterolemic effect of *Commiphora mukul*. Engl (Guggul) Indian J Med Res 57(10): 1950–1962

Schaaf O, Jarvis AP, van der Esch SA, Giagnacovo G, Oldham NJ (2000) Rapid and sensitive analysis of azadirachtin and related triterpenoids from neem (*Azadirachta indica*) by high-performance liquid chromatography-atmospheric pressure chemical ionization mass spectrometry. J Chromatogr A 886(1–2):89–97. https://doi.org/10.1016/s0021-9673(00)00492-1

Schönberg A, Sina A (1948) Xanthotoxin from the fruits of Ammi majus L. Nature 161:481–482. https://doi.org/10.1038/161481c0

Sebastian P (2006) Health sciences; Ayurvedic medicine: the principles of traditional practice, vol 2. Elsevier, New York, pp 135–136

Shah SMA, Naqvi SAR, Munir N, Zafar S, Akram M, Nisar J (2020) Antihypertensive and Antihyperlipidemic activity of aqueous Methanolic extract of *Rauwolfia Serpentina* in albino rats. Dose Response 18(3). https://doi.org/10.1177/1559325820942077

Shailajan S, Menon S, Singh A, Mhatre M, Sayed N, Joshi H, Tiwari B (2012) Estimation of psoralen from herbal formulations containing *Psoralea corylifolia* using the RP-HPLC-DAD method and its application to a pharmacokinetic study. Int J Green Pharm 6(3):217–223. https://doi.org/10.4103/0973-8258.104935

Shamsi Y, Kumar H, Tamanna SA, Khan EA (2006) Effect of a polyherbal Unani formulation on chronic urticaria. Ind J Trad Knowl 5(2):279–283

Sharma PC, Yelne MB, Dennis TJ (2001) Database on medicinal plants used in Ayurveda, vol 2. Central Council for Research in Ayurveda and Siddha, New Delhi, pp 89–93

Shen YH, Li RT, Xiao WL, Xu G, Lin ZW, Zhao QS, Sun HD (2006) ent-Labdane diterpenoids from *Andrographis paniculata*. J Nat Prod 69(3):319–322. https://doi.org/10.1021/np050160u

Shokeen P, Bala M, Singh M, Tandon V (2008) *In vitro* activity of eugenol, an active component from *Ocimum sanctum*, against multiresistant and susceptible strains of *Neisseria gonorrhoeae*. Int J Antimicrob Agents 32(2):174–179. https://doi.org/10.1016/j.ijantimicag.2008.03.018

Singh N, Sastry MS (1997) Antimicrobial activity of neem oil. Indian J Pharmacol 13:102–106

Singh DK, Kumar N, Sachan A, Lakhani P, Tutu S, Nath R, Sachan AK, Dixit RK (2017) Hypolipidaemic effects of *Gymnema sylvestre* on high fat diet induced Dyslipidaemia in Wistar rats. J Clin Diagn Res 11(5):FF01-FF05. https://doi.org/10.7860/JCDR/2017/27430.9859

Singhal M, Kansara N (2012) *Cassia tora* Linn cream inhibits ultraviolet-B-induced psoriasis in rats. ISRN Dermatol 2012:346510. https://doi.org/10.5402/2012/346510

Siriyong T, Srimanote P, Chusri S, Yingyongnarongkul BE, Suaisom C, Tipmanee V, Voravuthikunchai SP (2017) Conessine as a novel inhibitor of multidrug efflux pump systems in *Pseudomonas aeruginosa*. BMC Complement Altern Med 17(1):405. https://doi.org/10.1186/s12906-017-1913-y

Späth E, Holzen H (1933) Ber Dtsch Chem Ges 66:1137

Staniszewska I, Królicka A, Maliński E, Łojkowska E, Szafranek J (2003) Elicitation of secondary metabolites in in vitro cultures of *Ammi majus* L. Enzym Microb Technol 33:565–568. https://doi.org/10.1016/S0141-0229(03)00180-7

Sultana B, Anwar F, Przybylski R (2007) Antioxidant activity of phenolic components present in barks of *Azadirachta indica*, *Terminalia arjuna*, *Acacia nilotica*, and *Eugenia jambolana* Lam. trees. Food Chem 104(3):1106–1114. https://doi.org/10.1016/j.foodchem.2007.01.019

Tanaka M, Misawa E, Yamauchi K, Abe F, Ishizaki C (2015) Effects of plant sterols derived from *Aloe vera* gel on human dermal fibroblasts in vitro and on skin condition in Japanese women. Clin Cosmet Invest Dermat 8:95–104. https://doi.org/10.2147/CCID.S75441

Tanaka M, Yamamoto Y, Misawa E, Nabeshima K, Saito M, Yamauchi K, Furukawa F (2016) Aloe sterol supplementation improves skin elasticity in Japanese men with sunlight-exposed skin: a 12-week double-blind, randomized controlled trial. Clin Cosmet Invest Dermat 9:435–442. https://doi.org/10.2147/CCID.S118947

Tanvir EM, Hossen MS, Hossain MF, Afroz R, Gan SH, Khalil MI, Karim N (2017) Antioxidant properties of popular turmeric (*Curcuma longa*) varieties from Bangladesh. J Food Qual 2017: 1–8. https://doi.org/10.1155/2017/8471785

Thappa DM, Dogra J (1994) Nodulocystic acne: oral gugulipid versus tetracycline. J Dermatol 10: 729–731. https://doi.org/10.1111/j.1346-8138.1994.tb03277.x

Upton R, Axentiev MSP, Swisher MAD. (2012) *Aloe vera* leaf. American Herbal Pharmacopoeia, pp 1–52

Usmani QI, Jahan N, Aleem M, Hasan SA (2021) Aatrilal (*Ammi majus* L.), an important drug of Unani system of medicine: a review. J Ethnopharmacol 276:114144. https://doi.org/10.1016/j.jep.2021.114144

Uzzaman S (2020) Pharmacological activities of neem (*Azadirachta indica*): a review. Int J Pharma Life Sci 1(1):38–41. https://doi.org/10.33545/27072827.2020.v1.i1a.8

Vakil RJ (1949) A clinical trial of *Rauwolfia serpentina* in essential hypertension. Br Heart J 11(4): 350–355. https://doi.org/10.1136/hrt.11.4.350

Vakil RJ (1955) *Rauwolfia serpentina* in the treatment of high blood pressure: a review of the literature. Circulation 12(2):220–229

Vaughn AR, Branum A, Sivamani RK (2016) Effects of turmeric (*Curcuma longa*) on skin health: a systematic review of the clinical evidence. Phytother Res 30(8):1243–1264

Vijayalakshmi A, Madhira G (2014) Anti-psoriatic activity of flavonoids from *Cassia tora* leaves using the rat ultraviolet B ray photodermatitis model. Rev Bras Farma 24:322–329

Vollono L, Falconi M, Gaziano R, Iacovelli F, Dika E, Terracciano C, Bianchi L, Campione E (2019) Potential of curcumin in skin disorders. Nutrients 11(9):2169. https://doi.org/10.3390/nu11092169

Wahedi HM, Jeong M, Chae JK, Do SG, Yoon H, Kim SY (2017) Aloesin from *Aloe vera* accelerates skin wound healing by modulating MAPK/Rho and Smad signaling pathways *in vitro* and *in vivo*. Phytomedicine 28:19–26. https://doi.org/10.1016/j.phymed.2017.02.005

Wang XY, Wang JX (2007) Synergistic effect of psoralen cooperated with substrates on Tyrosinase activation. Nat Prod Res Dev 19:77–80

Wilson CO, Gisvold O, Doerge RF (1999) Textbook of organic medicinal and pharmaceutical chemistry, 7th edn, pp 956

Woodson RE, Yongken HW, Schlitter E, Schneider JA (1957) *Rauwolfia*: botany, pharmacognosy, chemistry and pharmacology. Little, Brown, Boston, p 150

Wu X, Ding W, Zhong J, Wan J, Xie Z (2013) Simultaneous qualitative and quantitative determination of phenolic compounds in *Aloe barbadensis* mill by liquid chromatography-mass spectrometry-ion trap-time-of-flight and high-performance liquid chromatography-diode array detector. J Pharm Biomed Anal 80:94–106. https://doi.org/10.1016/j.jpba.2013.02.034

Yadava RN, Verma V (2005) A new biologically active flavonol glycoside from *Psoralea corylifolia* (Linn.). J Asian Nat Prod Res 7(4):671–675. https://doi.org/10.1080/10286020310001608921

Yarnell E, Abascal K (2001) Treating hypertension botanically. Altern Complement Ther 7(5): 284–290

Yen GC, Chuang DY (2000) Antioxidant properties of water extracts from *Cassia tora* L. in relation to the degree of roasting. J Agric Food Chem 48(7):2760–2765. https://doi.org/10.1021/jf991010q

Zanwar AA, Badole SL, Menaa F (2013) *Curcuma longa*: use for skin disease care. In: Watson R, Zibadi S (eds) Bioactive dietary factors and plant extracts in dermatology. Nutrition and health. Humana Press, Totowa. https://doi.org/10.1007/978-1-62703-167-7_36

Chapter 10
Traditional Uses and Properties of Indian Medicinal Plants in the Treatment of Vitiligo

Mohd Kashif Husain, Goli Penchala Pratap, Mokhtar Alam, Ghazala Javed, Ahmed Minhajuddin, and Asim Ali Khan

Abstract Traditional health care practices have retained their importance for many centuries and are utilized by ethnic populations, based on traditional knowledge (TK). India is endowed with ethnic groups that have accumulated an extensive knowledge of traditional plant-based medicines. They have exploit these resources. India's contribution to the development of traditional health care systems, including Ayurveda, Siddha and Unani (ASU), continues to be exemplary. Vitiligo is an acquired chronic depigmentation, a skin disease of unknown etiology which usually occurs due to the selective destruction of melanocytes. Despite recent significant research achievements in developing viable therapies to combat this social stigma, there are still no approved therapies to treat it in modern systems.

In the Indian systems of medicine (ISM) plants provide both raw material for formulations and are used as single drugs to treat Vitiligo (Leucoderma). Plant species, like; *Ammi majus* (Atrilal), *Psoralea corylifolia* (Babchi), *Senna tora* (Panwad), *Senna absus* (Chaksu), *Eclipta prostrata* (Bhangra), *Ficus carica* (Anjeer) and *Plumbago zeylanica* (Sheetraj Hindi), provide a practical, productive, and affordable source of traditional medicine/drugs to cure stigmatic diseases, among them Vitiligo.

Keywords Indian medicinal plants · Traditional medicine · Unani · Vitiligo · Leucoderma · Babchi · Atrilal

M. K. Husain (✉) · G. P. Pratap · A. Minhajuddin
National Research Institute of Unani Medicine for Skin Disorders (NRIUMSD), Hyderabad, Telangana, India

M. Alam · G. Javed · A. A. Khan
Central Council for Research in Unani Medicine (CCRUM), New Delhi, India

Á. Máthé, I. A. Khan (eds.), *Medicinal and Aromatic Plants of India Vol. 2*,
Medicinal and Aromatic Plants of the World 9,
https://doi.org/10.1007/978-3-031-44914-7_10

10.1 Introduction

Traditional medicine has become more accepted in the treatment of many ailments because of the belief that it is safe, easily available and has less side-effects. A larger population of developing countries still depends on Medicinal plant (MP) derived drugs for their basic healthcare needs (Farnsworth et al. 1985).

Traditional knowledge (TK) refers to indigenous peoples' knowledge, innovations, know-how, skills and traditions. TK is often developed, maintained, and passed down from generation to generation, as part of a cultural or spiritual identity, and is adapted to the local culture and environment (UN 2019). Traditional knowledge is often transmitted orally and can be expressed in cultural values, folklore, proverbs, stories, songs, beliefs and religious ceremony. TK is also a source of traditional use and management of territory, land and resources, using local agricultural practices to think about the land without depleting its resources (WIPO 2021).

Medicinal and Aromatic Plants (MAPs) are essential components in rural people's healthcare, all around the world. Traditional medicine and medicinal plants have long been used as a normative basis for maintaining the good health of the people in most underdeveloped countries (Reddy et al. 2019). The resurgence of MAPs-use in high-income earning countries of the world has brought other types of its use in the set-up of complementary and alternative medicines (CAM). MAPs have now become "industrial products" with spanking new concepts like; nutraceuticals, phytotherapy, aromatherapy, cosmeceuticals and veterinary remedial uses, thus, widening the scope of its utilization. MAPs and their active chemical ingredients served as a lead and/or model materials for pharmaceutical investigations, research and production of useful medication.

There are numerous traditional medical systems, each with its own set of principles and cultural roots. TM has developed from the rich traditions and scientific heritage of ancient civilizations (WHO 2004).

The earliest written evidence of TM use extends back to the about five thousand (5000) years in Indian, Greek, Chinese, Roman, Syrian, and Egyptian classical texts/ manuscripts. The-'*Charaka Samhita*, *Sushruta Samhita* and *Rig-Veda* are only a few examples of ancient Indian literature that supports these beliefs and age-old traditions (Kamboj 2000).

10.2 Medicinal Plants and Indian Cultural Ethos

India, with its diverse ecosystems, presents a different tribal population representing a complex cultural mosaic. Indian society is divided into tribal, rural, and urban societies on the basis of geographical environment and socio-cultural characteristics. Tribal communities are an integral part of Indian society. Tribal populations are found in almost every part of the world. India has the second largest tribal population in the world, followed by Africa.

An anthropological survey of India published in the 'People of India Project', identified 4635 ethnic communities, of which 461 communities belonged to designated scheduled tribes. For centuries, the ethnic population have protected and preserved a rich cultural and spiritual heritage of living in harmony with nature (https://ansi.gov.in/people-ofindia\). Local and traditional communities continue to venerate nature and preserve biodiversity through traditional practices and institutional structures of religious importance, such as sacred groves.

10.3 Indian Traditional Systems of Medicine

The age-old Indian traditional systems of medicine (TSM) is one of the oldest systems of medicine exist in the world, played a vital role in providing health care services to human civilization, since thousands of years. The Ayurveda, Yoga and Naturopathy, Unani, Siddha, Homoeopathy and Sowa Rigpa system of India are officially well recognized TSM which has the exclusive peculiarity. These systems are based on the distinct concept and philosophies which promotes a healthy lifestyle, maintenance of health with well-established and conventional ideas. The basic approach to treatment of all these systems is holistic and pharmacological procedures are based on natural products, mainly those of MAPs origin followed by mineral and animal origin (Husain 2021a, b).

Among Indian TSM, Unani (also known as Greco-Arab System) originated from the ideas and philosophies of Hippocrates, which further refined and systematized by the physicians of Arab. In India, the Unani system was introduced during the Mughal period (eleventh to thirteenth century) and further spread through the practices and teachings of the eminent Unani physicians and scholars of Indian origin (Husain et al. 2017).

10.4 Vitiligo, Skin Depigmentation

Vitiligo is an acquired chronic depigmentation skin disease of unknown origin resulting from the particular destruction of the cells (melanocytes) in the skin and eyes. Some workers believed that skin depigmentation is the result of an autoimmune process that destructs the melanin producing cells. It is the utmost frequent reason of depigmentation (lack of melanin) at global level, with an assessed prevalence of 0.5–1% of the world population. This disorder can be psychologically disturbing and stigmatizing, especially in dusky skinned individuals, young girls and women. Vitiligo is clinically characterized by the appearance of white patches due to the loss of functional melanocytes in the skin, hair, or both. Vitiligo can affect any area of the skin, but it commonly occurs on the face, neck, arms, and skin folds.Two forms of the vitiligo are well recognised: segmental and non-segmental. Of these two, non-segmental Vitiligo is most common (Hussain 2021; Ezzedine et al. 2015).

Although various predisposing factors have been proposed for vitiligo, the exact cause of the disease remains unclear. The condition negatively affects the bodily image of the individuals, particularly females due to their major cosmetic concerns. This has a psychological and economic impact on affected persons as it increases the clinical visit for dermatological consultations (Khandalavala and Nirmalraj 2014).

10.4.1 Natural Products, Traditional Systems of Medicine and Vitiligo

Over the past few decades' the demand of natural products obtained from MAPs has increased and gained popularity among the masses due to its age-old use in TSM, specifically; Ayurveda and Unani. It is considered relatively safe, reliable, easy to use and affordable for most of the people. The use of natural herbal products (complementary/alternative medicines) for chronic disease conditions is common and individuals may find this approach to treat Vitiligo also (Hussain 2021). Studies have been reported involving the use of MAPs in the conditions of Vitiligo. TSM and natural health products are promoted for use in the conditions of Vitiligo. However, they have been found to be mentioned in old classic treatises, but there is limited scientific evidence to support their effectiveness for this purpose (Szczurko and Boon 2008; Fisk et al. 2014).

10.4.2 Concept of Vitiligo in Ayurveda System

'Shweta Kushta' in Ayurveda is grouped under skin disorders. '*Shweta*' means white, and '*Kushta*' means skin disease. It is caused by the amplification of "*Pitta*" in the body. *Pitta* is Ayurvedic humor that indicates fire elements and is manifested in the form of skin disorder. The *Pitta* dosha is of five types, in which '*Bhrajak Pitta'* gives discoloration to the skin (Shingadiya et al. 2018). As per the Ayurvedic concept, Vitiligo (*shwitra*) occupies only blood, skin, muscle tissues (*mansadhatu*) and adipose tissues (*medadhatu*). The Vitiligo disease is noncommunicable and it neither destroys the body tissues nor produces any contagious discharge (*vyadhiswabhava*) (Agnivesh et al. 1993).

10.4.3 Ayurvedic Medication for Vitiligo Based on Medicinal Plants

A short survey on the available literature based on the published reports revealed that the majority of Ayurvedic physicians used the drugs– '*Savaranakara Yoga*' internally and '*Savaranakara Lepa*' externally, followed by '*Arogyavardhini Rasa*' for

Fig. 10.1 (**a**). *Zingiber officinale*; (**b**). *Ammi majus*; (**c**). *Plumbago zeylanica*; (**d**). *Ipomea nil*

the treatment of Vitiligo (Fig. 10.1). The constituents of different Ayurvedic formulations used to treat Vitiligo are summarized in Table 10.1.

10.4.4 Vitiligo Concept in the Unani System

Ancient Unani physicians' defined Vitiligo (*Baraṣ*) as white discoloration of the skin which can appear anywhere in the body but mostly occurs on hands and feet. It is a skin disease. When it involves most of the body's skin it is known as extensive Vitiligo (*Baraṣ Muntashir*). According to Galen (Jalinoos, 130–200 AD) as mentioned in the script- *Moalijat-e-Buqratiya* (tenth century, AD), the *Baraṣ* is caused by the instability of transformative faculty (*Quwwat Mughayyira*), cold-impaired nature of organs,or it may be congenital (Anonymous 2012; Tabari 1997). Rabban Tabari (810–895 AD) while describing the etiology of *Baraṣ* in his well-known manuscript *Firdaus al-Hikmat* says: "impairment of blood (*Fasad-ud-dam*) and

Table 10.1 Ayurvedic formulations for Vitiligo,based on medicinal plants

S. no.	Ayurvedic formulations	Constituents	References
1.	Arogyavardhini	Haritaki (*Terminalia chebula*) dried fruit rind Bibhitaka (*Terminalia bellerica*) dried fruit rind Amalaki (*Phyllanthus emblica*) dried fruit rind Gum resin of guggul (*Commiphora wightii*) Chitrakamoola (*Plumbago zeylanica*) roots Kutaki (*Picrorhiza kurroa*) roots, Nimba (*Azadirachta indica*) leaves	Sapkota et al. (2016)
2.	Kanakabindvarishta	Khadira (*Acacia catechu*) heartwood Haritaki (*Terminalia chebula*) dried fruit rind Bahera (*Terminalia bellirica*) dried fruit rind Saunth (*Zingiber officinale*) dried rhizome, Maricha (*Piper nigrum*) fruits, Pippali (*Piper longum*) fruits, Vaividang (*Embelia ribes*) fruits, Haridra (*Curcuma longa*) dried rhizome Nagarmotha (*Cyperus rotundus*) dried rhizomes, Vasa (*Justicia adhatoda*) leaves Indrajau (*Holarrhena pubescens*) dried bark Guduchi (*Tinospora cordifolia*) dried stem	Shingadiya et al. (2015)
3.	Manjishthadi Kwatha	Manjishtha (*Rubia cordifolia*) root powder Haritaki (*Terminalia chebula)* dried fruit rind powder Bahera (*Terminalia bellirica*) dried fruit rind powder Amalaki (*Phyllanthus emblica*) dried fruit rind powder and Haridra (*Curcuma longa*) dried rhizome powder, Manjishtha root powder (*Rubia cordifolia*)	Rao et al. (2017)
4.	Savaranakara Lepa (Ointment)	Bakuchi (*Psoralea corylifolia*) seeds	Shingadiya et al. (2016, 2018)
5.	Savaranakara Yoga (Oral)	Bakuchi (*P. corylifolia*) seeds	Shingadiya et al. (2016, 2018)
6.	Shvitrahara Kashaya (Decoction)	Bakuchi (*P. corylifolia*) seed powder Khadira (*Acacia catechu*) heartwood powder Haridra (*Curcuma longa*) dried rhizome powder Sariva (*Hemidesmus indicus*) root powder Chakramarda (*Senna tora*) seed powder Chakshushya (*Senna absus*) seed powder	Dhanik et al. (2011)
7.	Shwitra Yoga (Oral)	Bakuchi (*P. corylifolia*) seed powder Khadira (*Acacia catechu*) heartwood powder Chitrakamoola (*Plumbago zeylanica*) root powder Jatamansi (*Nardostachys jatamansi*) root powder Amalaki (*Phyllanthus emblica*) dried fruit rind powder	Mulla (2018)

coldness of blood (*Burudat-ud-dam*) are the major reason of *Baraṣ*. The white patches (*Baraṣ*) appear when the digestive faculty of the body cannot properly digest the food and the blood throughout the body becomes impure due to phlegm (*Balgham*) or coldness" (Tabari 1996).

Rhazes (Zakariya-Al-Razi, 850–925, AD) has given a complete explanation of *Baraṣ* disease in his famous book *Kitab-ul-Hawi Fit Tib* as- "*Baraṣ* appears when flesh becomes phlegmatic. Thus the blood reaching to phlegmatic flesh turns the flesh become soft as that of the mollusk. The body part receiving such blood become undernourished and is transform into phlegmatic substance" (Razi 1970). *Baraṣ* is a chronic ailment and therefore, all Unani physicians believe that its treatment must begin with the removal of harmful substances from the body (*Tanquiyah-e-Badan*)) with concoctive and purgative (*Mundij* and *Mushil*) therapies.

10.4.5 Medicinal Plants (Unani Single Drugs) Used in the Treatment of Vitiligo

Traditional Unani system of medicine uses only natural resources as drug. The plant origin drugs are mostly obtained from the MAPs. The plants are used in the form of crude drug prepared of herbs, shrubs and trees. The herbs are used as whole plant or plant organs (parts). The parts (root, stem, branch, leaf, flower, fruit and seed) of shrubs and trees are used either as single drugs (*Mufrad*) or compound (*Murakkab*) formulations. The age-old wisdom of Unani Medicine is contained in its classical literature. The classical Unani formulations which have been in use for centuries are generally considered safe. The documentation of theories and philosophy of the ancient Unani scholars in the form of classical manuscripts and rare books, has clearly mentioned the cure of various skin diseases including Vitiligo (Khan 2014, Hakeem 2002) with the help of plant origin single drugs (Figs. 10.1 and 10.2) such as; *Ammi majus* (Atrilal), *Psoralea corylifolia* (Babchi), *Ipomoea nil* (Habbul-Neel), *Senna tora* (Panwar), and *Plumbago zeylanica* (Sheetraj). Some important single drugs of MAPs origin mentioned in Unani classical texts are briefly described in the followings:

10.4.5.1 *Ammi majus* L. (Apiaceae)

Unani Name Atrilal; **Part Used:** Seeds; dried powder is used as paste for topical application and for making tablets to treat Vitiligo. **Growth Habit:** An annual, erect, weak stemmed herbs, grows up to 1 m. tall; **Habitat:** Introduced herb, often cultivated,

Geographic Distribution In India; Delhi, Jammu, Garhwal region, Dehra Dun, and Tamil Nadu.

Fig. 10.2 (**a**) *Psoralea corylifolia*; (**b**). *Eclipta prostrata*; (**c**). *Terminalia chebula*; (**d**) *Operculina turpethum*

Morphological Description Annual herb with linear lanceolate leaflets with serrate margin; flowers small, white, in compound umbels; involucre bracts linear, trifid; involucel bracts linear; petals with unequal lobes; fruit about 2 mm.

Traditional Uses Dried fruit powder mix with honey is used topically to treat Vitiligo. It is diuretic and used to treat leprosy, renal calculi, and UTI-urinary tract infections.

Active Ingredients Isopimpinellin, xanthotoxin (8-methoxypsoralen), imperatorin and majudin (bergapten).

Pharmacological Activities and Uses Emmenagogue (to regulate menstruation), diuretic, and for the treatment of leprosy, kidney stones, and urinary tract infections. It is stimulatory, carminative and extremely helpful to treat Vitiligo and leucoderma. Anti-inflammatory, analgesic, and antipyretic effects.

10.4.5.2 *Cassia absus* L.; *Senna absus* L. (Caesalpiniaceae)

Unani Name Chaksu; **Part Used:** Seeds, dry powder is used externally to cure leucoderma and other skin ailments.

Growth Habit An annual, small, erect, sometimes procumbent, much-branched, pubescent herbs; **Habitat:** Abundant in open habitat, disturbed grounds in deciduous forests, also degraded forests and plains. **Geographic Distribution:** Within India, it is distributed almost throughout.

Morphological Description A small herb covered with viscid glandular hairs. Rachis with a gland between the lowest pair of the leaflets. Leaflets elliptic-obovate. Raceme short, terminal or axillary. Flowers yellow or reddish tinged. Pods flat covered with stiff glandular hairs.

Traditional Uses Astringent, detergent, diuretic, useful for vision and eye diseases such as eye pain, conjunctivitis and cataracts. It is used externally for vitiligo, ringworm, sexually transmitted diseases and other skin diseases (Ghani 2010).

Active Ingredients Sitosterol-beta-D-glucoside(seeds) and chaksine and isochaksine -alkaloids.

Pharmacological Activities and Uses Antihypertensive, antifertility, antifungal, anti-inflammatory, antibacterial, and antioxidant, antinicotinic, nonspecific muscle relaxant and curariform action.

10.4.5.3 *Cassia tora* L.; *Senna tora* (L.) Roxb. (Caesalpiniaceae)

Unani Name Panwad; **Part used:** Seeds; dried powder is used topically, leaves; taken internally to prevent skin diseases.

Growth Habit An annual, erect, shrubby, fetid, subglabrous herbs; **Habitat:** Moist deciduous forests, also low lying places and roadsides. **Geographic Distribution:** It is found as a weed throughout India in the monsoon season.

Morphological Description A tall, glabrous herb. Flowers axillary, yellow or reddish tinged. Pods subterete, 4-angled.

Traditional Uses Seeds, roots and leaves are used in paste and oil form to treat skin diseases and rheumatic diseases. Leaves- taken internally for the prevention of skin diseases; useful in treating eczema and ringworm.

Active Ingredients Seeds- naptho-pyrone glycosides anthraquinone glycosides, cassiaside and rubrofusarin-6-β-gentiobioside.

Pharmacological Activities and Uses Antioxidant and hepatoprotective.

10.4.5.4 *Cuscuta chinensis* Lam. (Cuscutaceae)

Unani Name Aftimoon vilayeti; **Part Used:** Stem; dried powder.

Growth Habit A rootless, leafless, thread-like yellowish perennial, obligate climbing parasitic vine; **Habitat:** Moist and dry deciduous forests, **Geographic Distribution:** Native to India, found at many places including Gujarat, Maharashtra, Rajasthan.

Morphological Description A leafless parasite, much branched. Stem golden yellow, filiform. Flowers yellow in solitary or shortly pedunculate cymes. Fruit a septifragal capsule, seeds small, 4 in each capsule.

Traditional Uses Beneficial for epilepsy, abnormal uterine bleeding, useful in the treatment of insanity, epilepsy and melancholy.

Active Ingredients Flavonoids (hyperoside, isorhamnetin, d-sesamin), saponins, sterols/triterpenes, b-sitosterol, d-sesamin, 9(R)-hydroxy-d-sesamin, d-pinoresinol, lignin glycosides.

Pharmacological Activities and Uses Anti-inflammatory, immunomodulatory, hepatoprotective and antioxidant.

10.4.5.5 *Eclipta prostrata* (L.) L. (Asteraceae)

Unani Name Bhangra; **Part Used:** Whole plant in form of dried powder, paste and fresh juice.

Geographic Distribution Almost throughout India. **Growth Habit:** An annual herb with erect, sub-erect or prostrate branches, often rooting at the nodes and branches are strigose; **Habitat:** Common weed in moist situations.

Morphological Description An annual herb with prostrate or erect stems. Leaves; sessile, opposite, oblong-lanceolate, or linear-oblong. Flower heads white.

Traditional Uses Leaf juice is prescribed for skin diseases, allergic urticaria, flatulence, colic and liver disease. The seeds are used for sexual weakness as a tonic and aphrodisiac. Externally, a paste of leaves is applied to swelling. It is used for hepatitis, splenomegaly, and chronic skin disease.

Active Ingredients Stigmasterol, α-tertienmethanol, wedelolactone, desmethylwedelolactone and desmethylwedelolactone-7-glucoside. Aerial part β-amyrin and luteolin-7-O-glucoside. Root (gentriacontanol and stigmasterol).

Pharmacological Activities and Uses Hepatoprotective, anticatarrhal, anticatarrhal, deobstruent, spasmogenic, hypotensive.

10.4.5.6 *Ficus carica* L. (Moraceae)

Unani Name Anjeer; **Part used:** Fruit and leaf (juice and paste), bark (dried powder).

Growth Habit A small deciduous tree with milky sap.

Habitat Cultivated; **Geographic Distribution:** Within India, it is found in Punjab, Uttar Pradesh and cultivated near Pune, Bellary, Mysore and Anantapur districts of South India.

Morphological Description A small deciduous tree with milky sap. Leaves alternate, cordate, 3–5 lobed. Fruits are pear shaped, green in color at first and turns deep reddish-purple at maturity.

Traditional Uses Leaves- used for vitiligo. Bark; useful in eczema and other skin ailments. Dried figs are carminative, laxative, nourishing and useful for back pain and urinary incontinence. Fresh fruits improve digestion having laxative and nutritional effects. Fruits are carminative, refrigerant, hematinic, fattening, strengthens the function of liver, beneficial in phlegmatic ailments such as; epilepsy, paralysis and palpitation.

Active Ingredients Phenolic compounds; protocatechuic acid, chlorogenic acid, vanillic acid, rutin, luteolin-3,7-di-O-glucoside. Psoralen and bergapten are the only important photoactive compounds present in significant amount in shoot sap and leaf. The leaves contain the furocoumarins psoralen, bergapten and coumarins umbelliferone, 4′,5′-dihydropsoralen and marmesin.

Pharmacological Activities and Uses Analgesic and anti-inflammatory, anticancer, hepatoprotective, hypoglycemic, hypolipidemic, antiradical, antiproliferative and antimicrobial activity.

10.4.5.7 *Terminalia chebula* Retz. (Combretaceae)

Unani Name Halela Siyah; **Part used:** Dried fruit powder (oral use).

Growth Habit A medium sized deciduous tree with umbrella-shaped crowded branchesTree; **Habitat:** Dry and moist deciduous forests; **Geographic Distribution:** It is distributed from the sub-Himalayan region to West Bengal and Assam. In South India, it is found in the states of Tamil Nadu and Karnataka.

Morphological Description A medium sized tree. The bark is dark brown and is often split vertically. Leaves are ovate or elliptical with a pair of large glands at the top of the petiole. Flowers are yellowish white and bloom in terminal raceme. The drupe is oval, obovate or ovoid, yellow to orange-brown, sometimes red or black, and hard when ripe. On drying the drupe become 5 ribbed; seeds hard and pale yellow.

Traditional Uses The fruit has a strong astringent taste and is more intense in the large intestine than in the small or stomach; used in diarrhea, piles, paralysis, enterorrhagia, metrorrhagia, leucorrhea headache, epilepsy and loss of memory, enlarged liver and spleen, indigestion, vomiting, bronchial asthma and for metabolic harmony. Unripe fruits have a laxative effect, while ripe fruits have a more astringent taste. Young unripe fruits have purgative effects, while ripe fruits have a more astringent taste. Bark is diuretic.

Active Ingredients Chebulic acid, chebulinic acid, 1,2,3,4,6-penta-O-galloyl-b-D-glucose gallic acid, punicalagin, corilagin, chebulanin.

Pharmacological Activities and Uses Immunostimulant, adaptogenic properties, anti-inflammatory, antidiarrheal, antimicrobial.

10.4.5.8 *Glycyrrhiza glabra* L. (Fabaceae)

Unani Name Asal-us-Soos; **Part used:** Dried root powder; topical preparations.

Growth Habit A perennial, stoloniferous undershrub with woody stolons; Habitat: Cultivated; **Geographic Distribution:** The plant is imported to India in significant quantities from Asia Minor, Iraq, Persia and other Central Asian countries. In India, it is grown in Punjab and south of the Himalayas. **Morphological Description:** An erect perennial plant. Leaflets are 4–7 pairs, oblong to elliptic-lanceolate, acute or obtuse. The racemes are loose, shorter or slightly longer than the leaves. The Pods are oblong to linear, 1–3 cm long. It is flat, straight, more or less densely echinate and glandular.

Traditional Uses Expectorant, nervine tonic, analgesic, resolvent, detergent, general tonic, laxative carminative, diuretic, emmenogogue, aphrodisiac, antacid, emetic, beneficial in chronic fever. It is specially used in peptic ulcer, arthritis, liver diseases, cough, tuberculosis, bronchial asthma burning micturition, gonorrhoea, hoarseness of voice, and corneal opacity. It is also recommended for the treatment of epilepsy and also used in early menopausal condition in women. It is also used for the common cold, viral hepatitis and viral infection.

Active Ingredients The main chemical constituent is Glycyrrhizin (glycyrrhizic acid or glycyrrhizinic acid). Saponins- glabranin-A and B.

Pharmacological Activities and Uses Antiallergic, antiviral, anti-inflammatory, spasmolytic, antistress, antidepressive, antiulcer, hepatoprotective, estrogenic and antidiabetic.

10.4.5.9 *Operculina turpethum* (L.) Silva Manso (Convolvulaceae)

Unani Name Turbud; **Part Used:** Dried root powder.

Growth Habit A perennial, herbaceous, hairy climber with milky sap and narrowly winged stems; **Habitat:** Degraded forests, moist areas; **Geographic Distribution:** common distribution in tropical areas of India.

Morphological Description A large perennial twinner with milky juice and fleshy, branched roots. Leaves vary greatly in shape; flowers are tubular-companulate, white, in a small number of flowered cymes. Capsule globose, with 3–4 dull black glabrous seeds.

Traditional Uses Purgative, anthelmintic, laxative and carminative, Useful in the treatment of phlegmatic and nervous ailments such as; cough, asthma, arthritis, gout, paralysis, palsy, sciatica, ulcers and tumors.

Active Ingredients Turpethinic acids, convolvulin (causes sneezing), operculinosides, turpethic acids, glycosidic acids, resin glycosides and turpethosides.

Pharmacological Activities and Uses Antioxidant, antibacterial, hepatoprotective and radical scavenging activities.

10.4.5.10 *Plumbago zeylanica* L. (Plumbaginaceae)

Unani Name Sheetraj Hindi; **Part used:** Dried root powder and leaves.

Growth Habit A perennial undershrub with long, glabrous and rambling branchlets; **Habitat:** Common in deciduous forests and plains.

Geographic Distribution In India, it has been recorded in the Western Himalayas of Jammu and Kashmir, Himachal Pradesh, and Uttar Pradesh.

Morphological Description A sub-scandent perennial herb. Leaves alternate, ovate, sub-acute, entire, glabrous. White flowers in spikes. Fruit capsule; enclosed in permanent calyx.

Traditional Uses The root is considered a cooling, nerve stimulant, aphrodisiac (stimulant), external cleansing and irritant. It is used for skin conditions such as Vitiligo (leucoderma), and discoloured patches of skin (bahaq). A paste made of vinegar, milk, salt and water is used for leprosy and other ailments.

Active Ingredients Coumarins (seselin, suberosin, xanthoxyletin), naphthaquinones (plumbagin, chitranone), and plumbagic acid glucosides.

Pharmacological Activities and Uses Anti-inflammatory, nephroprotective, antioxidant, antibacterial.

10.4.5.11 *Psoralea corylifolia* L. (Fabaceae)

Unani Name Babchi; **Part used:** Dried seed powder and leaves are useful as topical applications.

Growth Habit An erect, annual, pubescent undershrub with glandular hairs; **Habitat:** Occasional in plains and as weed in arable lands **Geographic Distribution:** Almost throughout India.

Morphological Description An erect, annual herb. Leaves unifoliolate, lanceolate; leaflets broadly ovate, obtuse at apex, crenate-dentate. Flowers sessile or subsessile, peduncled axillary subcapitate racemes. Pods black, glabrous, 1-seeded, ribbed.

Traditional Uses The seeds are used in the form of pills (qurs) and ointments (zimads) for vitiligo and leprous lesions. Also useful in psoriasis, pityriasis, phlegmatic fever, intestinal parasites and other inflammatory skin diseases.

Active Ingredients Monotrepenoid compounds of phenol-bakuchiol, raffinose and coumarin, namely; psoralen, isopsoralen, psoralidin, isopsolaridin, and corylifolin. Two therapeutically active ingredients; psoralen and isopsoralen are present in the seed.

Pharmacological Activities and Uses Antibacterial, antitumor, antioxidant, anti-inflammatory, antifungal and immunomodulatory activity.

Important MAPs origin Unani compound formulations (*Majoon-e-Atrilal*, *Safūf-i Baras, Roughan-e Bars,* and *Zimad-e-Bars*), which are used orally as well as topically in Vitiligo (bars) are summarized in Table 10.2 as;

10.5 Traditional and Ethnobotanical Uses of Medicinal Plants for Vitiligo

Based on traditional knowledge (TK) and experience, many local and indigenous populations in India still meet their elementary necessities through MAPs and its products. Most of the tribal and rural populations living in fairly remote areas are somewhat dependent on forest products, especially on MAPs. Tribal communities living close to the natural forest have acquired the knowledge of using medicinal plants to treat various disease conditions (Husain et al. 2015). The ethno-medicines derived from medicinal plants are considered much safer. It has been proven effective in the treatment of multiple diseases (Mitalaya et al. 2003).

Interest in TK on ethno-medicine is constantly growing. Recently, a number of ethnobotanical studies have been conducted and many researchers have reported and documented in detail the traditional knowledge of various tribal communities in Telangana State, Andhra Pradesh, and other regions of India. (Husain et al. 2015, 2016). The use of MAPs for the treatment of Vitiligo is summarized in Table 10.3.

Table 10.2 Unani formulations for Vitiligo, based on medicinal plants

S. no.	Unani Formulations	Constituents	References
1.	Habb-i-Baras	*Psoralea corylifolia* (Babchi) *Zingiber officinale* (Zanjabeel) (red soil- Geru and purified Sulphur –Gandak AmlaSar)	Hakim (2006)
2.	Habb-i-Hindi	*P. corylifolia* (Babchi) *Ficus hispida* L.f. (AnjeerDashti/ Katomari) *Azadirachta indica* A. Juss. (neem; leaf bark of stem and root) *Senegalia catechu* (L.f.) P.J.H. Hurter &Mabb. (Kher Heartwood)	Hakim (2006)
3.	Majoon Atrilaal	*Ammi majus* (Atrilal), *Anacyclus pyrethrum* (Aqraqaqrha), *Operculina turpetum* (Turbud Safaid), *Zingiber officinalis* (Sonth), Honey (Shehad)	Kabeeruddin (2002), Ghani (2010)
4.	Roghan e Babchi	*P. corylifolia* (Babchi) *Sesamum indicum* (Til oil)	Said (1997), Kabeeruddin (2010)
5.	Safūf-i-Baras	*P. corylifolia* (Babchi)- Seeds *S. absus* (Chaksu)- Seeds *S. tora* (Panwar)- Seeds *Ficus carica* (Anjeer-khushk) -Fruits	Standard Unani Treatment Guidelines for Common Diseases (2014)
6.	Zimad-i-bars	1. *Ficus carica* (Bekh-e-Anjeer Dashti-100 g.)- Roots 2. *P. corylifolia* (Babchi- 100 g.)- Seeds 3. *C. tora* (Tukhm-e-Panwar-100 g.)- Seeds 4. *Zingiber zerumbet* (L.) Roscoe ex Sm. (Zarambad/Narkachoor 100 g.) 5. *Citrus aurantifolia* (Christm. & Panz.) Swingle (Aab-e-Lemu)	NFUM (2007a, b)

A survey of the available literature on the 'treatment of Vitiligo by the use of traditional medicines' showed that *P. corylifolia* is the most reported plant and its seeds are most frequently used plant part (Figs. 10.3 and 10.4). Most of the reports consist of external/ topical application (62%) of the drugs/medicinal plants in the form of powder (24%) or extracts, gel or capsule (31%).

Table 10.3 Traditional MAPs used in Vitiligo

S. no.	Botanical name	Part used	Form of drug/ bioavailability	Mode of adminis-tration	References
1.	*Abrus precatorius* L.	Leaves	Paste	External application	Basha et al. (2014)
2.	*Aegle marmelos* (L.) Corrêa	Leaves and seeds	Paste	External application	Topno and Sinha (2018)
3.	*Albizia lebbeck* (L.) Benth.	Bark and leaves	Paste	External application	Basha et al. (2014), Navneet et al. (2012)
4.	*Ammi majus* L.	Seeds Fruits	Pastedecoction of the fresh fruit	External and oral	Usmani et al. (2021), Al-douri (2000)
5.	*Ammi visnaga* (L.) Lam.	Crystals produced from leaves and seeds	Crystals in the form of spray Topical ointment containing khellin (K), oral vitamin E Topical gel application and photo -chemotherapy	External application	Leeuw et al. (2011) Saraceno et al. (2009) Orecchia et al. (1998)
6.	*Carum carvi*	Seeds	Tablet	Oral	Abuduaini et al. (2021)
7.	*Cucumis melo* L.	Seeds	*Cucumis melo* extract, and ace-tyl cysteine gel	External	Gianfaldoni et al. (2018), Guarneri et al. (2021), Fahamiya et al. (2018), Rehman et al. (2018)
8.	*Cucumis sativus* L.	Fruit	Sulfur powder adhered to fresh cucumber slices to rub on vitiligo lesion	External application	Liu et al. (2019)
9.	*Curcuma longa* L.	Rhizome	Paste extract	External application	Kadam et al. (2020), Gianfaldoni et al. (2018), Gupta et al. (2013)
10.	*Cuscuta chinensis* Lam.	Whole plant	Extract in the form of capsule	Oral	Yang et al. (2017)
11.	*Ginkgo biloba* L.	Leaves and seeds	Powder; in the form tablet	Orally	Gianfaldoniet al. (2018), Szczurko et al. (2011), Ahmed et al. (2013), Parasad (2003), Zhang et al. (2019)
12.	*Launea aspleniifolia* (Willd.)Hook.f.	Whole plant	Paste	Applied externally	Singh and Narain (2010)

(continued)

Table 10.3 (continued)

S. no.	Botanical name	Part used	Form of drug/ bioavailability	Mode of adminstration	References
13.	*Melia azedarach* L.	Bark and seeds	Bark decoction, seed powder	Oral and external application	Vashistha (2015)
14.	*Nigella sativa* L.	Seeds	Paste	External application	Tahir et al. (2010)
15.	*Phyllanthus emblica* L.	Fruits	Extract in the form of tablet	Oral	Colucci et al. (2015)
16.	*Picrorhiza kurroa* Royle ex Benth.	Roots	Powder in the form of a tablet	Oral	Gianfaldoni et al. (2018), Beidi et al. (1989), Narayanaswamy and Ismail (2018)
17.	*Piper nigrum* L.	Fruits	Piperine extracted from fruits	External application	Mihăilă et al. (2019)
18.	*Polypodium leucatomos* Poir.	Whole plant	Extract in the form of a tablet Extract in the form of capsule	External application Oral	Gianfaldoni et al. (2018), Eduardo et al. (2005), Lucy and Martin (2012), Martiza et al. (2004), Mark et al. (2015)
19.	*Raphanus sativus* L.	Seeds	25 g powder of radish seeds with red vinegar	External application	Basha et al. (2014)
20.	*Psoralea corylifolia* L.	Seeds Seeds	Powder in the form of ointment Extract, seed oil	External External and Oral	Hussain et al. (2016), Sunil et al. (2019), Khushboo et al. (2010), Khan et al. (2015), Aruna et al. (2019), Khandekar et al. (2015), Pandey (2015), Hamzavi et al. (2012), Kumar et al. (2015)
21.	*Telosma pallida* (Roxb.) Craib.	Fruit	Latex of fruit and fruit paste	External application	Singh and Narain (2010)
22.	*Tribulus terrestris* L.	Fruits	Paste	External application	Lin et al. (1999), Chhatre et al. (2014)

10.6 Conclusions

Over the last few decades, a rapid development has been witnessed in the practice and application of Traditional System of Medicine to cure the diseases for which there is limited or no treatment available in the contemporary system. The recent resurgence of MAPs-derived remedies is the result of several factors, such as the

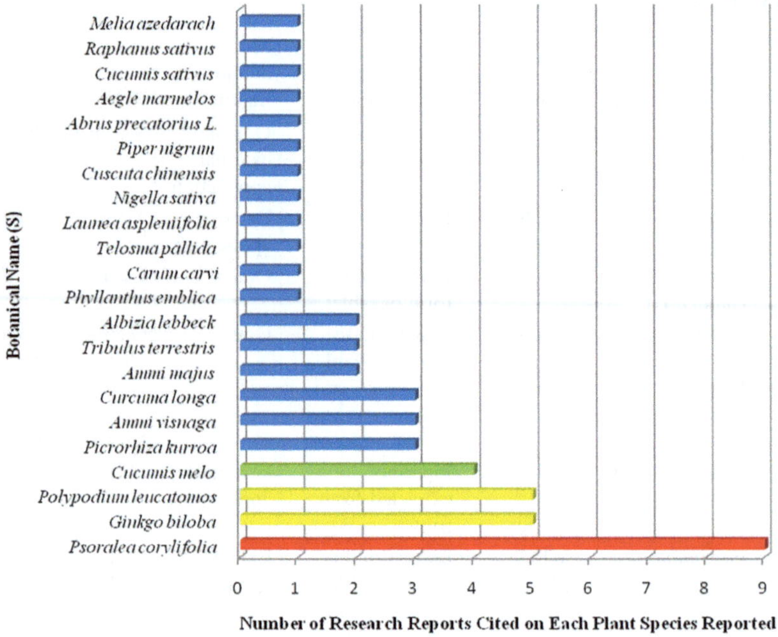

Fig. 10.3 Number of research reports documented on Traditional uses of MAPs in Vitiligo

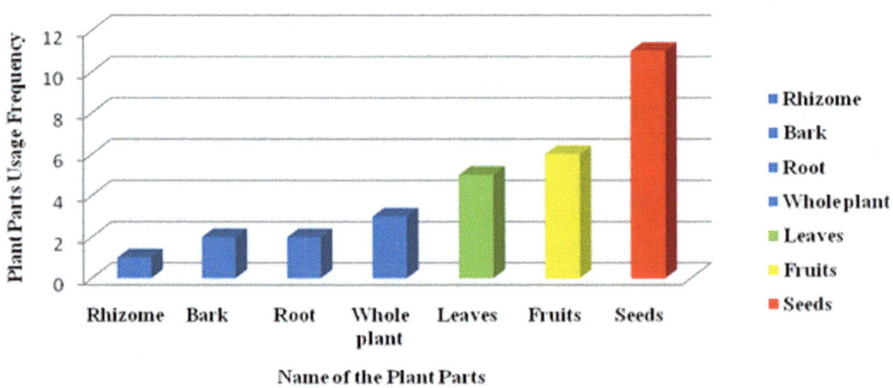

Fig. 10.4 Frequency and Use of Medicinal Plant (MAP) parts in Vitiligo

efficacy of plant origin drugs having fewer side effects compared to modern medicines. The age-old time-tested Indian traditional medications (Unani and Ayurveda) in the form of mono- or multi-component crude drugs obtained from the valuable MAPs including; *Ammi majus*, *Senna tora*, *Eclipta prostrata*, *Ficus carica*, *Psoralea corylifolia*, *Plumbago zeylanica* and *Zingiber officinale* have found their way to exhibit the desired curative and preventive management to treat stigmatic disease, like Vitiligo.

Sustainable use and conservation based on *ex situ* and *in situ* strategies should be undertaken to preserve the natural resources. The new protocols should serve multidimensional and interdisciplinary, integrative, collaborative, preclinical and clinical research programs, at both national and international level.

References

Abuduaini A, Lu X, Zang D, Tao W, Haji Akbar Aisa AG (2021) Effects of a traditional caraway formulation on experimental models of vitiligo and mechanisms of melanogenesis. Evid Based Complement Alternat Med:17

Agnivesh, Charaka, Dridhabala (1993) Charaksamhita. Chaukhambha Bharati Academy, Varanasi chikitsathana 7:162–177. Cited in: Mulla SM (2018) Vitiligo-Ayurvedic Treatment Approach. J Clin Cosmet Dermatol 2(3):1–7. https://doi.org/10.16966/2576-2826.129

Ahmed R, Abu R, Noor MA et al (2013) Evaluation of standardized extract of *Ginkgo biloba* in vitiligo remedy. Asian J Pharm Clin Res 6(5):127–130

Al-douri NA (2000) A survey of medicinal plants and their traditional uses in Iraq. Pharm Biol 38(1):74–79. https://doi.org/10.1076/1388-0209(200001)3811-BFT074

Anonymous (2012) Standard Unani medical terminology, central Council for Research in Unani Medicine (CCRUM) department of AYUSH. Ministry of Health & Family Welfare, Government of India

Aruna V, Amruthavalli GV, Gayathri R (2019) Safety profile of synthetic versus natural Psoralen. J Med Res 5(5):194–197

Basha SK, Anjaneyulu E, Krishna SG, Parveen DN, Sudarsanam G (2014) Plants used in the treatment of Leucoderma by the Tribals of Yerramalai Forest of Kurnool District, Andhra Pradesh, India. Photo-Dermatology 121:761–766

Beidi KL, Usha Z, Chopra CL (1989) *Picrorhiza kurroa*, an ayurvedic herb, may potentiate photochemotherapy in vitiligo. J Ethnopharmacol 27:347–352. https://doi.org/10.1016/0378-8741(89)90009-3

Chhatre S, Nesari T, Somani G, Kanchan D, Sathaye S (2014) Phytopharmacological overview of *Tribulus terrestris*. Pharmacogn Rev 8(15):45–51. https://doi.org/10.4103/0973-7847.125530

Colucci R, Dragoni F, Conti R (2015) Evaluation of an oral supplement containing *Phyllanthus emblica* fruit extracts, vitamin E, and carotenoids in vitiligo treatment. Dermatolog Treat 28: 17–21. https://doi.org/10.1111/dth.12172

Dhanik A, Sujatha N, Rai NP (2011) Clinical evaluation of the efficacy of Shvitraharakashaya and lepa in vitiligo. Ayu Jan 32(1):66–69. https://doi.org/10.4103/0974-8520.85731

Eduardo R, Pedro J, Elena H et al (2005) Systemic immunomodulatory effects of *Polypodium leucotomos* as an adjuvant to PUVA therapy in generalized vitiligo: a pilot study. J Dermatol Sci 41:213–216. https://doi.org/10.1016/j.jdermsci.2005.12.006

Ezzedine K, Eleftheriadou V, Whitton M, van Geel N (2015) Vitiligo. Lancet 4(386):74–84. https://doi.org/10.1016/S0140-6736(14)60763-7

Fahamiya N, Aslam M, Siddiqui A, Shiffa M (2018) Review on *Cucumis melo*: ethnobotany and Unani medicine. World J Pharm Pharmaceut Sci 5(12):1–18

Farnsworth NR, Akerele O, Bingel AS, Soejarto DD, Guo Z (1985) Medicinal plants in therapy. Bull World Health Organ 63(6):965–981

Fisk WA, Agbai O, Lev-Tov HA, Sivamani RK (2014) The use of botanically derived agents for hyperpigmentation: a systematic review. J Am Acad Dermatol 70(2):352–365. https://doi.org/10.1016/j.jaad.2013.09.048

Ghani N (2010) Khazain-ul-Advia. Reprinted by Idara Kitabus Shifa, Darya Ganj, and Central Council for Research in Unani Medicine (CCRUM): New Delhi

Gianfaldoni S, Wollina U, Tirant M, Tchernev G, Lotti J, Satolli F, Rovesti M, França K, Lotti T (2018) Herbal compounds for the treatment of vitiligo: a review. Open Access Maced J Med Sci 6(1):203–207

Guarneri F, Bertino L, Pioggia G, Casciaro M, Gangemi S (2021) Therapies with antioxidant potential in psoriasis, vitiligo, and lichen planus. Antioxidants 10:1087. https://doi.org/10.3390/antiox10071087

Gupta SC, Patchva S, Aggarwal BB (2013) Therapeutic roles of curcumin: lessons learned from clinical trials. AAPS J 15(1):195–218. https://doi.org/10.1208/s12248-012-9432-8

Hakeem MA (2002) Bustan al-mufradat. Idara Kitab-us-Shifa, New Delhi, pp 109.171, 249, 340, 417, 419, 431

Hakim K (2006) Al-Qarabadin, TabaSani. Central Council for Research in Unani Medicine, New Delhi

Hamzavi IH, Lim HW, Syed ZU (2012) Ultraviolet-based therapy for vitiligo: What's new? Indian J Dermatol Venereol Leprol 78:42–48

Husain MK (2021a) Herbs that heal: relevance of traditional natural remedies in promotion of health. Ann Phytomed 10(2):4–21. https://doi.org/10.21276/ap.2021.10.2.2

Husain MK (2021b) COVID-19 pandemic: a narrative review on possible prophylactic role of AYUSH Unani system and medicinal plants. Ann Phytomed 10(2):S116–S124. https://doi.org/10.21276/ap.covid19.2021.10.2.13

Husain MK, Pratap GP, Kazmi MH (2015) Ethnopharmacological uses of medicinal plants in Jannaram Forest division of Telangana, India. Hippocratic J Unani Med 10(4):122–133

Husain MK, Pratap GP, Kazmi MH, Rahman R (2016) Folk-claims on medicinal plants in Kammarpally Forest range of Nizamabad forest division of Telangana state Hippocratic. J Unani Med 11(4):113–129

Husain MK, Khalid M, Pratap GP, Kazmi MH (2017) Relevance of traditional Unani (Greco-Arab) system of medicine in cancer: an update. In: Akhtar M, Swamy M (eds) Anticancer plants: clinical trials and nanotechnology. Springer, Singapore, pp 273–302. https://doi.org/10.1007/978-981-10-8216-0_10

Hussain I (2021) The safety of medicinal plants used in the treatment of vitiligo and Hypermelanosis: a systematic review of use and reports of harm. Clin Cosmet Investig Dermatol 23(14):261–284. https://doi.org/10.2147/CCID.S298342

Hussain I, Hussain I, Manan A et al (2016) Fabrication of anti-vitiligo ointment containing *Psoralea corylifolia*: *in vitro* and *in vivo* characterization. Drug Des Devel Ther 10:3805–3816. https://doi.org/10.2147/DDDT.S114328

Kabeeruddin M (2002) Makhzan –ul-Mufaridat, IdaraKitab –us-Shifa, Kocchachelan, Darya Ganj, New Delhi, pp 45–83

Kabeeruddin A (2010) Bayaze Kabeer, vol 24. IdaraKitab -us -Shifa, New Delhi, p 30

Kadam SM, Chorage C, Singh K (2020) A review on turmeric: for the treatment of skin disease (vitiligo). Int J Food Sci Nutr 5(1):82–85

Kamboj VP (2000) Herbal medicine. Curr Sci 78(1):35–39

Khan MA (2014) MuheeteAzam, vol III. CCRUM, New Delhi, pp 107–419

Khan MS, Lari QH, Khan MA (2015) Babchi (*Psoralia corylifolia* Linn.) and it's therapeutic uses in Unani system of medicine – a review. Int J Pharm Phytopharmacol Res 5(1):41–45

Khandalavala BN, Nirmalraj MC (2014) Rapid partial repigmentation of vitiligo in a young female adult with a gluten-free diet. Clin Dermatol 6(3):283–287

Khandekar A, Jadhav JH, Danga SK (2015) Management of vitiligo: an ayurvedic perspective. Indian J Drugs Dermatol 1:41–43

Khushboo PS, Jadhav VM, Kadam VJ, Sathe NS (2010) *Psoralea corylifolia* Linn.- "Kushtanashini". Pharmacogn Rev 4(7):69–76. https://doi.org/10.4103/0973-7847.65331

Kumar GA, Mahendra P, Meena MS (2015) Effect of Bakuchi on vitiligo– a case study. Int Ayu Med J 3(1):193–196

Leeuw JD, Assen YJ, Van N et al (2011) Treatment of vitiligo with khellin liposomes, ultraviolet light and blister roof transplantation. J Eur Acad Dermatol Venereol 25(1):74–81. https://doi.org/10.1111/j.1468-3083.2010.03701.x

Lin ZX, Hoult JR, Raman A (1999) Sulphorhodamine B assay for measuring proliferation of a pigmented melanocyte cell line and its application to the evaluation of crude drugs used in the treatment of vitiligo. J Ethnopharmacol 66:141–150

Liu Z, Wang R, Zhang C, Guo S, Chen P (2019) A case of vitiligo cured with cucumber and sulfur. Phytother Res 33(4):1241–1242. https://doi.org/10.1002/ptr.6309

Lucy K, Martin M (2012) A randomized double-blind placebo controlled study evaluating the effectiveness and tolerability of oral polypodiumleucotomos in patients with melasma. J Am Acad Dermatol 2012(Ab21):4630

Mark S, Nestor BD, Nicole S (2015) Safety and efficacy of oral *Polypodium leucotomos* extract in healthy adult subjects. J Clin Aesthet Dermatol 8(2):19–23

Martiza A, Middelkamp H, Madhu A et al (2004) Orally administered *Polypodium leucotomos* extract decreases psoralen UVA induced phototoxicity, pigmentation, and damage of human skin. J Am Acad Dermatol 50:41–49. https://doi.org/10.1016/S0190-9622(03)02732-4

Mihăilă B, Dinică RM, Tatu AL, Buzia OD (2019) New insights in vitiligo treatments using bioactive compounds from *Piper nigrum*. Exp Ther Med 17(2):1039–1044. https://doi.org/10.3892/etm.2018.6977

Mitalaya KD, Bhatt DC, Patel NK, Didia SK (2003) Herbal remedies used for hair disorders by tribals & rural Folk in Gujarat. Indian J Tradit Knowl 2:389–392

Mulla SM (2018) Vitiligo-Ayurvedic treatment approach. J Clin Cosmet Dermatol 2(3):1–7. https://doi.org/10.16966/2576-2826.129

Narayanaswamy R, Ismail IS (2018) Role of herbal medicines in vitiligo treatment – current status and future perspectives. Asian J Pharm Clin Res 11(9):19–23

NFUM-National formulary of Unani Medicine (2007a) Ministry of Health and family welfare, Part I, vol II, Department of AYUSH, pp 11–12, 31–32, 85–86

NFUM-National formulary of Unani Medicine (2007b) Ministry of Health and family welfare, Part I, vol 1, Department of AYUSH, pp 13–14, 88–89

Navneet K, Sukhbir K, Sharma AK (2012) A review on leucoderma and reported herbs for its treatment. J Drug Deliv Ther 2(3):53–59

Orecchia G, Sangalli M, Gazzaniga A, Giordano F (1998) Topical photochemotherapy of vitiligo with a new Khellin formulation. J Dermatolog Treat 9(2):65–69. https://doi.org/10.3109/09546639809161375

Pandey RS (2015) Discovery of novel ayurvedic formulation for the treatment of vitiligo. Asian J Med Sci 6(1):1–2

Parasad P (2003) Effectiveness of oral *Ginkgo biloba* in treating limited, slowly spreading vitiligo. Clin Exp Dermatol 28:285–287. https://doi.org/10.1046/j.1365-2230.2003.01207

Rao R, Thakar AB, Patel KS, Bhatt NN, Gandhi R, Bhagiya S (2017) Certain works done on shwitra (vitiligo) by ayurvedic management: a brief review. Int J Adv Res 5(4):262–267

RaziABZ. (1970) *Kitab-ul-Hawi Fit Tib* (Arabic version), vol 23. Dairatul Marif, Osmania University, Hyderabad, pp 72–75

Reddy MT, Kalpana M, Sivaraj N, Kamala V, Pandravada SR, Sunil N (2019) Indigenous traditional knowledge on health and equitable benefits of oil palm (*Elaeis* spp.). Open Access Libr J 6:e5103. https://doi.org/10.4236/oalib.1105103

Rehman S et al (2018) Pharmacognostical and pharmacological review of *Cucumis melo* L. including Unani medicine perspective. Int J Pharmacogn Chinese Med 2(3):000140

Said H M (1997) Hamdard pharmacopoeia of eastern medicine. Published by Sri Satguru Publications, Delhi, pp 68, 133, 143, 155,186, 206, 261, 272, 278, 195

Sapkota YR, Bedarkar P, Shukla VJ, Prajapati PK (2016) Quality control parameters of Arogyavardhini rasa prepared by classical method. J Ayu Herb Med 2(4):104–111

Saraceno R, Nistico SP, Capriotti E, Chimenti S (2009) Monochromatic excimer light 308 nm in monotherapy and combined with topical khellin 4% in the treatment of vitiligo: a controlled study. Dermatolog Treat 25:391–394. https://doi.org/10.1111/j.1529-8019.2009.01252.x

Shingadiya R, Dhruve K, Shukla VJ, Prajapati PK (2015) Standard manufacturing procedure and quality parameters of Kanakbindvarishta. Int J Herbal Med 3(1):33–36

Shingadiya R, Joshi K, Shukla VJ, Prajapati PK (2016) Pharmaceutical and analytical profiles of Savarnakara yoga. J Ayu Herb Med 2(3):73–77

Shingadiya RK, Gohel JK, Chaudhary SA, Bedarkar P, Patgiri BJ, Prajapati PK (2018) Ayurvedic management of chronic vitiligo (Shvitra): a case study. J Ayu Herb Med 4(2):57–59

Singh U, Narain S (2010) Traditional treatment of leucoderma by Kol tribes of Vindhyan region of Uttar Pradesh. Indian J Tradit Knowl 9(1):173–174

Standard Unani Treatment Guidelines for Common Diseases (2014) Volume-I, Central Council for Research in Unani Medicine (CCRUM), New Delhi

Sunil G, Varsha S, Srivastava SN (2019) New hope in treatment of vitiligo (Switra) by Ayurvedic medicines (a case study). Int J Yogic Hum Mov Sports Scinces 4(1):1283–1286

Szczurko O, Boon HS (2008) A systematic review of natural health product treatment for vitiligo. BMC Dermatol 22(8):2. https://doi.org/10.1186/1471-5945-8-2

Szczurko O, Shear N, Taddio A, Boon H (2011) *Ginkgo biloba* for the treatment of vitilgo vulgaris: an open label pilot clinical trial. BMC Complement Altern Med Mar 11:21. https://doi.org/10.1186/1472-6882-11-21

Tabari ABM (1997) Moalijat-e-Buqratia (Urdu version), vol 2. Central Council for Research in Unani Medicine, New Delhi, pp 199–200

Tabari Abul Hasan Ali Bin Sahl Rabban. 1996 *Firdausul Hikmat* (Urdu version), vol 1. Idaria Tarjmantibb, Karachi, pp 825

Tahir MA, Pramod K, Ansari SH, Ali J (2010) Current remedies for vitiligo. Autoimmun Rev 9(7):516–520. https://doi.org/10.1016/j.autrev.2010.02.013

Topno SC, Sinha MR (2018) Study of medicinal plants used to heal skin diseases by tribes of west Singhbhum district of Jharkhand (India). J Pharmacogn Phytochem 7(1):371–376

United Nations-UN (2019). https://www.un.org/development/desa/indigenouspeoples/wp-ontent/uploads/sites/19/2019/04/Traditional-Knowledge-backgrounder-FINAL.pdf

Usmani QI, Jahan N, Aleem M, Hasan SA (2021) Aatrilal (*Ammi majus* L.), an important drug of Unani system of medicine: a review. J Ethnopharmacol 10:276. https://doi.org/10.1016/j.jep.2021.114144

Vashistha PBD (2015) An ethnobotanical study of plains of Yamuna Nagar district, Haryana, India. Int J Innov Res Sci Eng Technol 4(1):18600–18607

WHO (2004) WHO guidelines on safety monitoring of herbal medicines in pharmacovigilance systems. World Health Organization, Geneva

WIPO-World Intellectual Property Organization (2021). https://www.wipo.int/tk/en/tk/

Yang B, Yang Q, Yan H, Yang X, Lu Q-P (2017) Hyperoside elevates the melanin content and promotes the migration of human melanocytes. Int J Clin Exp Med 10(2):2953–2959

Zhang S, Yi X, Su X et al (2019) *Ginkgo biloba* extract protects human melanocytes from H2O2-induced oxidative stress by activating Nrf2. J Cell Mol Med 23:5193–5199. https://doi.org/10.1111/jcmm.14393

Chapter 11
Indian Herbs with Hepato-Protectant Potentials

Karishma Khan, Himanshu Dawda, and Usha Mukundan

Abstract India has a fine history of herbs used as medicine, based upon different therapeutic systems. Indian Ayurvedic pharmacopoeia has recorded around 540 species. Medicinal herbs are used directly or indirectly in current medicines. Several plant species and plant-based drugs play an important role in managing diseases related to the liver. To date not each plant species has been evaluated regarding its pharmacological or/and antiviral activity, even though described as a hepatoprotectant or against viral hepatitis. There is also a lack of the randomized controlled clinical trials and toxicological evaluation. The present literature review is aimed at compiling information on the so called "encouraging phytochemicals", i.e.: phytochemicals that have already been investigated in the hepatotoxicity models. Herb processing strategies and therapeutic applications including the status of global predominance, as well as research and advancement, have also been considered.

Keywords Ayurveda · Ethnobotany · Hepatoprotectant and liver

11.1 Introduction

The demands for plant-based drugs have substantially increased in recent scenarios, owing to parallel and speedy improvement of research on medicinal herbs efficacy. Ayurveda has a vivid history in the therapeutic usage of medicinal herbs. Several herbs are commonly used for therapeutic purposes and are specifically prepared by utilising distinct administering techniques (Jaiswal et al. 2016).

India has a fine history of herbs used as medicine, centred upon the six systems of medicine. Associated six systems of medicine are Ayurveda, Homeopathy, Unani, Siddha, Yoga and Naturopathy. Among these systems, ancient Ayurveda is most extensively practised and recognised. Frequently, in these systems, the same species

K. Khan · H. Dawda · U. Mukundan (✉)
Department of Botany, Ramniranjan Jhunjhunwala College, Mumbai, Maharashtra, India

Á. Máthé, I. A. Khan (eds.), *Medicinal and Aromatic Plants of India Vol. 2*,
Medicinal and Aromatic Plants of the World 9,
https://doi.org/10.1007/978-3-031-44914-7_11

are utilized for unique therapeutic functions depending on the region and knowledge and understanding of local people.

Recently on 13th August 2018, National Medicinal Plants Board, India launched a new section i.e., Bhuvan e-HERBS, a portal where the data of specific herbs will be documented for their medicinal application. The database may be one of the authentic prospective databases only designed for herb therapeutic materials employed in all traditional medicine of India in different groups of medicine.

Ayurveda uses numerous plant species in therapeutic remedies, which enhances human well-being through the usage of organic medicinal substances with a universal approach.

The Indian Ayurvedic Pharmacopoeia has recorded around 540 species of medicinal ingredients, in which only some part(s) of the plant are employed for medicine, based on their medicinal efficacies which are rarely unique. Studies encouraging the comprehensive reasonable scrutiny of selected attributes may assist investigators and conventional experts to achieve deeper insight into age-old medical writings and their primary principles. This can be of distinct importance in those cases where some herbs with similar therapeutic action, belonging to the same group but different species can be utilized as alternatives for each other (Jaiswal et al. 2016; Jain et al. 2013).

11.2 Hepatoprotective Herbs in Ethnobotany

A medicinal herb can contain ingredients that can be applied for therapeutic uses or serve as starting material for the synthesis of valuable drugs. Medicinal herbs can be used directly or indirectly in modern medicines. Herbal treatment is deeply rooted in Indian culture (Jaiswal et al. 2016; Sofowora et al. 2013).

Traditional medicine has evolved through the ages through the accumulation of skills of several generations with a blend of traces of Ayurvedic and Unani categories of medicine.

The WHO (World Health Organization) currently promotes, recommends, and supports low-cost, comparatively safe, and culturally acceptable traditional remedies in national healthcare plans worldwide. WHO, in several resolutions has highlighted the need to safeguard quality using modern techniques and employing its suitable criteria. The purpose to regulate a traditional drug is to make sure of its efficacy and safety (Jaiswal et al. 2016).

Established on the significance of the pharmacological activity, plants can be classified for their nutraceutical and phytochemical usage. There are several nutraceuticals and phytochemicals compounds which are foremost constituents of the herbal medicine of all ethnic groups in India and around the world (Sofowora et al. 2013).

Traditional herbal medicines are also essential resources for the development of hepatoprotective drugs. Presently, convincing evidence from large-scale randomised regulated trials is still scarce. According to a current survey an increase has been found in the traditional medication use by patients (around 20–30% in certain parts of Asian countries). Additional surveys also have revealed a similar percentage of herbal usage for curing persistent liver disorders, as a corresponding and alternate medicine (Yan et al. 2020).

Several herbal isolates have been developed to acquire hepatoprotective activities. Indian Medicine System contain 300 plus formulations which are known for medication for diseases like jaundice and chronic liver. Around 87 therapeutic plants are employed with combinations as an herbal drug(s) for liver diseases. Remarkably, only a modest number of plants have been assessed for their pharmacological and antiviral efficacy, even though numerous species were described as hepatoprotective and effective against viral hepatitis.

Several plants are being included in polyherbal formulations (mixtures of various herbal extracts) with assertions of hepatoprotective activities. From a total of 110 plants in 55 families, 170 phytoconstituents with liver protective action have been isolated. Nearly, 600 commercial herbal preparations having hepatoprotective activity are marketed across the world. Around 93 plus medicinal plants in India have been utilised in different mixtures, for preparations of around 40 polyherbal formulations which have been patented. These formulations are considered Ayurvedic medicines and are marketed in India for liver diseases and liver conditions. The formulations are systematically utilised for liver protection, promoters of appetite and growth, regulators for gastrointestinal and hepatic conditions, hepatic dysfunction, and regeneration, along with liver tonic acting as stimulants. Only a modest proportion of hepatoprotective herbal formulations are employed in traditional medicine and are pharmacologically assessed for safety and efficacy (Girish and Pradhan 2017).

The usage of natural treatments for diseases related to the liver has extended history from Ayurvedic treatments to European, Chinese, and various supplementary systems of standard medicines. Twenty-First century focuses on paradigm alteration concerning therapeutic and herbal evaluation for manufacturing goods for liver disease. These herbal products are sensibly synergising the benefit of medicine in traditional systems with the modern theory of evidence-based medicinal valuation. They have standardisation and randomised placebo-monitored clinical trials to provide for clinical efficacy.

The present review accumulates data, built on evaluated bioactive phytochemicals from therapeutic herbs and plants, which have been utilised in hepatotoxicity empirical models. We have tried to include all relevant evidence needed for insight application of some Indian herbs for their uses as hepatoprotectant.

11.3 Indian Herbs as Resources of Hepatoprotective Agents

Documentation and literature prove the usage of more than a hundred plant species in a wide-ranging condition of liver disorders. In the following review, we have compiled a concise description of 25 plant species that have been reported to be used and/or studied intensively as hepatoprotectant.

11.3.1 Achillea millefolium *L.*

Taxonomic Characteristics *Achillea millefolium* (Asteraceae), is commonly known as Yarrow in India.

Crude Drug Used Flowers, stem, and leaves are used as active constituents.

Major Chemical Constituents Achilleine, camphene, flavonoids and terpenoids.

Modern Medicine Based on Traditional Uses Yarrow is a herb well known for its capability of treating diseases. Normally as an essential oil-containing species, *Achillea millefolium* contains the alkaloid achilleine which is known for stopping bleeding. It delivers treatment to stomach spasms and pain; it also provides remedies to gastrointestinal complications involving dyspepsia and plays a role as an anti-haemorrhoid by interrupting external and internal blood loss. The extract of *A. millefolium* is hepatoprotectant, as the extract shows calcium channel blocking to reduce the assembly of macrophage-associated cytokines and lipid intermediaries and guards against their detrimental consequences. Water and alcohol extracts of *A. millefolium* revealed the hepatoprotective activity against D-Galactosamine/lipopolysaccharide, along with an antispasmodic effect facilitated through calcium channel blockade. The researcher also concluded its conventional use as a hepatoprotective and antispasmodic agent (Yaeesh et al. 2006).

Recent studies discovered the hepatoprotective activity against carbon tetrachloride-induced hepatotoxicity with the help of methanolic extract of *A. millefolium.* The documented work confirmed the utilisation of 100–200 mg per kg of active constituents for their significant activity to play a vital role in the treatment and its protective effects on the liver. The observed activity was associated with its highly bioactive compounds including flavonoids and other compounds (Al-Ezzy et al. 2017).

Yarrow is widely known for its ability to cure diseases. It usually grows in Western Himalayas exclusively from Kashmir to the area of Kumaon. Yarrow's Latin name *A. millefolium* is brought from a renowned character from Greek mythology known as Achilles. Achilles was the first to use the plant, Yarrow for treating wounds. Yarrow is approved by the German Commission E as a remedy for gastrointestinal tract complications and dyspeptic including moderate spasms.

11.3.2 Aloe barbadensis *Miller* (Aloe vera)

Taxonomic Characteristics *Aloe barbadensis* (Liliaceae) is usually known as Aloe vera or Aloe in India.

Crude Drug Used Shade-dried aerial *parts* of the plant are used for active constituents.

Major Chemical Constituents vitamins, enzymes, anthrones, Cardiac glycosides, and anthraquinones like aloe-emodin and aloesin.

Modern Medicine Based on Traditional Uses Hepatoprotective ability of *Aloe barbadensis* aqueous extracts significantly studied against hepatotoxin carbon tetrachloride. Aqueous extracts of *A. barbadensis* substantially reduced the hexobarbitone "sleep time" and zoxazolamine "paralysis time" as equated to carbon-tetrachloride-controlled treated animals. Aqueous extracts of *A. barbadensis* show the ability of hepatic drugs to metabolise enzymes. Prophylactic and curative therapy with aqueous extracts of *A. barbadensis* counteracts the carbon tetrachloride-induced decline of microsomal aniline hydroxylase enzyme and amidopyrine-*N*-demethylase enzyme actions which specifies efficient cells integrity. It was concluded by researchers that *Aloe barbadensis* aqueous extracts have antioxidant activity, a potential mechanism as hepatoprotectant. Researchers have also suggested about *A. barbadensis* aqueous extracts protect against increased lipid peroxidation and regulation with glutathione contents. In 2013, Cui et al., showed *Aloe vera's* bioactive component had protective sound effects and underlying mechanisms against alcoholic liver disease with the help of an enduring long-term (chronic) alcohol-feeding mouse model. The results of chronic alcohol ingestion ensured a decrease in body weight and degradation of the liver. This was evidenced by the elevation of activities of alanine aminotransferase and aspartate aminotransferase, total cholesterol, triglyceride and low-density lipoprotein levels, steatosis, and inflammatory response, reflecting early biochemical and pathological changes in alcoholic liver disease. *Aloe vera* utilised bioactive component offered a substantial defence to chronic alcohol-intoxicated mice as the bioactive compound facilitates by reversing levels of serum marker enzymes and lipids and restoring the body weight, indicating stabilisation of plasma membrane and repair of hepatic tissue. *Aloe vera's* bioactive component enhanced the hepatic histopathological changes induced by alcohol. Bioactive components of *Aloe vera* alleviate hepatotoxicity of chronic alcohol ingestion and the hepatoprotective effect is the same as a conventional commercial hepatoprotective drug. Several findings have verified fatty liver, which is the most general feature observed during alcohol hepatotoxicity and promotes the advancement of alcoholic liver disease. Chronic alcohol consumption increases the hepatic NADH/NAD+ ratio, resulting in the suppression of fatty acids mitochondrial β-oxidation and stimulating conventional lipogenesis, thus causing lipid accumulation within hepatocytes. Activated protein kinase, a key regulator of lipid metabolism, is liable for the inactivation of acetyl-CoA carboxylase, its deactivation results in supplementary oxidation of fatty acid in the liver (Chandan et al. 2007 and Cui et al. 2014).

11.3.3 Andrographis paniculata *Burm f*

Taxonomic Characteristics *Andrographis paniculata* (Acanthaceae): in India, it is commonly known as Green Chirayata.

Crude Drug Used Especially roots and whole plants are utilised for the active constituent.

Major Chemical Constituents Diterpenoids, Flavonoids, Polyphenols, Andrographolide and β-sitosterol.

Andrographis paniculata alcoholic extracts are utilised to check the presence of secondary metabolites by phytochemical screening. Qualitative tests detected the presence of alkaloids, saponins, flavonoids and steroid constituents in alcoholic extracts. Tannin was absent in alcoholic extracts.

Modern Medicine Based on Traditional Uses Hepatoprotective effect has been stated as quite significant against acute hepatitis induced by the carbon tetrachloride, paracetamol, and D- galactosamine as in noticeable clinical studies established on biochemical, and functional parameters (Nagalekshmi et al. 2011).

11.3.4 Berberis aristata *DC*

Taxonomic Characteristics *Berberis aristata* (Berberidaceae) is commonly called Chitra in India.

Crude Drug Used Root, stem and leaves are used for active constituents.

Major Chemical Constituents Berberine and berbamine, flavonoids like quercetin and rutin and polyphenols like caffeic acid.

Modern Medicine Based on Traditional Uses Root extract of *Berberis aristata* is commonly used hepatotoxin in a rat model of acute hepatotoxicity induced with carbon tetrachloride. Liver damage induced by carbon tetrachloride concerns biotransformation of free radical derivatives, increase in lipid peroxidation and excessive cell death. The root extract of *B. aristata* has berberine chloride; a pharmacological bioactive possessing antimicrobial, antiviral, anti-inflammatory, cholesterol-lowering, anticancer and antioxidant effects. Hepatoprotective activities of berberine have been seen to chemically induce hepatotoxicity in rats. The preventive and curative effect of berberine is confirmed by its hepatoprotective effects against carbon tetrachloride-induced hepatotoxicity. Berberines have the prospective for developing an additional drug to treat liver toxicity (Dehar et al. 2013). Berberine reduces the excretion of excessive formation of bile pigments facilitating the reduction of serum enzyme levels in the blood and decreasing inflammation in liver cells. Berberine properties of antioxidative, anti-inflammatory, anticancer, hepatoprotective and immunomodulatory are useful in treating many disorders like dysentery, anorexia, problems related to the gallbladder and also in hepatitis (Kiran et al. 2021).

11.3.5 Boerhavia diffusa *L.*

Taxonomic Characteristics *Boerhavia diffusa* (Nyctaginaceae) is commonly called Punarnava in India.

Crude Drug Used Roots are used for the active constituents. The alcoholic extract of *Boerhavia diffusa* whole plant appears to be a potent and safe antihepatotoxic drug (Chandan et al. 1991).

Major Chemical Constituents Boeravinone, boerhavia acid, some isoflavonoids, punarnavine, sitosterol, palmitic acid, lignan glycosides, steroids, and esters of sitosterol.

Modern Medicine Based on Traditional Uses Hepatoprotective activity of *Boerhavia diffusa* L. roots is confirmed by serum constraints in toxicity of thioacetamide in experimental mice. Aqueous extract of thin roots collected in the summertime has supplementary activity, which suggests a selection of specific periods and types of roots to be collected for maximum results. The exploration also validates the use of roots in hepatic ailments by several tribes in India (Rawat et al. 1997). *B. diffusa* extract protects the liver against ethanol-induced hepatotoxicity. The mechanism is confirmed by inducing acceleration of liver cell regeneration by minimising leakage of marker enzymes into the blood and thereby lowering their values in serum. Hydroalcoholic extract of *B. diffusa* L. has hepatoprotective and antioxidant activity, as it shows a lowering of serum hepatic marker enzyme actions and increased antioxidant enzyme levels. Hydroalcoholic extract of *B. diffusa* L. also showed promising hepatoprotective and antioxidant activity against D-Galactosamine-induced Hepatotoxicity (Devaki et al. 2004; Nalini et al. 2018).

Boerhavia diffusa is abundantly found throughout India and is considered a weed. Methanolic and chloroform extract of root and aerial part of *B. diffusa* demonstrated hepatoprotective activity against carbon tetrachloride, paracetamol and D-galactosamine intoxication in investigational rats. Roots of *B. diffusa* L. usually known as '*Punarnava*', are utilised by Indian tribes for hepatic illnesses therapy. Clinical data confirmed the effectiveness of *B. diffusa* L. in cases related to oedema caused by initial liver cirrhosis and persistent peritonitis. One of the findings suggests that alcoholic extract of *B. diffusa* L. roots induces hepatotoxicity, and it was analysed in albino rats by Chandan et al. (1991). Findings of histopathological tests reveal a decrease in fat deposits after receiving *B. diffusa* L. country-made liquor. It proves that *B. diffusa* L. shielded rats from hepatotoxic activity by diminishing alanine aminotransferase enzyme, triglycerides, cholesterol, and total lipid amount equally in serum and tissues. *B. diffusa* L. or *Punarnava* embraces alkaloids like punarnavine and punarnavoside showing antifibrinolytic and hepatoprotective activity. Findings of ursolic acid isolates from *B. diffusa* L. leaves showed a dose-dependent (5–20 mg/kg) hepatoprotective activity (21–100%) against galactosamine, thioacetamide, and carbon tetrachloride-induced hepatotoxicity in rats. Pre-treatment with ursolic acid improved the viability of rat hepatocytes significantly (Devaki et al. 2004; Nalini et al. 2018).

11.3.6 Capparis spinosa *L.*

Taxonomic Characteristics *Capparis spinosa* (Capparidaceae) is commonly known as Kabra and Himsra in India. Himsra is a bushy plant commonly found in areas of the Mediterranean. Himsra bush has large white to pinkish white flowers, thick and curved leaves, and a persistent climber, or shrub as it is equipped with light yellow strong thorns. It mostly matures in rocky coastal regions or on walls.

Crude Drug Used Fruits, roots, and root bark are used.

Major Chemical Constituents Alkaloids, glycosides, tannins, phenolics, flavonoids, triterpenoids steroids and saponins.

Modern Medicine Based on Traditional Uses Root and bark of *Capparis spinosa* L. contain rutic acid, a volatile component, and stachydrine. Caper is utilised as an expectorant and diuretic for relieving problems of the liver, kidney, and spleen. Caper berry is used in pickled food. Caper is highly popular because its bud and fruit are edible, commonly called caper berry. Caper Bush is a potent hepatoprotective known to inhibit the increase of malondialdehyde (biomarker for oxidative stress) levels in plasma and hepatic cells. Caper inhibits alanine aminotransferase and AST enzyme levels and improves the functional efficiency of the liver. Flavonoids of Caper Bush display considerable antioxidant properties. Hydroalcoholic extract of the root bark revealed hepatoprotective effects against carbon tetrachloride-stimulated liver damage in mice. *C. spinosa* can cause nephrotoxicity and hepatotoxicity based on dosage (Aghel et al. 2007; Fanoudi et al. 2017). Protective effects against acute liver injury are observed as caper shows a higher free radical scavenging effect and inhibition of lipid peroxidation with increased antioxidant activity. Corresponding to the results obtained, abundant flavonoid compounds such as quercetin are considered the main hepatoprotective factor in the hydroalcoholic extract of *C. spinosa.* (Kalantari et al. 2018).

11.3.7 Cassia occidentalis *or* Senna occidentalis *(L.)*

Taxonomic Characteristics *Cassia occidentalis* (Fabaceae) is generally known as Senna in India.

Crude Drug Used Roots are used for active constituents. *Cassia occidentalis* or coffee senna are different from normal coffee plants. But as it is utilised as a substitute for coffee.

Major Chemical Constituents Alkaloids, flavonoids, glycosides, tannins, phlobatannins, chrysophanol, emodin and tetrahydroanthracene.

Modern Medicine Based on Traditional Uses The use of coffee Senna for regulating the peristalsis of the intestines is well-known in traditional medicine.

It works likewise to the Senna species which have more purgative actions. Seeds of coffee senna serve as a form of diuretic which helps in the purification or detoxification of blood. They are used to cure heart diseases, whooping cough, and convulsions. Senna seeds, leaves, and roots contain volatile oils, which can fight against fungi and bacteria. The seeds provide relief for diseases in the respiratory system (Nwaehujor et al. 2011).

11.3.8 Cichorium intybus *L.*

Taxonomic Characteristics *Cichorium intybus* (Asteraceae) is usually known as *Chicory* (Kasani) in India.

Crude Drug Used Roots of plants are used for the active constituent.

Major Chemical Constituents Alkaloids, coumarins, flavonoids, sesquiterpenoids, steroids, organic acid and beta-sitosterol.

Modern Medicine Based on Traditional Uses *Chicory* (Kasani) is a potent antioxidant having free radical scavenging and hepatoprotective properties. *Cichorium intybus* extract protects liver damage from carbon tetrachloride with a dose of 50–100 mg/kg, but concentrations higher than 200 mg/kg are not effective. *C. intybus* extract protects the cells against carbon tetrachloride-induced cytotoxicity, but concentrations of *C. intybus* around 1.5 mg/mL and higher favour the carbon tetrachloride-induced cytotoxicity. The hepatoprotective effect of *C. intybus* extracts is well documented: high concentrations were hepatotoxic (Jamshidzadeh et al. 2006). A decrease in total bilirubin, serum glutamic-oxaloacetic transaminase, serum glutamic pyruvic transaminase and alkaline phosphatase activities was observed when associated with thioacetamide treated group (Madani et al. 2008). A hepatoprotective effect of leaf extracts was demonstrated, by decreasing aspartate aminotransferase and alanine aminotransferase amounts thus preventing histopathological changes in rat liver with carbon tetrachloride-induced hepatotoxicity. Leaf extracts attenuated the reduction of glutathione and superoxide dismutase and decreased malondialdehyde levels in rats with carbon tetrachloride-induced hepatotoxicity. Cytoplasmic transaminase declines the structural integrity of the liver, confirmed by a rise in serum enzymes level. Carbon tetrachloride toxicity is equivalent to the cleavage of the carbon-chlorine bonds to create a trichloromethyl free radical. Released free radical responds promptly with oxygen-forming trichloromethyl peroxy radical contributing to hepatotoxicity and an increase in the hepatic enzymes. Serum levels concerning hepatic enzymes aspartate aminotransferase and alanine aminotransferase were enhanced, showing hepatocellular injury in the carbon tetrachloride-induced injury model. However, therapy with *C. intybus* leaf extracts lowered the aspartate aminotransferase and alanine aminotransferase levels of carbon tetrachloride exposed animals. Liver index and hexadecenoic acid serum levels, laminin and hydroxyproline were enhanced in rats exposed to carbon tetrachloride. *C. intybus* leaf extract reduced the hexadecenoic acid serum levels, laminin, hydroxyproline and the liver index, implying that the extract has hepatoprotective effects (Li et al. 2014).

11.3.9 Eclipta alba *(L.) Hassk*

Taxonomic Characteristics *Eclipta alba* (Asteraceae) is generally known as Bhringraj in India.

Crude Drug Used Roots and leaves are used for active constituents.

Major Chemical Constituents Alkaloids, coumarins, thiophene, flavonoids, polyacetylenes, triterpenes and glycosides.

Modern Medicine Based on Traditional Uses *Eclipta alba* Hassk. mostly known as *Bhringraj,* belongs to the Asteraceae family. *Eclipta alba* leaf extract is considered a powerful liver tonic in Ayurvedic medicine. It possesses a wide range of biological actions and is used for the medication of hepatitis and cirrhosis (Wagner 1986). *E. alba* exhibited antihepatotoxic activity against carbon tetrachloride with the help of D-galactosamine and phalloidin which induce toxicity against rat hepatocytes. Studies carried out by Ahirwar and Saxena (2008) revealed that hepatoprotective activity is related to the regulation of hepatic microsomal drug metabolising enzymes. Extract of *E. alba* was competent to lower the major promoted biochemical parameter changes related to carbon tetrachloride-stimulated hepatic damage in experimental rats. The concentration of total protein was at a normal level after the treatment (Ahirwar and Saxena 2008).

The effectiveness of *Eclipta alba* as a hepatoprotectant in counter to paracetamol stimulated hepatotoxicity, by regulating serum enzymes alkaline, acid phosphatase & 5′-nucleotidase which are known to facilitate hepatotoxic i.e., hepatic necrosis in rats. A substantial decline in serum concentration and rise in cholesterol level are observed in paracetamol-exposed experimental rats, which are regulated by therapy with *E. alba*. Its role as a hepatoprotective agent has been proven in reducing the side effects and confirming the lowering of hepatotoxicity (Indhuleka and Jeyaraj 2019). A study conducted by Murugaian et al. (2008) on whole plant reveals that *E. alba* extract exhibited protective activity against carbon tetrachloride-induced liver injury and its alkaloid ecliptine shows choleretic action (extract augmented in the bile flow of rats resulting in stimulation of liver secretory capacity).

11.3.10 Embelia Ribes

Taxonomic Characteristics *Embelia ribes* (Primulaceae) are usually known as Vidanga in India.

Crude Drug Used Fruits, roots, stem, and leaves are used for the active constituents.

Major Chemical Constituents Embelin, alkaloids, coumarins, quercetin, polyacetylenes, glycosides, tannins and terpenoids.

Modern Medicine Based on Traditional Uses Embelin restores carbon tetrachloride-induced abnormality in AST, serum alanine aminotransferase, ALP, LDH, GGT, and total bilirubin in serum. Antioxidative and immunoregulatory activities of embelin in acute liver injury are well documented. Hence, embelin can be an influential therapeutic bioactive against acute hepatic injury and the effects of its underlying mechanisms can be investigated for the development of beneficial pharmaceutical products. Embelin substantially reduced thioacetamide and induced liver cell necrosis. Massive inflammatory cell infiltration is consistent with hepatic fibrogenesis (a healing process), and it occurs in embelin treated recovery group. Moreover, an increase in macrophage activities is also observed with embelin treatment. Embelin oral administration increases enzyme activities produced by carbon tetrachloride favouring the consequent recovery towards enzymes normalisation suggesting the possibility of embelin to improve the condition of liver cells and cause accelerated renewal of parenchymal cells. Stabilisation of total serum bilirubin and total protein level with the oral embelin and silymarin in rats indicates improvement of the functional status of the hepatic cells. Findings suggest that the hepatic antioxidant perspective of embelin is approved due to the hydroxyl group at the ortho position. The O-hydroxyl group has resonance property which simply donates free radicals and efficiently neutralises them. The presence of a hydroxyl group in the ortho position increases its antioxidant potential through intermolecular hydrogen bonding involving the -SH group of non-protein thiols and enzymes resulting in conserving the antioxidant system against peroxidative damage in liver tissue (Singh et al. 2009; Jain et al. 2013; Wang et al. 2019).

11.3.11 Fumaria officinalis *L.*

Taxonomic Characteristics *Fumaria officinalis* (Papaveraceae) is frequently known as Papara in India.

Crude Drug Used Aerial parts and leaves are used for active constituents.

Major Chemical Constituents Alkaloid, phenol, flavonoid, glycoside, terpenoid, phytosterol, saponin, steroid and tannin.

Modern Medicine Based on Traditional Uses The *Fumaria officinalis* ethanolic extract with a dose of 200–500 mg has significant hepatoprotective action by reducing the serum markers enzymes like serum glutamic pyruvate transaminase, serum glutamic-oxaloacetic transaminase and alkaline phosphatase. *F. officinalis* ethanolic extract declines the high concentration of total serum bilirubin, cholesterol, and triglycerides. Ascorbic acid estimation in rat urine and histopathological studies conform to the hepatoprotective activity of *F. officinalis* when assessed with carbon tetrachloride-treated control groups. Authors have also compared the activity with commercially available Silymarin, the standard drug; the ethanolic extract of *F. officinalis* showed significant hepatoprotective activity (Sharma et al. 2012; Orhan et al. 2012).

11.3.12 Hygrophila auriculata *(Schumach.) Heine*

Taxonomic Characteristics *Hygrophila auriculata* (Acanthaceae), is commonly known in India as kokilaksha or Gokul Kant. *H. auriculata* is a herbaceous medicinal plant of the Acanthus family growing in marshy places and native regions of tropical Asia and Africa.

Crude Drug Used Roots and seeds of plants are used for the active constituent. *H. auriculata* aqueous extract does not show the existence of various phytoconstituents, like non-reducing and reducing sugar, saponin, polyphenol, flavonoid, tannin, alkaloid, triterpenes, protein, and amino acids; for which hydroalcoholic aqueous extract of *H. auriculata* is preferred.

Major Chemical Constituents Alkaloids, flavonoids, terpenoids, saponins, steroids, tannins, and triterpenoids.

Modern Medicine Based on Traditional Uses In an animal clinical trial as discussed by Shanmugasundaram and Venkataraman (2006), it was observed that *Hygrophila auriculata* extract in the experimentally treated clusters exhibited regeneration of liver cells, normalisation of fatty changes and liver cell necrosis when compared with histopathological studies of control rats. Results confirmed by researchers based on the antioxidant activity of aqueous extract *H. auriculata* in vitro, by Ferric thiocyanate and Thiobarbituric acid methods (Shanmugasundaram and Venkataraman 2006).

11.3.13 Phyllanthus Amarus/Niruri

Taxonomic Characteristics *Phyllanthus niruri* (*Phyllanthaceae*) is known as Indian gooseberry and *Bhui Amla* in India.

Crude Drug Used Fruits and aerial parts are used for active constituents.

Major Chemical Constituents Phenolic and polyphenols, tannins, flavonoids, triterpenoids, stigmasterol and β-sitosterol.

Modern Medicine Based on Traditional Uses *Phyllanthus niruri* and *amarus* are acknowledged to acquire hepatoprotective activity. High potency is exhibited by crude extracts of *P. niruri* via free radical scavenging, inhibition of reactive species (ROS) and lipid peroxidation is well documented. *P. niruri* aqueous and methanolic extracts are effective inhibitors of microsomal lipid peroxidation induced by Fe2+ and ascorbate when employed *in vitro*. Leaf and fruit extract of *P. niruri* has antioxidant activity. Results of inhibition of the superoxide (ROS) *in vitro* revealed more potency of aqueous extracts of leaf and fruit than methanolic extracts. All parts of *P. niruri* extract when examined with DPPH radical scavenging activity, exhibited very high potency because of the free radical quenching properties with the IC50 values. Phenolic chemical constituents such as flavonoids and tannins are showing antioxidant activity. *P. niruri* Linn. stem decoctions protect the liver from oxidative damage; hence it is utilised as an effective protector in carbon tetrachloride-induced damage. *P. niruri* Linn. is a proven antiviral medicine against Hepatitis-B in human subjects. Preclinical studies demonstrate that *P. niruri* Linn. extract limits the functioning of endogenous DNA polymerase of the hepatitis B virus and binds to the surface antigen. Pre-treatment with extract reduces the paracetamol-induced acute hepatic damage in rats when monitored with estimating serum glutamic-oxaloacetic transaminase. In the *in vitro* study, it decreased the release of AST and serum alanine aminotransferase in rat primary cultured hepatocytes treated with ethanol (Kodoli et al. 2021; Sattar et al. 2015).

11.3.14 Picrorhiza kurroa *Royle ex Benth*

Taxonomic Characteristics *Picrorhiza kurroa* (Plantaginaceae), is known as Anjani and Kutki in India.

Crude Drug Used Roots and rhizomes are used for active constituents.

Major Chemical Constituents Phenolic compounds, flavonoids, triterpenoids, glycosides, aromatic esters, bis-iridoid, phenyl propenoids, kutkiol, kutki sterol and apocynin.

Modern Medicine Based on Traditional Uses *Picrorhiza kurroa* Royle ex Benth. Is a renowned herb in Ayurveda. It has traditionally been utilised to treat disorders of the liver, and upper respiratory tract, reduce fevers, and treat dyspepsia, chronic diarrhoea, and scorpion sting. Kutkin or *P. kurroa* contains kutkoside and iridoid glycosides, picrosides I, II, and III as a bioactive compound. The hepatoprotective action of *P. kurroa* is recognised by its ability to hinder ROS generation, scavenge free radicals, strong antioxidant activity and inhibit lipid peroxidation significantly. Cells treated only with *P. kurroa* extract alone are explicit to the detoxification of reactive metabolites generated from ethanol metabolism in hepatocytes. *Picrorhiza kurroa* hepatoprotective action is confirmed by an increase in Hb, PCV, TEC, lymphocytes, total protein, albumin, and globulin levels while it decreases glucose, total cholesterol, bilirubin, AST, serum alanine aminotransferase, ALP and LDH values in blood level normalcy. *P. kurroa* extract inhibits the production of oxygen anions and scavenges free radicals. Its antioxidant effect is the same as that of superoxide dismutase, metal ion chelators, and xanthine oxidase inhibitors. Animal studies signal that the components demonstrate a strong anticholestatic activity against a range of liver toxic substances when compared to the commercial silymarin effect. The dose-dependent choleretic activity of *Picrorhiza* is evidenced by an increase in bile salts and acids, and bile flow (Sinha et al. 2011; Kumar and Shukla 2017).

11.3.15 Piper longum *L.*

Taxonomic Characteristics *Piper longum* (Piperaceae) is known as Pippali in India.

Crude Drug Used Fruits are used for active constituents.

Major Chemical Constituents Alkaloids, phenolic compounds, flavonoids, triterpenoids, glycosides, coumaperine, sesquiterpene hydrocarbons and ethers.

Modern Medicine Based on Traditional Uses *Piper longum* possess antiasthmatic, anti-inflammatory, hypocholestremic, hepatoprotective, and immunomodulatory activities. *P. longum* contains various alkaloid like piperine, piperlongumine, piperlonguminine, etc. which helps in the regeneration and development of liver cells. The hepatoprotective action of fruits of *P. longum* induces microsomal enzymes accelerating the excretion of carbon tetrachloride or inhibiting lipid peroxidation by carbon tetrachloride. A decrease in the activity of SGOT and SGPT enzymes shows a reversal of the induced toxicity of the liver. Biochemical and histopathological observation reveals that ethanolic extract and butanol fraction of fruits of *P. longum* exhibits significant hepatoprotective activity. Ethanolic extract and butanol fraction of the fruits exerts a clear protective action against carbon tetrachloride-induced liver cell damage. An evident decline in serum enzyme level, total bilirubin and direct bilirubin were observed. Histopathological findings reveal the administration of *P. longum* L. milk extract proposed protection for the liver cell damage induced by Carbon tetrachloride with mild fatty changes in liver parenchymal cells. It also showed the regeneration of liver cells near the area of the necrotic zone (Jalalpure et al. 2003; Patel 2009).

11.3.16 Plumbago zeylanica *L.*

Taxonomic Characteristics *Plumbago zeylanica* (Plumbaginaceae) is commonly known as Chitrak in India.

Crude Drug Used Roots and aerial parts are used for an active constituent.

Major Chemical Constituents Alkaloids, phenolic compounds, flavonoids, triterpenoids, glycosides, plumbagin, plumbagic acid, beta-sitosterol, naphthoquinone, trans-cinnamic acid and vanillic acid.

Modern Medicine Based on Traditional Uses The qualitative phytochemical investigation of *Plumbago zeylanica* using different extracts showed positive tests for carbohydrates, triterpenes, steroids, tannins, and flavonoids. Only the methanolic *P. zeylanica* extract has a higher content of triterpenes, tannins, and flavonoids which shows significant hepatoprotective activity. Crude powder of roots showed the hepatoprotective effect related to carbon tetrachloride-induced liver destruction in Wistar rats. The antioxidant effect of *P. zeylanica* exerts its protective effect against carbon tetrachloride by induced alteration in its active ingredient of crude root powder. Additional work is necessary to elucidate the constituent responsible for hepatoprotective activity along with their mechanism of action (Kumar et al. 2009; Tayubi et al. 2018; Akhilraj et al. 2021).

11.3.17 Raphanus sativus *L.*

Taxonomic Characteristics *Raphanus sativus* (Brassicaceae) commonly known as Mulak in India.

Crude Drug Used Roots are used for an active constituent.

Major Chemical Constituents Phenolic compounds, flavonoids, triterpenoids, linoleic acid, glucosinolates and isothiocyanate.

Modern Medicine Based on Traditional Uses *Raphanus sativus* enzyme extract prevents hepatic damage by inhibiting the inflammation in carbon tetrachloride-induced rats with hepatic fibrosis/cirrhosis. But it is not revealed that, whether *R. sativus* enzyme extract plays a central role in the mechanism or not. Enzyme extract of *R. sativus* has a defending consequence on tacrine-induced hepatotoxicity in human hepatoma cell line. Administration of *R. sativus* enzyme extract also prevented biochemical and histomorphological alteration induced by carbon tetrachloride (Lee et al. 2012). A phytochemical screening study revealed the presence of flavonoids, terpenoids, alkaloids, saponin and sterol. Findings suggest a dose-dependent rise in the oxidative potential of an extract with phenolic substance. Ethanolic extract of *R. sativus* increased the levels of aspartate aminotransferase, alanine aminotransferase and alkaline phosphatase as related to the negative control. The glutathione and catalase in ethanolic and aqueous extract of *R. sativus* considerably increased, while malondialdehyde levels were decreased when compared to the negative control. The ethanol and aqueous extract of *R. sativus* have partial hepatoprotection against carbon tetrachloride toxicity (Syed et al. 2014).

11.3.18 Ricinus communis *L.*

Taxonomic Characteristics *Ricinus communis* (Euphorbiaceae), is commonly known as castor in India.

Crude Drug Used Leaves, roots, and bark are used as the active constituent.

Major Chemical Constituents Alkaloids, flavonoids, triterpenoids, glycosides, ricinoleic acid, stearic, linoleic, palmitic acid, sitosterol, squalene, tocopherols, stearic acid, ricinine and ricin.

Modern Medicine Based on Traditional Uses Leaves extract of *Ricinus communis* exhibited dose-associated hepatoprotective choleretic and anticholestatic activity. *R. communis* ethanolic extract has hepatoprotective outcomes against carbon tetrachloride-induced hepatotoxicity in rats. The positive effect is due to flavonoids as it exhibits membrane stabilisation and antiperoxidative effects. Flavonoids and tannins of *R. communis* ethanol extracts have effectively enhanced the regenerative and reparative capacity of the liver cells. Although *R. communis* alcohol extract has an equivalent hepatoprotective effect with the silymarin, the researcher is trying to clarify the hepatoprotective mechanism and constituents of effective components from *R. communis* extract. The hepatoprotective effect is due to high amounts of rutin and other bioactive compounds. The extract is potent, as a reversal in the pathological challenges occurs after D-galactosamine administration, which promotes hepatocytes regeneration, scavenging excess free radicals, mostly by activating antioxidant resistance system and membrane stabilisation through inhibition of peroxidation of lipid, glycoprotein biosynthesis enhancement and stabilisation of cell membrane. The hepatoprotective property of the *R. communis* extract is attributed to strong antioxidants like rutin, gentisic acid, quercetin, gallic acid, ellagic acid and epicatechin bioactive compounds. The presence of high amounts of rutin compounds may be responsible for the hepatoprotective effect (Visen et al. 1992; Prince et al. 2011; Babu et al. 2017).

11.3.19 Silybum marianum *(L.) Gaertn*

Taxonomic Characteristics *Silybum marianum* (Asteraceae)- is known as milk thistle in India.

Crude Drug Used Above-ground parts and seeds are used for an active constituent.

Major Chemical Constituents Phenolic compounds, flavonoids, flavolignans, triterpenoids, glycosides, silymarin, isosilychristin, silychristin, silydianin, silybin and isosilybin.

Modern Medicine Based on Traditional Uses The *Silybum marianum* polyphenolic extracts have a shielding effect against hepatic cell injury induced mostly by thioacetamide. Polyphenolic compounds act as antioxidants and anti-prooxidants by showing scavenging reactive oxygen species via enzymatic and non-enzymatic pathways. Silymarin has multiple mechanisms of action, and acts as a hepatoprotectant, which includes antioxidant activity by flavonoid. Silymarin is 10 times more potent than vitamin E; it has anti-lipid prooxidative activity; the induced detoxification system includes protection of cells against employed glutathione; lessening of lukoterin formation from unsaturated free acid; enhanced synthesis of protein; and stabilisation of mast cell and regulation of immuno functions. *S. marianum* bioactive component inhibit the cyt-P450 detoxification system and prevents the metabolism of toxic compounds such as thioacetamide, tetrachloride and acetaminophen (Madani et al. 2008). Strong antioxidant activity is exhibited by Silymarin and its shows protective effects against hepatic toxicity induced by a wide variety of agents by inhibiting lipid peroxidation. Higher total phenolic content is known to contribute to the antioxidant activity of extracts, while antioxidant activity is linked to the hepatoprotective effect of *S. marianum* extracts. These findings corroborate the researcher's results on the ability of Silymarin to exert a hepatoprotective activity (Freitag et al. 2015).

11.3.20 Solanum nigrum *L.*

Taxonomic Characteristics *Solanum nigrum* (Solanaceae), is commonly known as black nightshade in India.

Crude Drug Used Berries, leaves, and roots are used for active constituents.

Major Chemical Constituents Phenolic compounds, flavonoids, triterpenoids, anthocyanidin, gentisic acid, luteolin, apigenin, kaempferol, and m-coumaric acid.

Modern Medicine Based on Traditional Uses Two extracts, water and methanolic, of *Solanum nigrum* have shown hepatoprotective effects in opposition to carbon tetrachloride intoxicated rats. Water extract appears to have a more potent hepatoprotective influence than the methanolic one, due to more polar phytoconstituent in *S. nigrum.* The aqueous extract is well-known as a potent hepatoprotective agent. But the mechanism of hepatoprotective activity of the aqueous extract is not known yet. It is assumed that the effect of *S. nigrum* extract on liver protection may be due to glutathione-mediated detoxification. *S. nigrum* may have enhanced GSH status in cells and thereby afforded protection to hepatic cells from toxic damage (Elhag et al. 2011; Kumar et al. 2013a, b). As *S. nigrum* is a potent hepatoprotective plant due to its high flavonoid content it prevents carbon tetrachloride-induced adduct formations in tissue macromolecules (Krithika and Verma 2019). Hepatoprotective effects against ethanol-induced injury have been described under both *in vitro* and *in vivo* conditions. *S. nigrum* possesses different types of alkaloids which help to regulate in improving skin, bladder, liver, and kidney conditions. This plant may also be used as a diuretic, laxative, emollient, and antiseptic and cures liver cirrhosis (Liu et al. 2016).

11.3.21 Taraxacum officinale *F.H. Wigg*

Taxonomic Characteristics *Taraxacum officinale* (Asteraceae), is commonly known as Dandelion in India.

Crude Drug Used Roots and leaves are used for active constituents.

Major Chemical Constituents Phenolic compounds, flavonoids, triterpenoids, glycosides, sesquiterpene lactones, taraxasterol, taraxerol and chlorogenic acid.

Modern Medicine Based on Traditional Uses Dandelions are utilised as a folklore medicine for the medication of disorders related to the liver and kidney. *Taraxacum officinale* root aqueous extract shows a protective effect when compared to ethanol-induced liver damage. Reactive oxygen species facilitate hepatic tissue harm induced by ethanol. The induction of cytochrome P-450 2E1 due to ethanol is believed to play a major role in enhancing the generation of reactive oxygen species like reactive oxygen and hydrogen peroxide. It normally causes a prooxidative state in the cellular environment resulting in oxidative damage. Irreversible liver damage stimulated by the consumption of excessive ethanol is undoubtedly associated with oxidative stress via enhanced lipid peroxidation and reactive oxygen species production. *T. officinale* root aqueous extract is implicated as a potentially functional radical scavenger for the host. *T. officinale* root hot water extract supplementation antagonises ethanol-induced hepatic damage by changing hepatic antioxidant status and lipid peroxidation. The antioxidant scavenging action of the aqueous extracts was found to be accountable for the amelioration of oxidative stress during alcohol toxicity. Hence, the development of dietary supplementation by using *T. officinale* could be helpful to protect against alcoholic liver damage mediated by oxidative stress (You et al. 2010). Dandelion hot water extract declines lipid peroxidation and increases antioxidant enzyme activity. It was found that alcohol-induced oxidative stress is linked with a reduction in cell viability, but by using *T. officinale* aqueous extract increase in cell viability is observed in the presence of ethanol. The hepatoprotective effects of *T. officinale* root extracts are associated with antioxidant activities. The root extract of Dandelion increases the antioxidant enzymes and ameliorated the liver enzymes which protects the liver against oxidative stress induced by ethanol (Mahboubi and Mahboubi 2020).

11.3.22 Terminalia arjuna *(Roxb. ex DC.) Wight & Arn.*

Taxonomic Characteristics *Terminalia arjuna* (Combretaceae), is commonly known as arjuna in India.

Crude Drug Used Root and root bark are used as active constituents.

Major Chemical Constituents Flavonoids, glycosides, lactones, polyphenols, phytosterol, saponins, tannins, triterpenoids, sterols, and minerals.

Modern Medicine Based on Traditional Uses Arjuna tree consists of Co-enzyme Q-10 in large amounts. A bioactive substance i.e., Co-enzyme Q-10 is known to prevent the occurrence of heart attacks. People showing coronary heart disease problems are given *Terminalia arjuna* extract and powder prepared from its bark. *T. arjuna* is taken directly as it has properties related to cardioprotective which makes cardiac muscles much stronger and enhances the pumping action of the heart and many other functions of the cardiovascular system. Moreover, it stipulates a better prognosis for persons with the condition of liver cirrhosis, and it also gives people relief from nervousness and other forms of stress. *Terminalia* is extremely valuable in delivering treatment for diseases related to the coronary artery, hypercholesterolemia, and heart failure. It blocks gastric ulcers in people who are on non-steroidal anti-inflammatory drugs. Since 2500 B.C., *T. arjuna* is used by Ayurvedic doctors only due to its cardioprotective properties. Aqueous extracts of the bark have been shown to possess potent antioxidative activity that protects the kidney and liver against carbon tetrachloride-induced oxidative stress via modification of cytochrome P-450. The aqueous extract of *T. arjuna* protects liver and kidney tissues against oxidative damage (Manna et al. 2006; Subasini et al. 2007; Haidry and Malik 2014).

11.3.23 Terminalia chebula *(Gaertn.) Retz.*

Taxonomic Characteristics *Terminalia chebula* (Combretaceae), is commonly known as Harda in India.

Crude Drug Used Fruit and root are used for active constituent.

Major Chemical Constituents Polyphenols, flavonoids, tannins, triterpenoids, saponins, phytosterol, glycosides, sterols, chebulagic acid and eugenol.

Modern Medicine Based on Traditional Uses *Terminalia chebula* water extract positively affects on acute and severe liver injury. *T. chebula* water extract clarifies its mechanisms by involving the inhibition of oxidative stress and inflammatory cytokines. The hepatoprotective role of ethanolic fruit extract of *T. chebula* fruit is due to the antioxidant, as it shows potential mechanism suggesting that *T. chebula* water extract is useful to counteract the ethanol-induced hepatotoxicity. Chebulic acid prevented oxidative stress which was induced by t-BHP in L-02 cells through upregulation of expression levels (HO-1 and NQO1), which was mediated by Nrf2 via the MAPK signalling pathway (Choi et al. 2015; Balakrishna and Lakshmi 2017; Feng et al. 2021).

11.3.24 Tinospora cordifolia *(Thunb.) Miers*

Taxonomic Characteristics *Tinospora cordifolia* (Menispermaceae), is commonly known as *Guduchi, Amrita* in India.

Crude Drug Used Stem and leaves are used as the active constituent.

Major Chemical Constituents Phenolic compounds, flavonoids, terpenoid, alkaloid, lignans, steroids and tinosporide.

Modern Medicine Based on Traditional Uses *Tinospora cordifolia* (Wild.) Miers., *Guduchi, Amrita* Ayurveda precious medicinal herbs. The word *'Amrita'* means youthfulness, vitality, and longevity. Current modern medicine acknowledged *Guduchi* as an immunomodulatory, adaptogenic, hepatoprotective, and antifibrinolytic activity. The efficient source of tinosporin is known to adjust immunosuppression correlated with the disordered function of hepatic cells (Kashaw et al. 2011). Kupffer cells are good indicators of the liver injuries. Antihepatotoxic action was also explored in albino mice with carbon tetrachloride activity. Nagarkatti et al. (1994) conducted a study on *T. cordifolia* (Wild.) Miers. which indicated a reduction in fibrosis in experimental rats, induced by carbon tetrachloride and substantially improved oppressed function of Kupffer cells in an additional rat model of the chronic liver damage mostly induced by heterologous serum. The study elevates the prospect of the antifibrotic effect of *T. cordifolia* being intervened through activation of the Kupffer cells. Hence, an aqueous extract of *T. cordifolia* is a potent hepatoprotective agent. Aqueous extract effectively controls serum alanine aminotransferase, ALP, and total bilirubin levels during the study. The histopathological studies supported the hepatoprotective activity of the aqueous extract. *T. sinensis* has been found to be comparatively more potent than *T. cordifolia*, although both formulations have considerable protection against paracetamol-induced hepatic toxicity. In Ayurvedic literature both species are utilised as *guduchi* and have been described as similar, therefore further comparative characterisation of the chemical constituents of each species should identify the potent Hepatoprotective components along with their proportionate combination (Kumar et al. 2013a, b; Nagarkar et al. 2013; Singh et al. 2015).

11.3.25 Tamarix dioica and T. species

Taxonomic Characteristics *Tamarix dioica* (Tamaricaceae), is commonly known as Farash in India.

Crude Drug Used Aerial parts are used for active constituent.

Major Chemical Constituents Phenolic compounds, flavonoids, and triterpenoids like hexadecane, octadecane, dodecanoic acid, docosane, hexadecane, nonanal, and nonanoic acid.

Modern Medicine Based on Traditional Uses *Tamarix dioica* and its species have the presence of antioxidant activity (both *in-vitro* and *in-vivo)* mostly because of the phenolic constituent involved in the hepatoprotective potent property. *T. dioica* extracts decline the acute carbon tetrachloride-mediated damage of the liver *in-vivo* and ameliorates histopathological and biochemical parameters of the mice. *T. dioica* could alleviate the seriousness of liver damage produced by carbon tetrachloride. Hence, it is presumed that *T. dioica* may be used as a supportive cure for the medication of drug-induced and other oxidative stress-mediated hepatoxicity in future (AbouZid and Sleem 2011; Komal et al. 2021).

Table 11.1 provides a tabulated overview of the above-discussed species; with the aim to facilitate orientation, they are ranked according to plant families. The table also contains both the Latin and common/vernacular names of species, as well as the most important sources of reference on their hepatoprotective properties.

11.4 Conclusions

Herbal, plant-based remedies for liver disorders have been utilised for a very long period. They are commercialised worldwide by major pharmaceutical industries. Despite of their significant popularity, a various array of herbal medicines is still objectionable for the medication of liver disorders and conditions. These constraining factors are mostly due to the lack of herbal drug standardisation, identification as well as the insufficient knowledge about the mechanism of action of active ingredients or their bioactive principles (including the lack of statistics treated randomised controlled clinical trials with toxicological evaluation). Traditional medicinal research materials of Ayurveda are worldwide used for pharmaceutical research and are increasing day by day. The recent upgrading of these ayurvedic systems has had a positive impact by relatively improving the reliability of their utilisation. Recent data from the scientific literature suggest a range of phytochemicals from Indian medicinal plants, that can be analysed in hepatotoxicity model studies. Elaboration on managing strategies for Indian medicinal plants for therapeutic purposes is encouraging, and the status of their international prevalence is promising for their future.

Table 11.1 Indian medicinal plants with hepatoprotective properties

Botanical name	Family	Common name	References
Achillea millefolium	Asteraceae	Puthkanda	Yaeesh et al. (2006) and Al-Ezzy et al. (2017)
Aloe vera	Liliaceae	Aloe vera	Chandan et al. (2007) and Cui et al. (2014)
Andrographis paniculata	Acanthaceae	Green Chirayata	Nagalekshmi et al. (2011)
Berberis aristata	Berberidaceae	Chitra	Dehar et al. (2013) and Kiran et al. (2021)
Boerhavia diffusa	Nyctaginaceae	Punarnava	Rawat et al. (1997), Devaki et al. (2004) and Nalini et al. (2018)
Capparis spinosa	Capparidaceae	Kabra	Aghel et al. (2007), Fanoudi et al. (2017) and Kalantari et al. (2018)
Cichorium intybus	Asteraceae	Kasni	Jamshidzadeh et al. (2006) and Li et al. (2014)
Eclipta alba	Asteraceae	Bhringraj	Ahirwar an Saxena (2008), Indhuleka and Jeyaraj (2019)
Embelia ribes	Primulaceae	Vidanga	Singh et al. (2009), Jain et al. (2013), and Wang et al. (2019)
Fumaria officinalis	Papaveraceae	Papara	Sharma et al. (2012) and Orhan et al. (2012)
Hygrophila auriculata	Acanthaceae	Tamilkhana marsh barbel	Shanmugasundaram and Venkataraman (2006)
Phyllanthus amarus	*Phyllanthaceae*	*Bhui Amla*	Kodoli et al. (2021) and Sattar et al. (2015)
Plumbago zeylanica	Plumbaginaceae	Chitrak	Tayubi et al. (2018) and Akhilraj et al. (2021)
Piper longum	Piperaceae	Pippali	Jalalpure et al. (2003) and Patel (2009)
Picrorhiza kurroa	Plantaginaceae	Anjani, Kutki	Sinha et al. (2011) and Kumar et al. (2017)
Raphanus sativus	Brassicaceae	Mulak	Syed et al. (2014) and Lee et al. (2012)
Ricinus communis	Euphorbiaceae	Castor	Visen et al. (1992), Prince et al. (2011) and Babu et al. (2017)
Senna occidentalis	Caesalpiniaceae	Kasunda	Nwaehujor et al. (2011)
Silybum marianum	Asteraceae	Milk thistle	Madani et al. (2008) and Freitag et al. (2015)
Solanum nigrum	Solanaceae	Black night-shade, makoi	Elhag et al. (2011), Kumar et al. (2013a, b), Liu et al. (2016) and Krithika and Verma (2019)
Tamarix dioica	Tamaricaceae	Farash	AbouZid and Sleem (2011) and Komal et al. (2021)
Taraxacum officinale	Asteraceae	*Dandelion*	You et al. (2010), Mahboubi and Mahboubi (2020)

(continued)

Table 11.1 (continued)

Botanical name	Family	Common name	References
Tinospora cordifolia	Menispermaceae	Amruta, Gudvel, Guduchi	Kumar et al. (2013a, b), Nagarkar et al. (2013) and Singh et al. (2015)
Terminalia arjuna	Combretaceae	Arjuna	Manna et al. (2006), Subasini et al. (2007) and Haidry and Malik (2014)
Terminalia chebula	Combretaceae	Harda	Choi et al. (2015), Balakrishna and Lakshmi (2017) and Feng et al. (2021)

References

AbouZid S, Sleem A (2011) Hepatoprotective and antioxidant activities of *Tamarix nilotica* flowers. Pharm Biol 49(4):392–395

Aghel N, Rashidi I, Mombeini A (2007) Hepatoprotective activity of *Capparis spinosa* root bark against carbon tetrachloride induced hepatic damage in mice. Iran J Pharm Res 6(4):285–290

Ahirwar DK, Saxena RC (2008) Hepatoprotective activity of ethanolic extract of *Eclipta alba* in albino rats. Biomed Pharma J 1(1):235–238

Akhilraj AR, Bhat S, Priyalatha B, Vimala KS (2021) Comparative hepatoprotective activity of detoxified roots of *Plumbago zeylanica* L. and *Plumbago rosea* L. in Wistar rats. J Ayurveda Integr Med 12(3):452–457

Al-Ezzy RM, Al Anee RS, Kathum OA (2017) Hepatoprotective effects of *Achillea millefolium* methanolic extract on carbon tetrachloride induced hepatotoxicity on albino male mice. Int J Adv Res Biol Science 4(8):98–109

Babu PR, Bhuvaneswar C, Sandeep G, Ramaiah CV, Rajendra W (2017) Hepatoprotective role of *Ricinus communis* leaf extract against d-galactosamine induced acute hepatitis in albino rats. Biomed Pharmacother 88:658–666

Balakrishna V, Lakshmi T (2017) Hepatoprotective activity of ethanolic extract of *Terminalia chebula* fruit against ethanol-induced hepatotoxicity in rats. Asian J Pharm Clin Res 10(11): 55–58

Chandan BK, Sharma AK, Anand KK (1991) *Boerhaavia diffusa*: a study of its hepatoprotective activity. J Ethnopharmacol 31(3):299–307

Chandan BK, Saxena AK, Shukla S, Sharma N, Gupta DK, Suri KA, Suri J, Bhadauria M, Singh B (2007) Hepatoprotective potential of *Aloe barbadensis* mill. Against carbon tetrachloride induced hepatotoxicity. J Ethnopharmacol 111(3):560–566

Choi MK, Kim HG, Han JM, Lee JS, Lee JS, Chung SH, Son CG (2015) Hepatoprotective effect of *Terminalia chebula* against t-BHP-induced acute liver injury in C57/BL6 mice. Evid Based Complement Alternat Med:1–11

Cui Y, Ye Q, Wang H, Li Y, Yao W, Qian H (2014) Hepatoprotective potential of Aloe vera polysaccharides against chronic alcohol-induced hepatotoxicity in mice. J Sci Food Agric 94(9): 1764–1771

Dehar N, Walia R, Verma RB, Pandey P (2013) Hepatoprotective activity of *Berberis aristata* root extract against chemical induced acute hepatotoxicity in rats. Asian J Pharm Clin Res:53–56

Devaki T, Shivashangari KS, Ravikumar V, Govindaraju P (2004) Hepatoprotective activity of *Boerhaavia diffusa* on ethanol-induced liver damage in rats. J Nat Remed 1(2):109–115

Elhag RA, El Badwi SM, Bakhiet AO, Galal M (2011) Hepatoprotective activity of *Solanum nigrum* extracts on chemically induced liver damage in rats. J Vet Med Anim Health 3(4):45–50

Fanoudi S, Rakhshandeh H, Afshari AR, Mollazadeh H, Taher M (2017) Nephrotoxicity and hepatotoxicity of *Capparis spinosa* hydro-alcoholic extract in mice. JOJ Uro Nephron 4(2): 555–638

Feng XH, Xu HY, Wang JY, Duan S, Wang YC, Ma CM (2021) *In vivo* hepatoprotective activity and the underlying mechanism of chebulinic acid from *Terminalia chebula* fruit. Phytomedicine 1(83):153–479

Freitag AF, Cardia GF, da Rocha BA, Aguiar RP, Silva-Comar FM, Spironello RA, Grespan R, Caparroz-Assef SM, Bersani-Amado CA, Cuman RK (2015) Hepatoprotective effect of silymarin (*Silybum marianum*) on hepatotoxicity induced by acetaminophen in spontaneously hypertensive rats. Evid Based Complement Alternat Med:1–8

Girish C, Pradhan SC (2017) Herbal drugs on the liver. In: Liver pathophysiology. Academic, pp 605–620

Haidry MT, Malik A (2014) Hepatoprotective and antioxidative effects of *Terminalia arjuna* against cadmium provoked toxicity in albino rats (*Ratus norvigicus*). Biochem Pharmacol 3(1):1–4

Indhuleka A, Jeyaraj M (2019) Hepatoprotective effect of *Eclipta Alba* on membrane marker enzymes against paracetamol induced liver damage. Orient J Chem 35(3):1215–1219

Jain SK, Rajvaidy S, Desai P, Singh GK, Nagori BP (2013) Herbal extract as hepatoprotective-a review. J Pharmacogn Phytochem 2(3):170–175

Jaiswal Y, Liang Z, Zhao Z (2016) Botanical drugs in Ayurveda and traditional Chinese medicine. J Ethnopharmacol 194:245–259

Jalalpure SS, Patil MB, Prakash NS, Hemalata K, Manvi FV (2003) Hepatoprotective activity of the fruits of *Piper longum* Linn. Indian J Pharm Sci 65(4):363–366

Jamshidzadeh A, Khoshnoud MJ, Dehghani Z, Niknahad H (2006) Hepatoprotective activity of *Cichorium intybus* L. leaves extract against carbon tetrachloride induced toxicity. Iran J Pharm Res 1:41–46

Kalantari H, Foruozandeh H, Khodayar MJ, Siahpoosh A, Saki N, Kheradmand P (2018) Antioxidant and hepatoprotective effects of *Capparis spinosa* L. fractions and quercetin on tert-butyl hydroperoxide-induced acute liver damage in mice. J Tradit Complement Med 8(1):120–127

Kashaw V, Nema AK, Agarwal A (2011) Hepatoprotective prospective of herbal drugs and their vesicular carriers – a review. Int J Res Pharmaceut Biomed Sci (2):360–374

Kiran P, Aku R, Gaurav S (2021) Review on Hepatoprotective Effect of *Berberis Aristata* Dc, 190–199.

Kodoli RS, Galatage ST, Killedar SG, Pishwikar SA, Habbu PV, Bhagwat DA (2021) Hepatoprotective activity of *Phyllanthus niruri* Linn. Endophytes. Future J Pharm Sci 7(1):1

Komal S, Malik A, Akhtar N, Kazmi SA, Anjum F, Rida A (2021) *Tamarix dioica* (Ghaz) protective potential in the carbon tetrachloride-induced hepatotoxicity animal model. PRO 35(3):37–43

Krithika R, Verma RJ (2019) *Solanum nigrum* confers protection against carbon tetrachloride -induced experimental hepatotoxicity by increasing hepatic protein synthesis and regulation of energy metabolism. Clin Phytosci 5(1):1–8

Kumar P, Shukla SK (2017) Hepatoprotective efficacy of *Picrorhiza kurroa* in experimentally induced hepatotoxicity in cockerels. Int J Curr Microbiol Appl Science 6:2614–2622

Kumar R, Kumar S, Patra A, Jayalakshmi S (2009) Hepatoprotective activity of aerial parts of *Plumbago zeylanica* Linn against carbon tetrachloride-induced hepatotoxicity in rats. Int J Pharm Pharm Science 1(1):171–175

Kumar V, Modi PK, Saxena KK (2013a) Exploration of hepatoprotective activity of aqueous extract of *Tinospora cordifolia*-an experimental study. Asian J Pharm Clin Research 6(1):87–91

Kumar V, Sharma S, Modi PK (2013b) Exploration of hepatoprotective activity of aqueous extract of *Solanum nigrum* – an experimental study. Int J Pharm Sci Res 4(1):464–470

Kumar V, Chauhan RS, Tandon C (2017) Biosynthesis and therapeutic implications of iridoid glycosides from Picrorhiza genus: the road ahead. J Plant Biochem Biotechnol 26:1–13

Lee SW, Yang KM, Kim JK, Nam BH, Lee CM, Jeong MH, Seo SY, Kim GY, Jo WS (2012 Sep) Effects of white radish (*Raphanus sativus*) enzyme extract on hepatotoxicity. Toxicol Res 28(3): 165–172

Li GY, Gao HY, Huang J, Lu J, Gu JK, Wang JH (2014) Hepatoprotective effect of *Cichorium intybus* L., a traditional Uighur medicine, against carbon tetrachloride-induced hepatic fibrosis in rats. World J Gastroenterol 20(16):4753–4760

Liu FP, Ma X, Li MM, Li Z, Han Q, Li R, Li CW, Chang YC, Zhao CW, Lin YX (2016) Hepatoprotective effects of *Solanum nigrum* against ethanol-induced injury in primary hepatocytes and mice with analysis of glutathione S-transferase A1. J Chin Med Assoc 79(2):65–71

Madani H, Talebolhosseini M, Asgary S, Naderi GH (2008) Hepatoprotective activity of *Silybum marianum* and *Cichorium intybus* against thioacetamide in rat. Pak J Nutr 7(1):172–176

Mahboubi M, Mahboubi M (2020) Hepatoprotection by dandelion (*Taraxacum officinale*) and mechanisms. Asian Pac J Trop Biomed 10(1):1–10

Manna P, Sinha M, Sil PC (2006) Aqueous extract of *Terminalia arjuna* prevents carbon tetrachloride induced hepatic and renal disorders. BMC Complement Altern Med 6(1):1

Murugaian P, Ramamurthy V, Karmegam N (2008) Hepatoprotective activity of Eclipta alba L. against acute hepatotoxicity in rats. Res J Agri Bio Sci 4(6):685–687

Nagalekshmi R, Menon A, Chandrasekharan DK, Nair CK (2011) Hepatoprotective activity of *Andrographis paniculata* and Swertia chirayita. Food Chem Toxicol 49(12):3367–3373

Nagarkar B, Kulkarni R, Bhondave P, Kasote D, Kulkarni O, Harsulkar A, Jagtap S (2013) Comparative hepatoprotective potential of *Tinospora cordifolia, Tinospora sinensis* and neem-guduchi. J Pharm Res Int:906–916

Nagarkatti DS, Rege NN, Desai NK, Dahanukar SA (1994) Modulation of Kupffer cell activity by Tinospora cordifolia in liver damage. J Postgrad Med (40):65–67

Nalini G, Ajithadasaruna CN, Jegan N (2018) Hepatoprotective activity of hydroalcoholic extract of *Boerhaavia diffusa* linn. Against d-galactosamine induced hepatotoxicity in mice. Int J Pharm Sci Res 9(8):3367–3372

Nwaehujor CO, Ode OJ, Okoyie DN (2011) The hepatoprotective effect of *Senna occidentalis* methanol leaf extract against acetaminophen induced hepatic damage in rats. J Pharmacol Toxicol 6(7):637–646

Orhan IE, Şener B, Musharraf SG (2012) Antioxidant and hepatoprotective activity appraisal of four selected *Fumaria* species and their total phenol and flavonoid quantities. Exp Toxicol Pathol 64(3):205–209

Patel JA (2009) Hepatoprotective activity of *Piper longum* traditional milk extract on carbon tetrachloride induced liver toxicity in Wistar rats. Boletín Latinoamericano y del Caribe de Plantas Medicinales y Aromáticas 8(2):121–129

Prince ES, Parameswari P, Mahaboob KR (2011) Protective effect of *Ricinus communis* leaves extract on carbon tetrachloride induced hepatotoxicity in albino rats. Iran J Pharm Sci 7(4): 269–278

Rawat AK, Mehrotra S, Tripathi SC, Shome U (1997) Hepatoprotective activity of *Boerhaavia diffusa* L. roots – a popular Indian ethnomedicine. J Ethnopharmacol 56(1):61–66

Sattar MA, Iqbal MM, Asad BS, Ibrahim M (2015) Pharmacological evaluation of *P. amarus* seeds and L. aspera leaves for its hepatoprotective and nephroprotective activities. Res J Pharm, Biol Chem Sci 6(1):1017–1025

Shanmugasundaram P, Venkataraman S (2006) Hepatoprotective and antioxidant effects of *Hygrophila auriculata* (K. Schum) Heine Acanthaceae root extract. J Ethnopharmacol 104(1–2):124–128

Sharma UR, Prakash T, Surendra V, Roopakarki N, Rao NR, Goli D (2012) Hepatoprotective activity of *Fumaria officinalis* against carbon tetrachloride-induced liver damage in rats. Pharmacologia 3(1):9–14

Singh D, Singh R, Singh P, Gupta RS (2009) Effects of embelin on lipid peroxidation and free radical scavenging activity against liver damage in rats. Basic Clin Pharmacol Toxicol 105(4): 243–248

Singh DP, Awasthi H, Luqman S, Singh S, Mani D (2015) Hepatoprotective effect of a polyherbal extract containing *Andrographis paniculata, Tinospora cordifolia* and *Solanum nigrum* against paracetamol induced hepatotoxicity. Pharmacogn Mag 11(3):375–379

Sinha S, Bhat J, Joshi M, Sinkar V, Ghaskadbi S (2011) Hepatoprotective activity of *Picrorhiza kurroa* Royle ex. Benth extract against alcohol cytotoxicity in mouse liver slice culture. Int J Green Pharm 5(3):244–253

Sofowora A, Ogunbodede E, Onayade A (2013) The role and place of medicinal plants in the strategies for disease prevention. Afr J Tradit Complement Altern Med 10(5):210–229

Subasini U, Rajamanickam GV, Dubey GP, Prabu PC, Sahayam CS, Shabi MM, Gayathri K, Agrawal A (2007) Hydroalcoholic extract of *Terminalia arjuna* a potential hepatoprotective herb. J Biol Sci 7:255–262

Syed SN, Rizvi W, Kumar A, Khan AA, Moin S, Ahsan A (2014) In vitro antioxidant and in vivo hepatoprotective activity of leave extract of *Raphanus sativus* in rats using carbon tetrachloride model. Afr J Tradit Complement Altern Med 11(3):102–106

Tayubi IA, Desai A, Madar IH, Al Ssadh H (2018) Hepatoprotective activity of *Plumbago zeylanica* linn. against carbon tetrachloride induced hepatotoxicity in rats. Int J Sci Innov 5(2):94–98

Visen PK, Shukla B, Patnaik GK, Tripathi SC, Kulshreshtha DK, Srimal RC, Dhawan BN (1992) Hepatoprotective activity of *Ricinus communis* leaves. Int J Pharmacogn 30(4):241–250

Wagner H, Geyer B, Kiso Y, Hikino H, Rao GS (1986) Coumestans as the main active principles of the liver drugs *Eclipta alba* and *Wedelia calendulacea*. Planta Medica 52(05):370–374

Wang H, Zhang H, Wang Y, Yang L, Wang D (2019) Embelin can protect mice from thioacetamide-induced acute liver injury. Biomed Pharmacother 118(109360):1–7

Yaeesh S, Jamal Q, Khan AU, Gilani AH (2006) Studies on hepatoprotective, antispasmodic and calcium antagonist activities of the aqueous-methanol extract of *Achillea millefolium*. Phytothe Res Int J Dev Pharmacol Toxicol Eval Nat Prod Derivat 20(7):546–551

Yan T, Yan N, Wang P, Xia Y, Hao H, Wang G, Gonzalez FJ (2020) Herbal drug discovery for the treatment of nonalcoholic fatty liver disease. Acta Pharm Sin B 10(1):3–18

You Y, Yoo S, Yoon HG, Park J, Lee YH, Kim S, Oh KT, Lee J, Cho HY, Jun W (2010) *In vitro* and *in vivo* hepatoprotective effects of the aqueous extract from *Taraxacum officinale* (dandelion) root against alcohol-induced oxidative stress. Food Chem Toxicol 48(6):1632–1637

Chapter 12
Selected Medicinal Plants for Treatment of Mucormycosis, in India

Komal Lata, Priyanka, Ishan Sen, and Wamik Azmi

Abstract Fungal infections, that are becoming more common by the day, can mean a life-threatening menace to people who are immunological or medically compromised. Fungi are eukaryotic creatures that have a similarity with humans at molecular level and have many biochemical targets which are also found in eukaryotic cells. Due to this similarity, it is difficult to design drugs that target only fungal cells without affecting human cells. As a result, many antifungals have side effect. Therefore, such antifungals are required which target fungal cells but are nontoxic to human cells. To date, only a few numbers of antifungal medications, such as polyenes, azoles, echinocandins, and flucytosine, are available for the treatment of invasive fungal infections. Furthermore, toxicity, drug interaction and the emergence of fungal resistance to different fungicides limits the use of current antifungal drugs. This results in significant morbidity and mortality rates. Natural compounds with fungistatic and fungicidal activity to treat fungal infections can be found in plants. Secondary bioactive metabolites such as terpenoids, phenolics, flavonoids, saponins, and alkaloids, which have antifungal properties, are abundant in plants. Traditional medicine systems cite several medicinal plants for the treatment of both animal and human mycoses, thus offering them a promising future as a potential source for antifungal medicines.

Keywords Mucormycosis · Black fungus · Medicinal plants · Antifungals · Secondary metabolites

12.1 Introduction

Most fungi are ubiquitous and are not known to cause any invasive or life-threatening infections, however patients with a weakened immune system are at risk for fungal infection. Incidence of human fungal infections in immune

K. Lata · Priyanka · I. Sen · W. Azmi (✉)
Department of Biotechnology, Himachal Pradesh University, Shimla, Himachal Pradesh, India

Á. Máthé, I. A. Khan (eds.), *Medicinal and Aromatic Plants of India Vol. 2*,
Medicinal and Aromatic Plants of the World 9,
https://doi.org/10.1007/978-3-031-44914-7_12

compromised patients has been increased since last 20 years (Baddley et al. 2001). Mycosis is an uncommon and occasionally chronic disease caused by a fungus invading the tissues and producing systemic, superficial, or subcutaneous diseases that extend into tissues and nearby areas such as bones, organs and so on. Mucormycosis (also known as black fungus and Zygomycosis), is a serious major fungal disease caused by a group of microorganisms. It affects persons of all ages, but affects especially the immunocompromised, and is spread by inhaling and ingesting spores. It affects several body organs, including the brain, lungs, orbit, and sinuses, and can be fatal in some cases if left untreated.

Aspergillus fumigatus, Candida albicans and *Cryptococcus neoformans* are by far the most important causes of life-threatening invasive mycoses. Endemic mycoses, on the other hand, are caused by a thermophilic dimorphic heterogeneous group of fungi that occupy a specific ecological niche in the environment and can be found as a mould in the environment as well as in the human body as yeast. These are classified as primary pathogens because they are the primary cause of disease in both healthy and immunocompromised hosts. Dimorphic fungi are fungi that live in the environment as a saprophyte mould (i.e. filamentous like). They convert into pathogenic yeast after being inhaled. *Blastomyces dermatitidis, Histoplasma capsulatum, Coccidiodesimmitis, Paracoccidiodesimmitis*, and *Sporothrixschenckii* are examples of dimorphic fungi. Because of under-diagnosis and under-reporting, dimorphic fungus infection data are frequently inaccurate. Endemic mycosis is thought to cause roughly 65,000 life-threatening cases per year around the world (Goughenour and Rappleye 2017).

There has been a minimal success in the development of new antifungal drugs when compared to development of antibacterial drugs. Only a few antifungal medicines are now approved for use in the treatment of fungal infections (Borgers 1980; Barchiesi et al. 2000). There are only three types of antifungal therapy for systemic mycoses: polyenes, azoles, and echinocandins.

12.2 Antifungal Drugs

An antifungal drug is a drug that selectively eliminates fungal pathogens from a host with minimal toxicity to the host. Therapeutically allowed antifungals are categorized into four groups: (1) the polyenes, (2) the pyrimidine analogues, (3) the azoles and (4) the echinocandins (Carmona and Limper 2017). The development of antifungal agents in comparison with that of antibacterial agents is lagging behind, corely because of the cellular structure of the organisms, in addition to that fungi generally grow slowly and often in multicellular forms, they are more difficult to quantify than bacteria. This leads to complications in the experiments to check properties of a potential antifungal agent in *vitro* or *in vivo*. Despite of which, most important antifungal agents and their mode of action, drugs, side effects are summarised in Table 12.1.

Table 12.1 Classification of antifungal drugs

Antifungal drugs	Mode of action	Drugs	Side effects
Azole	Disrupt ergosterol biosynthesis pathway. Inhibit cytochrome P450 dependent C14-αdemethylase which is responsible for the conversion of lanosterol to ergosterol.	Fluconazole, Voriconazole	Hepatotoxicity and hormone-related effects, including gynecomastia, alopecia, decreased libido, oligospermia, azoospermia, impotence, hypokalemia, hyponatremia, and (rarely) adrenal insufficiency.
Echinocandins	Non-competitive inhibition of 1, 3-beta-D-glucan synthase, an enzyme which is responsible for fungal cell wall glucan synthesis. Beta-glucan destruction prevents resistance against osmotic pressure and leads to cell death.	Caspofungin, Andidulafungin	Facial flushing, swelling, rash and fever.
Polyenes	Forms complex with ergosterol and disrupts the integrity of fungal cell plasma membrane, which increase membrane permeability which leads to leakage of cytoplasmic contents into cell and causes the death of fungal cell.	Amphotericin B Nystatin	Infusion-related reactions including fever, shaking, chills, hypotension, tachypnea and renal toxicity.
Pyrimidine analogues	Inhibitor of DNA/RNA/Protein Synthesis by interfering with purine and pyrimidine metabolism. It is transferred into its active metabolite by fungal cells (Caromona et al. 2017). 5-FC act as antifungal after its conversion into 5fluorouracil.	Flucocytosine	Hepatotoxicity, bone marrow suppression (thrombocytopenia), enterocolitis, loss of appetite, diarrhoea and vomiting. Anaphylaxis and allergic reactions also occur sometimes.

12.3 Mechanism of Multidrug Resistance in Fungi

Currently, therapeutically licensed antifungal agents have several limitations. Most antifungal agents have toxicity and are immunosuppressive in nature. Many pathogenic fungal species are developing resistance against antifungal agents and multidrug resistance against available antifungal agents (Cowen et al. 2014).

There are generally four mechanisms for developing resistance against antifungal agents:

1. Decreased drug accumulation inside fungal cell.
2. Target for drugs changed by amino acid substitution.
3. Modification in metabolism to counterbalance the effects of antifungal drugs and
4. Overproduction of targeted enzyme (Kanafani and Perfect 2008).

Exposure of fungus to suboptimum level of drug is also a cause of antifungal resistance (Cowen et al. 2014).

12.4 Benefits of Using Medicinal Plant Drugs for Fungal Treatments

Medicinal plants are known to contain chemicals that can be utilised for therapeutic reasons or are precursors for the manufacture of valuable pharmaceuticals. They are frequently used to make traditional medicines with high therapeutic potential. Plant extracts of leaves, stems, flowers, roots, and essential oils, as well as their essential oils, provide unlimited opportunities for new drug leads because of their normally matchless chemical diversity. Many bioactive secondary metabolites found in plants, such as saponins, alkaloids, and terpenoids, have antifungal properties. These plants could be a promising source of antifungal medications in the future (Table 12.1). The growth of multidrug-resistant fungus strains, as well as the limited number of medications available, mandates the development of new antifungal classes from natural resources, such as medicinal plants. Traditional systems of medicine have also reported medicinal herbs for the treatment of both human and animal mycoses, and they are thought to be a valuable source for the discovery of new antifungal medicines (Mishra et al. 2020). The optimal amount of active ingredients must reach the target tissues for any herbal medication to display expected therapeutic activity. Herbal medications are easily destroyed by first-pass metabolism and the pH differential in the gastrointestinal tract. Various innovative drug delivery technologies, such as nanoparticles, nano-emulsions, phytosomes, transferosomes, and liposomes, bypass acidic pH and first-pass metabolism to deliver the maximum number of pharmaceuticals to target tissues. Nano carriers are smaller in size and have a faster beginning of action (Indalkar et al. 2015).

Herbal medications present themselves as a viable candidate for drug delivery employing an unique drug delivery system for the following reasons:

- Natural substances do not have the same negative effects as pharmaceutical medications.
- When multifunctional molecules are present in natural products, they have a synergistic impact.

- Natural substances have a long history of support for their action and safety, whereas contemporary drugs are more dangerous, even if they have been scientifically demonstrated to work (Sharma et al. 2011). The World Health Organization lists 21,000 plants that are widely utilised for medical purposes around the world. About 2500 species have been found in India, with 150 of them being commercially employed on a big scale by biopharmaceutical companies as mainstream medicine. India is the world's largest producer of medicinal plants and is known as the world's botanical garden (Yuan et al. 2016). Herbal medicine supports the body's natural healing process and aids in the elimination of infections, whereas synthetic medications target specific disorders. A symptom relief strategy comes under medicinal plant therapeutic strategies.

The Indian flora is reputed for its medicinal plants used in Traditional Medicine. Based on the special literature Table 12.2 contains a list of species that have been reported and/or have a potential to treat fungal infections (mainly Mucormycosis).

12.5 Five Most Frequently Used Species for Treatment of Mucormycosis, in India

The use of synthetic drugs as antifungal has shown toxic effects and its cost is high, as it is a therapy lasting a long time period. Further, it has been observed that due to its prolonged use resistant strains have developed (Sepahvand et al. 2017). Thus, the exploration of MAP's for their potential antifungal properties, especially against Mucorales, one of the most prevalent fungal infections in India. These plants due to the diversity of their phytoconstituents are a rich source of antimicrobial agents with the potential to prevent or treat various fungal infections. In the followings the five most effective Indian Medicinal plants will be reviewed for their anti-mucormycosis activities.

12.5.1 Ocimum Species

Species of the *Ocimum* genus, in the family Labiate, are known to have many therapeutical uses. *Ocimum sanctum* is commonly known as Tulsi or Holy Basil and is worshipped by the people of hindu religion. In traditional medicine, most frequently the leaves and seeds are used (Fig. 12.1).

The essential oil from *Ocimum sanctum* is shown to have significant antifungal properties. Eugenol has been reported to have the highest fungicidal properties (Zhao et al. 2021). According to Zabka et al. (2021) monoterpenoid alcohols like linalool make up about 90% of the essential oil from the *Ocimum sanctum* and another 10% is composed of estragole-methyl-cinnamate and eugenol. Other components includeeucalyptol, eugenol, *c*-methyl cinnamate, ocimene, terpinen-4-ol, bornyl acetate and camphor.

Table 12.2 Indian medicinal plants and herbs used for the treatment of fungal infection, mainly Mucormycosis

Botanical name of plant	Common name	Family	Characteristics of medicinal plants
Ocimum sanctum	Holy basil, tulsi plant	Laminaceae	Aromatic plant rich in vitamin C and contains oleanolic acid, rosamarinic acid, eugenol, carvacrol, linalool and beta-caryophyllene. Have antifungal and antibacterial properties (Balakumar et al. 2011).
Azadirachta indica	Neem, nimtree or Indian lilac	Meliaceae	Best known for antiaging properties and leaves is used for fungal infection. Used to treat foot fungi (Leonardelli et al. 2016).
Capparis spinosa	Caper brush, flinder rose	Capparaceae	Best known plant against fungal infections and *in vitro* against *Aspergillusniger*, *Penicillium* spp., and *Trichoderma viridiae* (Ali-Shtayeh and Abu Ghdeib 1999).
Anagallis arvensis	Scarlet pimpernel, red pimpernel, poor man barometer, poor man weather glass	Primulaceae	Used to treat fungal infections like *Candidaalbicans* (Taye et al. 2011).
Juglans regia	Persian walnut, madiera walnut, English walnut	Juglandaceae	Contains four extract fractions i.e., methanolic, ethyl acetate, alkaloid and hydrolysed methanolic that are derived from *Juglans regia* leaves for fungal infections like *Candida albicans* (Ali-Shtayeh and Abu Ghdeib 1999).
Inula viscosa	*Ditrrichiaviscosa*, false yellow head, woody fleabane, sticky fleabane	Asteraceae	Effective against dermatophytes, *aspergillus* spp. Contain tannin (TCT), phenol (TPC), flavonoid (TFC) and caffeoylquinic acid (CQC) derived from *Ditrrichia viscosa* were assessed for anti-candida activities (Ali-Shtayeh and Abu Ghdeib 1999).

(continued)

Table 12.2 (continued)

Botanical name of plant	Common name	Family	Characteristics of medicinal plants
Artemisia judaica	Mugwort, wormwood and sage brush	Asteraceae	Hardy herbaceous plants which are known for powerful constituents in their essential oils. Utilized for the treatment of fungal infections of tinea, tympanitis and thrush.
Mentha spicata	Spearm mint, garden mint, common mint	Lamiaceae	Aromatic herb of the mint family. Carvone is the major component of the essential oils and show effectiveness against *Cryptococcus neoformans*, the dermatophytes, *Trichophyton rubrum*, *Trichophyton verrucosum* (Mugnaini et al. 2012).
Heracleum persicum	Persian hogweed	Apiaceae	Perennial flowering plant, their fruits and essential oils is used as herbal medicine for assessing anti-Candida activity (Chowdhary et al. 2014).
Nigella sativa	Black cumin, kalonji, fennel flower, roman coriander	Ranuculaceae	Used as a medicinal plant: their essential oil is effective against *Trichophyton*, *Microsporum canis*, *Microsporum gypsum* and dermatophytes (Liu et al. 2016).
Rosamarinus officinalis	Saliviarosamarius and rosemary	Lamiaceae	Contains main constituents of essential oils like linalool, gamma-terpinene, thymol, beta-pinene, p-cymeme and eucalyptol are effective against *Alternaria alternate*, *Botrytis cinerea* and *Fusarium oxysporum* (Mugnaini et al. 2012).
Salvia fruticosa	Garden sage, culinary sage	Lamiaceae	Attractive and aromatic culinaryherb. Oil extracts includes cineole, borneol, ledene, beta-pinene, alpha-humulene and trans-caryophyllene are the natural alternative against antifungal diseases like *Candida albicans*, *Candida glabrata* (Adam et al. 1998).

(continued)

Table 12.2 (continued)

Botanical name of plant	Common name	Family	Characteristics of medicinal plants
Foeniculumvulgarae	Common fennel	Apiaceae	Typically grown in vegetable and herb gardens. Essential oil contains estragole, limonene, fenchone, alpha-pinene are effective against *Candida albicans*, *Aspergillus spp.* (*A. flavus*, *A. niger*), *Rhizopus* spp., *Trichophyton rubrum*, *Trichophyton mentagrophytes* and *Rhodotorula* spp. (Patra et al. 2002).

Fig. 12.1 *Ocimum sanctum*

In another study conducted by Khan et al. (2010) *Ocimum sanctum* was shown to have 44% chavicol and 21% linalool. They suggest that the essential oils of *Ocimum sanctum* causes the development of lesions on the plasma membrane of the fungi and cause a reduction in the ergosterol content resulting in their fungicidal property.

12.5.2 *Azadirachta indica* A. Juss.

This species of the family Meliaceae has been used in Indian ayurvedic practice. It contains a significant amount of antioxidants and has been known to the Chinese and

Fig. 12.2 *Azadirachta indica* A. Juss.

Unani medicine for many years (Alzohairy 2016). The plant is reported to have immunomodulatory, anti-inflammatory, anti-hyperglycaemic, anti-malarial, antifungal, antibacterial, antioxidant, anti-mutagenic and anti-carcinogenic properties (Subapriya and Nagini 2005). Common names for the plant are Neem, Nim tree, Nimb, Margosa tree, Nimbh and Indian lilac (Fig. 12.2).

The plant parts used in traditional medicine are the leaves, seeds, roots and the bark of the tree. Polyphenolic flavonoids (e.g. Quercitin and beta sitosterol) isolated from neem extracts have been shown to have significant antifungal properties (Govindachari et al. 1998). Further bioactive compounds found in the neem tree include Azadirachtin, Meliacin, nimbin, nimbolide and solanine (Hashmat et al. 2012).

12.5.3 Nigella sativa

The species belongs to the family Ranunculaceae. *Nigella sativa* is a shrub which is about 20–90 cm tall. It is native to southeast Asia, southern Europe and north Africa. The plant has also been cultivated in India for many centuries. The plant is known to survive even in the harshest of environments with short life span (Agradi et al. 2002). The seeds of the ripe fruit are dark in color and have been used in ayurvedic, unani, Arabic and Chinese medicine since ancient times (Islam et al. 2017). Crude extracts and essential oils from this plant have been shown to have several properties like antioxidants, anti-tumor, antibacterial, anti-parasitic, anti-inflammatory and anti fungal properties (Shokri 2016). Other common names for the plant *Nigella sativa* include black seed, black cumin, Roman coriander, kalonji and fennel flower (Fig. 12.3).

Fig. 12.3 *Nigella sativa* L.

Several important phytochemical components of the essential oil obtained from the *Nigella sativa* include thymoquinone, p-cymene, trans–anethole, terpinene, camphor, linalool, estragole etc. Thymol, thymoquinone and thymo-hydroquinone are shown to have an antifungal effect on various filamentous fungi (Shokri 2016). Against the dermatophytic fungi thymoquinone was shown to have a strong antifungal effect (Ali and Blunden 2003).

Shokri (2016) in their article mentioned that on treatment with the essential oil from *Nigella sativa* the fungal hyphae and membrane bound organelles like nuclei and mitochondrion were degraded which resulted in the fungicidal activity of the plant. The antifungal property of the seed oil from *Nigella satvia* has been recorded and confirmed in various researches which make it a fine candidate for the treatment of various fungal infections.

12.5.4 Mentha spicata L.

Mentha spicata also known as spearmint is a member of Lamiaceae family and is one of the most commonly grown herb in the world. The plant is grown due to its delightful aroma and high demand of its essential oils. The extracts of this plant are shown to have remedial effects on various human diseases and ailments (Ali-Shtayeh et al. 2019). The plant is a perennial herb and is grown worldwide. Mentha is used to treat bad breath, gastrointestinal, and some respiratory problems. The aerial parts of the plants have been used in teas from many years and are also useful in treating sore throat and some skin related problems. Other names for the plant include names like spearmint, English mint, pudina, common mint, garden mint and lamb mint (Mahendran et al. 2021).

The parts of the plant used to produce essential oils are the aerial parts. The essential oils of *Mentha spicata* contain terpenes and terpenoids. A study by Mahendran et al. (2021) reported 35 phytoconstituents obtained from the various parts of *Mentha spicata.* These include flavonoids, phenolic acids and lignans. The main components of the essential oils from *Mentha spicata* include carvone, carvacrol, trans-carveol, piperitone oxide, limonene, 1,8-cinéole, camphene, p-cymene, dihydrocarvone, pulegone, β-caryophyllene, germacrene D, menthone, α-pinene, and linalool (El Menyiy et al. 2022).

Several studies have been conducted to reveal the antifungal properties of essential oils obtained from *Mentha spicata*. The good antifungal property of this species is supposedly due to the presence of carvone which is known to have a very strong anti-fungal property. This is due to the high solubility of carvone in water (Soković et al. 2009). It can be said that the essential oils from *Mentha spicata* can be used to produce better and nontoxic antifungal compounds and can be used to replace the synthetic and harmful fungicides produced in the market.

12.5.5 Inula viscosa (L.) Aiton

The genus *Inula* belongs to the family Astereaceae and has about 100 species. The species *Inula viscosa* has been grown since ancient times, in Africa, Asia and Europe. It is known for its antioxidant, antiviral, anti-inflammatory, anti-tumor and anti-fungal activities. It is a well known in ancient folk medicine in the treatment for skin inflammation and scabies among other ailments (Mahmoudi et al. 2016).

Inula viscosa is a perennial herb which is native to the Mediterranean region and grows in hillsides, moist areas and on the sides of roads. It has sticky leaves with bright yellow flowers and blooming period is in August and November (Danino et al. 2009). Common names of the plant are false yellowhead, woody fleabane, sticky fleabane and yellow fleabane (Fig. 12.4).

The most commonly used parts of the plant are aerial parts and the roots. Phytochemical assay of essential oil of *Inula viscosa* showed that the two main compound found in the essential oil were Eudesma-3,11(13)-dien-12-oic acid (ESA) and of 1,2-Benzenedicarboxylic acid, di-isooctyl ester (BAE). The essential oil is reported to have shown remarkable anti fungal properties (Haoui et al. 2016).

Several phenolic and flavonoid compound were also found in the extracts of *Viscosa*p-coumaric acid, chromogenic acid, luteoin etc. Fungicidal activity of the plant extract were confirmed by Qasem et al. (1995). It can be said that the plant is a good and natural source of antifungal compounds and can be used to produce better antifungal remedies and medicines.

12.6 Conclusions

The use of synthetic drugs, as antifungal agents, has frequently shown toxic effects and high costs. Further, it has been observed that due to their prolonged use resistant strains have developed. Thus, the exploitation of MAP's for their potential

Fig. 12.4 *Inula viscosa* (L.) Aiton

antifungal properties seems promising, especially against Mucorales, one of the most prevalent fungal infections in India. These species, due to the diversity of their phytoconstituents, are a rich source of antimicrobial agents with the potential to prevent or treat various fungal infections. Some of the most effective Indian medicinal plants known for their anti-mucormycosis activities are: *Ocimum sanctum*, *Azadirachta indica*, *Capparis spinosa*, *Anagallis arvensis*, *Juglans regia*, *Inula viscosa*, *Artemisia judaica*, *Mentha spicata*, *Heracleum persicum*, *Nigella sativa*, *Rosmarinus officinalis*, *Salvia fruticosa* and *Foeniculum vulgare*. Despite the increasing number of reports on the use of antifungal therapy, the number of cases of infection and resistance against antifungals is still high. Protection against fungal diseases is far from being accomplished.

It is necessary to improve the efficiency of already existing molecules. New formulations and alternative therapies are needed for the prevention and treatment of fungal infections to improve life quality.

Research on the mechanisms of fungal pathogenesis – particularly during pathogenic period -may contribute to the progress of novel anti virulence. It may also serve as the basis for the substituent approaches for the modification of fungal infections caused by opportunistic fungi.

In the interim period to a successful breakthrough in research, medicinal plants remain a useful and efficient alternative in the fight against fungal infections.

References

Adam K, Sivropoulou A, Kokkini S, Lanaras T, Arsenakis M (1998) Antifungal activities of *Origanum vulgare* subsp. hirtum, *Mentha spicata*, *Lavandula angustifolia*, and *Salvia fruticosa* essential oils against human pathogenic fungi. J Agric Food Chem 46(5):1739–1745

Agradi E, Fico G, Cillo F, Francisci C, Tome F (2002) Estrogenic activity of *Nigella damascena* extracts, evaluated using a recombinant yeast screen. Phytother Res 16(5):414–416

Ali-Shtayeh M, Abu Ghdeib S (1999) Antifungal activity of plant extracts against dermatophytes. Mycoses 42(11–12):665–672

Ali BH, Blunden G (2003) Pharmacological and toxicological properties of *Nigella sativa*. Phytother Res 17(4):299–305

Ali-Shtayeh MS, Jamous RM, Abu-Zaitoun SY, Khasati AI, Kalbouneh SR (2019) Biological properties and bioactive components of *Mentha spicata* L. essential oil: Focus on potential benefits in the treatment of obesity, Alzheimer's disease, dermatophytosis, and drug-resistant infections. Evid Based Complement Alternat Med 3834265. https://doi.org/10.1155/2019/3834265

Alzohairy MA (2016) Therapeutics role of *Azadirachta indica* (Neem) and their active constituents in diseases prevention and treatment. Evid Based Complement Alternat Med 7382506. https://doi.org/10.1155/2016/7382506

Baddley JW, Stroud TP, Salzman D, Pappas PG (2001) Invasive mold infections in allogeneic bone marrow transplant recipients. Clin Infect Dis 32:1319–1324

Balakumar S, Rajan S, Thirunalasundari T, Jeeva S (2011) Antifungal activity of Ocimum sanctum Linn. (Lamiaceae) on clinically isolated dermatophytic fungi. Asian Pac J Trop Med 4(8): 654–657

Barchiesi F, Schimizzi A, Caselli F, Novelli A, Fallani S, Giannini D, Arzeni D, Di Cesare S, Di Francesco L, Fortuna M, Giacometti A, Carle F, Mazzei T, Scalise G (2000) Interactions between Triazoles and amphotericin B against *Cryptococcus neoformans*. Antimicrob Agents Chemother 44(9):2435–2441

Borgers M (1980) Mechanism of action of antifungal drugs, with special reference to the imidazole derivatives. Clin Infect Dis 2(4):520–534

Carmona E, d Limper A (2017) Overview of treatment approaches for fungal infections. Clin Chest Med 38(3):393–402

Chowdhary A, Sharma C, Hagen F, Meis J (2014) Exploring azole antifungal drug resistance in *Aspergillus fumigates* with special reference to resistance mechanisms. Future Microbiol 9(5): 697–711

Cowen LE, Sanglard D, Howard SJ, Rogers PD, Perlin DS (2014) Mechanisms of antifungal drug resistance. Cold Spring Harb Perspect Med 5:a019752

Danino O, Gottlieb HE, Grossman S, Bergman M (2009) Antioxidant activity of 1, 3-dicaffeoylquinic acid isolated from *Inula viscosa*. Food Res Int 42(9):1273–1280

El Menyiy N, Mrabti HN, El Omari N, Bakili AE, Bakrim S, Mekkaoui M, Balahbib A, Amiri-Ardekani E, Ullah R, Alqahtani AS, Shahat AA, Bouyahya A (2022) Medicinal Uses, Phytochemistry, Pharmacology, and Toxicology of Mentha spicata. Evid Based Complement Alternat Med 7990508. https://doi.org/10.1155/2022/7990508

Goughenour KD, Rappleye CA (2017) Antifungal therapeutics for dimorphic fungal pathogens. Virulence 8:211–221. https://doi.org/10.1080/21505594.2016.1235653

Govindachari TR, Suresh G, Gopalakrishnan G, Banumathy B, Masilamani S (1998) Identification of antifungal compounds from the seed oil of *Azadirachta Indica*. Phytoparasitica 26(2): 109–116

Haoui IE, Derriche R, Madani L, Oukali Z (2016) Extraction of essential oil from *Inula viscosa* (L.) leaves: composition, antifungal activity and kinetic data. J Essent Oil Bear Pl 19(1):108–118

Hashmat I, Azad H, Ahmed A (2012) Neem (Azadirachta indica A. Juss)-A nature's drugstore: an overview. Int Res J Biol Sci 1(6):76–79

Indalkar YR, Pimpodkar NV, Godase AS, Gaikwad PS (2015) A compressive review on the study of nanotechnology for herbal drugs. Asian J Pharma Res 5(4):203–207

Islam MT, Guha B, Hosen S, Riaz TA, Shahadat S, Sousa LD, Santos JV, Junior JJ, Lima RM, Braga AL, Reis AC (2017) Nigellalogy: a review on *Nigella sativa*. MOJ Bioequiv Availab 3(6):00056

Kanafani ZA, Perfect JR (2008) Resistance to antifungal agents: mechanisms and clinical impact. Clin Infect Dis 46:120–128

Khan A, Ahmad A, Manzoor N, Khan LA (2010) Antifungal activities of *Ocimum sanctum* essential oil and its lead molecules. Nat prod commun 5(2):345–349

Leonardelli F, Macedo D, Dudiuk C, Cabeza M, Gamarra S, Garcia-Effron G (2016) *Aspergillus fumigatus* intrinsic fluconazole resistance is due to the naturally occurring T301I substitution in Cyp51Ap. Antimicrob Agents Chemother 60(9):5420–5426

Liu N, Wang C, Su H, Zhang W, Sheng C (2016) Strategies in the discovery of novel antifungal scaffolds. Future Med Chem 8(12):1435–1454

Mahendran G, Verma SK, Rahman LU (2021). The traditional uses, phytochemistry and pharmacology of spearmint (Mentha spicata L.): A review. J Ethnopharmacol 278:114266

Mahmoudi H, Hosni K, Zaouali W, Amri I, Zargouni H, Hamida NB, Ouerghi Z (2016) Comprehensive phytochemical analysis, antioxidant and antifungal activities of *Inula viscosa* Aiton leaves. J Food Saf 36(1):77–88

Mishra KK, Kaur CD, Sahu AK, Panik R, Kashyap P, Mishra SP, Dutta S (2020) Medicinal plants having antifungal properties. In: Medicinal plants – use in prevention and treatment of diseases. IntechOpen

Mugnaini L, Nardoni S, Pinto L, Pistelli L, Leonardi M, Pisseri F, Mancianti F (2012) *In vitro* and *in vivo* antifungal activity of some essential oils against feline isolates of *Microsporum canis*. J de Mycol Méd 22(2):179–184

Patra M, Shahi S, Midgely G, Dikshit A (2002) Utilization of essential oil as natural antifungal against nail-infective fungi. Flavour Fragr J 17(2):91–94

Qasem JR, Al-Abed AS, Abu-Blan HA (1995) Antifungal activity of clammy inula (*Inula viscosa*) on *Helminthosporium sativum* and *Fusarium oxysporum* f. sp. lycopersici. Phytopathol Mediterr 34:7–14

Sepahvand A, Ezatpour B, Tarkhan F, Mahmoud Bahmani AK, Razi MRK (2017) Phytotherapy in fungi and fungal disease: a review of effective medicinal plants on important fungal strains and diseases. Int J Pharm Sci Res 8(11):4473–4495

Sharma AT, Mitkare SS, Moon RS (2011) Multicomponent herbal therapy: a review. Int J Pharm Sci Rev Res 6:185–197

Shokri H (2016) A review on the inhibitory potential of *Nigella sativa* against pathogenic and toxigenic fungi. Avicenna J Phytomed 6(1):21–33

Soković MD, Vukojević J, Marin PD, Brkić DD, Vajs V, Van Griensven LJ (2009) Chemical composition of essential oils of *Thymus* and *Mentha* species and their antifungal activities. Molecules 14(1):238–249

Subapriya R, Nagini S (2005) Medicinal properties of neem leaves: a review. Curr Med Chem Anticancer Agents 5(2):149–156

Taye B, Giday M, Animut A, Seid J (2011) Antibacterial activities of selected medicinal plants in traditional treatment of human wounds in Ethiopia. Asian Pac J Trop Biomed 1(5):370–375

Yuan H, Ma Q, Ye L, Piao G (2016) The traditional medicine and modern medicine from natural products. Molecules 21(5):559

Zhao Y, Wang Q, Wu X, Jiang M, Jin H, Tao K, Hou T (2021) Unraveling the polypharmacology of a natural antifungal product, eugenol, against *Rhizoctonia solani*. Pest Manag Sci 77(7): 3469–3483

Chapter 13
Selected Indian Medicinal and Aromatic Plants for Prophylactic Therapy

Nupur Mehrotra, Sara Anees Khan, and Kaustubh Jadhav

Abstract Ancient Indian medical practioners had extensive knowledge of medicinal properties of indigenous plants and used them to cure ailments Many of these MAPs are ingrained in our socio-cultural practices and over 700 species are being utilized for preparation of remedies. These herbal medicines are prepared from dried plant components such as barks, stems, leaves, flowers, fruits, seeds, and roots, while at times the entire plant is used. Over the years, the biologically active phytoconstituents, which are mainly secondary metabolites, have been isolated and characterized, and considered as lead molecules with potential for drug development. Several drugs have been found to facilitate protection against bacteria, viruses as well as parasites and boost immunity in many disorders, diseases, in both animals and human models.Their therapeutic efficacy and potency is variable. This work presents an insight into traditional Indian medicinal aromatic plants which are used for their healing properties and maintenance of good health.

Keywords Indian medicinal aromatic plants · Phytoconstituents · Disease defending

Abbreviations

ACE	Angiotensin Converting Enzyme
ADMET	Absorption, distribution, metabolism, excretion, and toxicity
ALT	Alanine transaminase
CAD	Coronary artery disease
CAGR	Compound Annual Growth Rate
CAT	Catalase

N. Mehrotra (✉) · S. A. Khan · K. Jadhav
Department of Biochemistry, SVKM's Mithibai College of Arts, Chauhan Institute of Science & Amrutben Jivanlal College of Commerce and Economics (Autonomous), Mumbai, Maharashtra, India
e-mail: nupur.mehrotra@mithibai.ac.in

Á. Máthé, I. A. Khan (eds.), *Medicinal and Aromatic Plants of India Vol. 2*, Medicinal and Aromatic Plants of the World 9,
https://doi.org/10.1007/978-3-031-44914-7_13

CYP	Cytochrome
CYP	Cytochrome P
DPPH	2,2-diphenyl-1-picryl-hydrazyl-hydrate
FGF	Fibroblast Growth Factor
GPx	Glutathione peroxidase
GSH	Glutathione
HbA1c	Glycosylated hemoglobin
HDL	High density lipoprotein
HIV	Human immunodeficiency virus
HSV-1	Herpes simplex virus type 1
IDDM	Insulin Dependent Diabetes Mellitis
IFN	Interferon
IL	Interleukin
LDL	Low density lipoprotein
LPS	Lipopolysaccharide
MAPK	Mitogen Activated Protein Kinase
MAPs	Medicinal and Aromatic Plants
MDA	Malondialdehyde
MMP-1	Matrix Metalloproteinase-1
ROS	Reactive oxygen species
SOD	Superoxide dismutase
TGF	Transforming Growth Factor
TGF	Transforming growth factor
TNF	Tumor necrosis factor
VLDL	Very low-density lipoprotein

13.1 Introduction

Ancient history across the globe, is a testament to the use of plants for medicinal advantage, documented in civilizations in Egypt, India, China, Rome, Thailand, etc. The earliest records discovered carved on clay plates date back to 2600 BC in ancient Mesopotamia, mentioning the use of close to 250 plants as herbal recipes for therapy of common ailments (Petrovska 2012). The first documented records on herbalism are the Eber Papyrus, written by Egyptians around 1500 BC listing about 850 medicines (Abou El-Soud 2010). Indian system of medicine Ayurveda is at least 5000-year-old and the texts like Charaka Samhita, Sushruta Samhita, Rig and Atharva Veda find mention of over 300 herbs used as medicines (Mehrotra 2021).

Amongst the traditional Indian non-conventional medicine systems, it has been suggested that 1200 species are used in Ayurveda, 900 in Siddha and around 700 in Unani sciences. 70% of these species are distributed in the tropical forests on the Western Ghats. Amongst the leading producers of herbal plants in India are the states of Kashmir, Uttarakhand, Gujarat, Rajasthan, Haryana, Tamil Nadu and Andhra

Pradesh and amongst the plants used for their medicinal properties 90% are collected from Indian forests. The country is rated amongst the top 12 mega diverse countries in the world and this diversity is spread over 10 biogeographic regions (Máthé and Khan 2021).

An array of medicinal forms is derived from MAPs in various systems of medicine. Medicinal plants, whose one or more parts (e.g.: bark, roots, leaves flower or fruits) have shown potential for therapeutic purposes, have been also frequently observed to posess prophylactic modality for specific diseases/disorders or used as precursor molecules for prophylactic drugs.

Plants, as sources of traditional medicines, also frequently offer an alternative to allopathic medicine. Medicinal plants are increasingly used in the treatment of several pathologic conditions. Additionally, many drugs currently used in allopathic medicine have their origin in medicinal plants.Medicinal and Aromatic Plants are, therefore a current topic of interest in medical research.

According to the 2019 WHO report, almost 25% of the prescribed medicines worldwide comprise of either a plant or a phytoconstituent. Medicinal and aromatic plants are the 'spine' of traditional medicine. Importantly, a staggering number of 3.3 billion people, mostly in the developing and less developed countries consume medicinal plants for therapy. Many of these MAPs are also employed as prophylactica or supplements to maintain good health and nutrition.

One of the main reasons leading to the extensive use of these plants is that they are attributed better adaptability with the human physiological system. In addition, they are associated with safety and fewer side effects in use.

Plant extracts and vegetal compounds used in monotherapy are important sources of therapeutic and prophylactic agents. They are also used successfully either alone or in combination with chemotherapeutic agents. Nowadays, oral products with vegetal compounds included are an important part of modern therapy (Milutinovici et al. 2021).

Medicinal plants are a topic of interest for current research in the field of medicine, being increasingly used for the treatment of a large number of pathologies. Additionally, many drugs currently used in allopathic medicine have their origin in medicinal plants. So, plants are both sources of traditional medicines and an alternative to them (Taheri et al. 2011).

Alternative therapy, including the use MAPs with prophylactic properties can play an important role in times when due to global ailments, like the COVID-19 pandemic, affect the world economy and health sector and there is no existing proven treatment for them (Parida et al. 2020).

13.2 Selected Prophylactic Plants of India

The diversity of MAPs found and used in India is immense. In the current chapter, we present selected Indian MAPs regarding their prophylactic use. The use of these plants holds relevance also for our times, as they have been already cited since time immemorial for prevention as well as therapy of common ailments.

13.2.1 Aloe vera *(L.) Burm. f*

Botanical Description
Aloe vera (L.). Burm. f., is an herb belonging to Family Liliaceae (Milutinovici et al. 2021), Xanthorrhoeaceae (ITIS 2022). Aloe is an Arabic word for 'Allaeh' that stands for 'shining bitter substances' and 'Vera' is a Latin word for 'true'. native to tropical regions. In India, it is predominantly found in the western and southern-western belt. *Aloe vera* is a perennial, shrubby and succulent plant with triangular shaped green colored leaves (Farha and Kumar 2021). Pharmacological and cosmetic applications are attributed to the sticky mucilaginous gel present in the leaves.

Major Bioactive Constituents
Many *in vitro* experiments have demonstrated the protective effect of *Aloe vera* and its phytoconstituents. According to Teplicki et al. (2018) and De Oliveira et al. (2018), the major constituents of *Aloe barbadensis* such as aloin, aloesin and emodin exhibited their defensive effects largely through their anti-inflammatory and antioxidant mechanisms. The active phytoconstituents (depicted in Fig. 13.1) are vitamins, sterols, polysaccharides, anthraquinones phytosterols, pyrones, flavonoids, proteins, tannins, carbohydrates, chromones, coumarins, alkaloids, organic compounds, mineral constituents and anthrones (Nalimu et al. 2021). Amongst the multiple phtoconstituents the structure of the major contributor- aloe-emodin, a lignin and is depicted in Fig. 13.2.

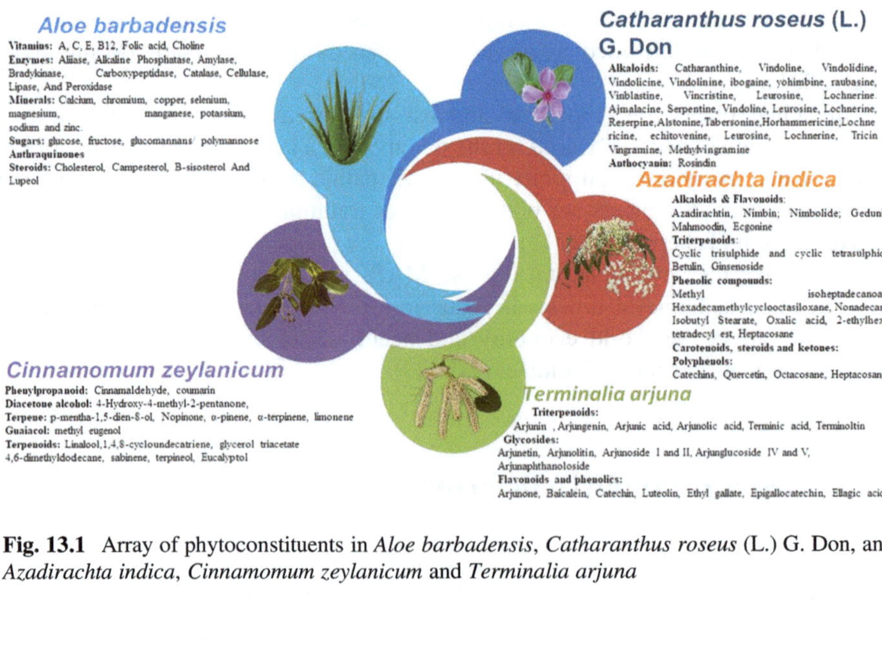

Fig. 13.1 Array of phytoconstituents in *Aloe barbadensis*, *Catharanthus roseus* (L.) G. Don, and *Azadirachta indica*, *Cinnamomum zeylanicum* and *Terminalia arjuna*

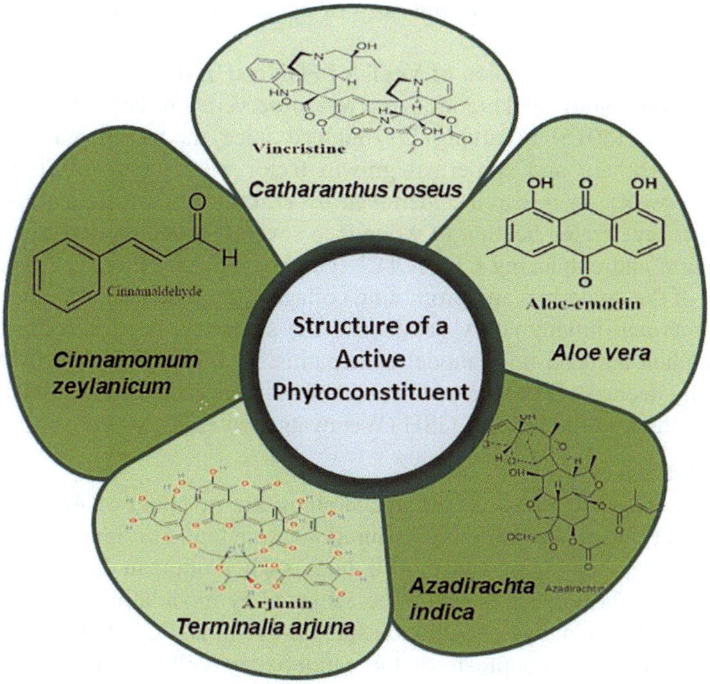

Fig. 13.2 Structures of the main phytoconstituents from the MAPs described

Medicinal Uses

Traditionally, this medicinal plant has been employed to treat skin problems (burns, wounds, and anti-inflammatory processes). *Aloe vera* gel is used extensively in several cosmetic and pharmacological preparations. Not only this, *Aloe vera* possesses a myriad of therapeutic properties like antihyperlipidemic, antioxidant, anticancer and antidiabetic to name a few (Sánchez et al. 2020). Periodontitis is a dental condition that involves inflammation of gums leading to bone, dental plaque and tissue deterioration. The early stage of periodontitis is gingivitis and clinical research findings of Kumar et al. (2014), suggest that the use of *Aloe vera* gel and mouthwash helped in the management and therapy of gingivitis as well as periodontitis by significantly decreasing plaque and gingival index.

Aloe vera syrup in a dosage of 10 mL/day for 4 weeks has been found to be an effective remedy in the treatment and management of gastroesophageal reflux disease. It also helped reduce the associated clinical manifestations like food regurgitation, flatulence, heartburn, belching, dysphagia and nausea (Panahi et al. 2015). In an *in vivo* study on a Balb/c mouse model with alcohol-induced acute gastritis, *Aloe vera* gel was found to protect against gastritis by promoting the matrix metalloproteinase-9 inhibitory activity (Park et al. 2017). Also, in a randomized, double-blind, placebo-controlled study by Størsrud et al. (2015), the extract of *Aloe vera* -AVH200®-decreased the intensity of symptoms in patients suffering from irritable bowel syndrome.

In fibroblasts, the expression of Vegf-A bFGF and TGFβ1 was up-regulated and enhanced proliferation of keratinocytes was observed. Furthermore, in the study of Saito et al. (2016) involving UV-induced mice model, increased levels of hyaluronan synthase and epidermal growth factor were found on application of *Aloe vera* powder. In research undertaken by Prakoso (2018), topical use of *Aloe vera* extract accelerated healing of wounds in several animal models by decreasing inflammation and enhancing CD^{4+}/CD^{8+} lymphocyte ratio, thereby increasing the thickness of the epidermis and promoting collagen accumulation. *Aloe barbadensis* also exhibits anti-inflammatory action and the same has been investigated in an acetaminophen-induced mice model of hepatitis. A 150 mg/kg dose of *Aloe vera* significantly decreased the levels of hepatic biomarkers such as IL-12, MDA, IL-18, ALT and enhanced the level of GSH (Werawatganon et al. 2014).

Current Advances

In a study performed to assess the anticarcinogenic potential of *Aloe vera*, crude extracts in varying concentrations and durations were found to significantly decrease the replicative potential of cancer cell lines such as human breast MCF-7 and cervical HeLa. The same was said to be achieved by modulating the expression of genes that mediate cell cycle regulation such as cyclin D1, CYP1A1 and CYP1A2. The extract promoted apoptosis of the carcinogenic cells by inducing apoptosis involving chromatin condensation and fragmentation (Trybus et al. 2018). Many *in vitro* and *in vivo* studies have explored the anticancer potential, especially in colon cancer. The active constituent aloe-emodin exhibited cyto-toxic effects at varying concentration between 10 and 40μM by activating cellular free radical production (Chen et al. 2018).

Researchers have found that *Aloe vera* reduced blood glucose levels in streptozotocin induced diabetic rats. Also, the functionality of pancreatic islet cells was found to be improved (Kim et al. 2018). The pathophysiology of diabetes involves generation of free radicals. However, *Aloe vera* extract was found to protect against oxidative stress and thus ameliorate the complications associated with diabetes, such as nephropathy and retinopathy to name a few. As per the *in vitro* study of Kim et al. (2018), the polysaccharides present in *Aloe vera* were found to elicit an anti-diabetic effect by decreasing apoptosis and the stress signaling associated with endoplasmic reticulum. *Aloe vera*, aloe-emodin, at a concentration of 20μM was found to protect RIN-5F (rat pancreatic β-cells) cells against glucotoxicity.

Aloe vera is a powerhouse of anti-oxidants. The antioxidant ability of crude methanolic extracts was examined using different *in vitro* methods from six different climatic zones of India (Kumar et al. 2017). It was observed that the antioxidant activity was higher in the species collected from North India as compared to South India. This was largely attributed to the higher content of flavonoids, phenolic compounds, alkaloids and saponin glycosides. As per the study of Ji et al. (2018), aloe-emodin when administered for 42 days in a dose of 50 and 100 mg/kg in male

Wistar rats, was found to reduce elevated lipid levels, total cholesterol and LDL. Aloe-emodin is also credited with possessing antibacterial activity against *Staphylococcus aureus*. Antibacterial activity is mediated by preventing biofilm formation and the synthesis of extracellular proteins that are required for bacterial growth and reproduction (Xiang et al. 2017). Jain et al. (2016) studied the antiviral activity in influenza virus subtype H1NI and herpes simplex virus type 1 (HSV 1). *Aloe vera* gel in concentrations of 0.2% to 5% exhibited antiviral effects against HSV 1. According to the study of Narkhede et al. (2020), anthraquinone- rhein and berberine, demonstrate strong binding affinities against the 3CLpro receptor of SARS-CoV-2 with a $\Delta G = -8.9$ kcal/mol.

13.2.2 *Azadirachta Indica*

Botanical Description

Azadirachta indica or Neem, is an evergreen wonder tree that has been cultivated for many centuries in several parts of India. Neem is not only cultivated in tropical and subtropical countries but also distributed worldwide (Hashmat et al. 2012). *Azadirachta indica* is classified under order Rutales and family Meliaceae. It is a 40 feet tall tree with upright trunk and long scattering branches. It has dark brown bark with compound leaves containing about 15 leaflets. The petals are white, oblanceolate and scented. It produces green fruits which on ripening turn yellow exhibiting aromatic odour (Hashmat et al. 2012).

Major Bioactive Constituents

The different parts of the tree including its flower, fruit, leaves and bark along with the gums are found to have therapeutic potential in the treatment of various disease conditions (Islas et al. 2020). Abundant types of active ingredients are present in diverse parts of the tree. The major ingredient in the bark is azadiaractin (Fig. 13.2) along with others are nimbolinin, nimbidol, nimbidin, gedunin, sodium nimbinate, quercetin and salannin. The leaves contain components like nimbin, nimbandiol, 6-desaacetylnimbinene, nimbanene, n-hexacosanol, ascorbic acid, nimbolide, and amino acid etc. (Hossain et al. 2014). The seed holds important constituents such as azadirachtin and gedunin whereas the bark possesses gallic acid and catechin in it (Gupta 2019). The active phyoconstituents are depicted in Fig. 13.1.

Medicinal Uses

All the parts like leaves, fruit, seeds, flowers, and bark extract show strong antioxidant activity (Ghimeray et al. 2009). When a methanolic extract was tested in rat models with induced intestinal ischemic reperfusion injury, it was reported that treated groups showed reduced level of inflammatory markers and increased level of glutathione that resulted in recovery of G6PD enzyme. This led to the conclusion that the extract aids in boosting the natural defenses of the body (Omóbòwálé et al. 2016). Experiments performed in diabetic rats demonstrated release of glucose 6-phosphate dehydrogenase on treatment with the plant extracts. This study was supported by retardation in kidney and liver damage when leaf and bark extract was used in the treatment (Upreti et al. 2013).

According to Barman et al. (2009), gram positive and negative organisms along with acid fast mycobacterium were highly sensitive to oil extracted from neem bark, leaves and seeds. Antimicrobial activity of the aqueous, methanolic and petroleum ether extracts of leaves was studied against four gram negative and two gram positive organism by Aditi et al. (2011) and they observed that with methanolic extract the highest susceptibility was noted for *B. subtilis.* Antifungal activity against *Aspergillus* and *Rhizopus* has been found in aqueous and alcoholic extracts from oil, leaf, bark and the other parts (Mahmoud et al. 2011).

Chattopadhyay et al. (2004) proved that the leaves exert antiulcer effect by inhibiting hydrogen potassium ATPase pump thereby blocking acid secretion and also by reducing oxidative damage and programmed cell death. The extract prepared using the bark has been documented to exhibit reduction in gastric acid hypersecretion thereby helping in recovery from gastroduodenal ulcers. Neem extract was also reported to be hepatoprotective (Chattopadhyay 2003) and Sithisarn et al. (2005) documented that leaf extract have potent antioxidant capacity. Anti-dermatophytic potential of leaf extract was reported against diverse dermatophyte species (Natarajan et al. 2002). A study of leaf extract on adhesion, cell surface hydrophobicity and biofilm formation found negative impact on *C. albicans*. The results proposed the anti-adhesive properties of neem *in vitro* (Polaquini et al. 2006).

Kumar et al. (2012) experimented with seed oil to evaluate it as an analgesic using albino rats and displayed significant analgesic effect at 1–2 mL/kg dose. An easy to develop animal model to check anti-inflammatory potential uses edema induced by carrageenan in the hind paw. Naik et al. (2014) studied the effect of seed oil and reported concentration dependent increased inhibition of edema with dose in range of 0.25–2 mL per kg body weight. At 4-h post injecting the highest dose of carrageenan, the inhibition of edema was 53.14%.

Dental gel using extract of leaves has been found to reduce plaque index along with bacterial count of *Streptococcus* (Pai et al. 2004). Prashant et al. (2007) verified that a combination of chewing sticks having neem and mango is found to be helpful in eradicating organisms causing dental caries. The bark of the tree has been an ingredient of products to maintain oral hygiene as it is antibacterial and an effective deodorant (Sharma et al. 2011). Nayak et al. (2011) reported that the plant was effective against periodontal pathogens as well as acidogenic bacteria found in the oral cavity that result in dental caries plagues.

Studies done by Yerima et al. (2012) elucidated the anti-HSV-1 activity of bark extract at concentration between 50 and 100μg/mL. Inhibitory potential was observed against dengue virus-2 (Parida et al. 2002) as well as against *A. parasiticus* produced aflatoxin (Allameh et al. 2002). In AIDS patients, oral administration of extract over 3 months exhibited a noteworthy influence on CD4 cells without cytotoxicity (Mbah et al. 2007).

Studies have documented the inhibitory activities of *Azadirachta* plant and its constituents against growth of malignant cells by modifying proliferation of cells, apoptosis, regulating tumor suppressor gene and other metabolic pathways

(Rahmani et al. 2014). It has been documented that the extract of leaves in ethanol is used for upregulating proapoptotic genes and proteins like p53, Bax, Bcl-2 etc. (Arumugam et al. 2014). Study by Uppuluri et al. (2015) on embryos of zebra fish revealed angiogenesis inhibitory activity of imatinib extracted from roots.

An active component -limonoids, show cytotoxic ability by acting as a modulator of inhibitory factors to macrophage migration that inhibits its tautomeric activity and prevents the release of proinflammatory cytokines like interleukins-1α, 1β, 6 and TNF-α (Shilpa et al. 2017). Pramanik et al. (2016) conducted a study wherein chemo protective activity of neem was tested on buccal carcinogenesis in hamsters and their results depict an up regulation of p21, cyclin D1 glutathione S-transferase confirming anticancer activity.

It is known that cisplatin leads to histological damage, especially to the kidney and also induces apoptosis. The methanolic leaf extract improved this damage suggesting that it holds therapeutic benefit when used with cisplatin. Further, through PCR studies they found that down regulation was seen for caspase-3, caspase-9 as well as Bax genes indicative that the damage induced by cisplatin was negated by the plant (Moneim 2014).

Current Advances

The neem leaf extract was found to be effective in increasing the white blood cells and blood platelets counts in turn used in treating the ZIKA virus and strengthening the patients' immune system. (Dohroo et al. 2016). Therefore, it can be concluded that *Azadirachta indica* would play an important character as an adjuvant to the antibiotics or may replace currently used antibiotics for treating opportunistic infections.

When concentrations of CD8+ and CD4+ lymphocytes and the CD4+/CD8+ ratio in plasma of HIV positive patients in age group 18–52 years without any comorbidities were determined after 6 months treatment with Azadirachta extract, there was significant rise in CD4+ cells observed. Further, the treatment of AIDS with antiretroviral therapy in combination with *Azadirachta* decoction was proved to be more efficient than antiretroviral therapy alone. (Hamadama et al. 2021).

The human prostate cancer DU145 cells when treated with doxorubicin (6.37μg/mL) and *A. indica* extract (IC50 41.78μg/mL) showed higher activity of the drug than the plant extract, But Azadirachta silver nanopracticals were found to exhibit good antiproliferative activities against the cancerous cells *in vitro*. The silver nanoparticles synthesized from Azadirachta were found to be selective towards only cancerous cells sparing normal cells. This could concur the potential anticancer activity of *Azadirachta indica* (Kitimu et al. 2022).

A pilot, randomized, placebo-controlled, double-blind study on hospital workers coming in contact with COVID-19 patients were given Neem capsules in the doses of 125, 250, and 500 mg. The study observed insignificant changes in hematological, vital, hepatic and renal function concluded the safety of use of neem capsules instead of hydroxychloroquine tablets suggesting potential use of prophylactic treatment against COVID-19. (Nesari et al. 2021).

An extensive study by Sarkar et al. (2022) checked the effect of *Azadirachta indica* bark extract and (NBE) administered orally and intranasally in Vero E6 cells and Murine Neuro-2A neuroblastoma cells infected with m-CoV-RSA5. The triterpenoids isolated from extract found to be inhibitory against m-CoV infection with docking energy between 8.5 to −10.5 kcal/mol for MHV Spike and − 8.4 to −10.5 kcal/mol for Spike protein of m-CoV-RSA5 suggesting such compounds are key intermediaries of *Azadirachta* induced antiviral properties. Molecular docking studies also confirmed the high affinity binding of Nimbin/4-Epinimbin to SARS-CoV-2 Spike protein and to RNA-dependent RNA polymerase thereby reducing the infectivity of virus.

13.2.3 Catharanthus roseus *(L.) G. Don*

Botanical Description

Species of the Apocynaceae family, *Catharanthus roseus* (L.) G. Don has been used since time immemorial for its therapeutic and pharmacological properties. It is also known as 'The Madagascar periwinkle'. A native of Madagascar, it is common throughout India and is cultivated widely in tropical and subtropical areas worldwide. It has earned the name 'Sadabahar' in Hindi as the flowers are literally ever blooming. *Catharanthus roseus* is an annual evergreen dicot herb. Plant leaves range in size from two to 10 cm approximately and are 1–3.5 cm wide. The shape of the leaf varies anywhere from oval to oblong. The plant boasts of pinkish flowers. The size of the basal tube is approximately 3 cm in length with the diameter of the corolla close to 2–5 cm having 5 petals like lobes (Lahare et al. 2020).

Major Bioactive Constituents

It is rich in several phytoconstituents that exhibit pharmacological properties. These include flavonoid, tannin, saponin, glycoside, terpenoid, protein, phenol and alkaloids (Tamizhazhagan et al. 2017). It is a reservoir of alkaloids and boasts of more than 400 alkaloids, Vincristine being the major one (Fig. 13.2). Majority of the characteristic properties like flavor and fragrance, role as a food additive and therapeutic efficacy associated with *C. roseus* is attributed to its alkaloids. It has been suggested that the different plant organs produce more than 100 monoterpenoid indole alkaloids (TIA). The shoot of the plant is rich in alkaloids whereas the underground plant parts have exhaustive reserve of reserpine vinceine, ajmalicine and vineamine. The flowers of the plant show an abundance of alkaloids, triterpenoids and tannins and anthocyanin (Tamizhazhagan et al. 2017) (Fig. 13.1).

Medicinal Uses

Periwinkle plant is rich in antioxidants. It has a substantial concentration of phenolic and volatile compounds like caffeoylquinic acids and flavonal glycosides to name a few (Kumar et al. 2015). These compounds also demonstrate anti-allergic, cardioprotective, anti-microbial, vasodilatory, anti-inflammatory and antithrombotic properties (Lahare et al. 2020). Terpenes or terpenoids indole alkaloids have been identified as active anti-cancer, anti-inflammatory and anti-bacterial anti-protozoal and anti-malarial agents in many pharmacological studies. In an experimental study on Wistar rats, involving diarrhea induction with castor oil, the ethanolic extract of *C. roseus* was found to exhibit significant anti-diarrheal property (Hassan et al. 2011). Another alkaloid, vindoline, has been reported to exhibit antiulcer property as observed in an experimental rat model (Babulova et al. 2003). In the research conducted by Zheng and Wang (2001), the antioxidant potential of pink and white flowered *C. roseus* and *C. alba*, respectively, was compared and it was elucidated that the antioxidant content in *C. roseus* was found to be more than *C. alba*. The juice was found to contain a compound similar to vinpocetine that exhibited properties comparable with flavonoids (Patel et al. 2011). According to the study of Ahmad et al. (2010), vinculin isolated from *C.roseus* was found to lower blood glucose levels. It was also reported that fresh juice from leaves significantly reduced glucose levels in alloxan and streptozocin induced diabetic rats. The methanolic, ethanolic, acetone and chloroform extract showed antimicrobial activity against many pathogenic organisms such as *S. faecalis, E. coli*, *S. aureus* and *V. cholera* as well as antiviral activity against HSV-1. One of the active constituents found to be present in *C. roseus* leaf, yohimbine, demonstrates effective protection against *C. albicans* infection (Patil and Ghosh 2010).

Current Advances

The enhanced antioxidant activity of *Catharanthus roseus* is attributed to elevated concentration of phenolic compounds (Lahare et al. 2020). *C. roseus* is also said to be used in the treatment and management of stomach disorders, kidney, liver and cardiovascular diseases as well as diabetes. (Kumar et al. 2022). According to the study of Espejel-Nava et al. (2018) the aqueous alkaloid-free extract of *C. roseus* was found to be rich in gallic and chlorogenic acid and exhibited a glucose lowering effect in both normal and diabetic mice.

One of the most distinguished properties of *C. roseus* is its anticancer activity. *C. roseus* has been well documented as an effective anticancer remedy attributed largely to the presence of alkaloids-vinblastine, vincristine and vindesine. According to the study of Jordan (2002), the alkaloids present act as inhibitors of mitotic spindle formation thus inhibiting mitosis, which prevents the division of metastatic cells. The vinca alkaloids-vincristine and vinblastine-function exhibits anticancer activity by binding to α/β-tubulin, thus interfering in its association into microtubules. Vinblastine has been successfully employed for the treatment of varied cancers like carcinoma of breast and lung, lymphosarcoma, neuroblastoma, lymphocytic leukemia and choriocarcinoma (Lahare et al. 2020). In a study done by Wei et al. (2017) for glioma treatment, it has been observed that vincristine along with procarbazine and lomustine supplemented with radiotherapy was a more effective treatment against glioma when compared to radiotherapy alone (Wei et al. 2017). A chemical vinpocetine, obtained from the *C.roseus* alkaloid vincamine has been reported to be improve memory and other neurological symptoms, which will be extremely beneficial for patients suffering from Alzheimer's or Parkinson's (Nayak and Pereira 2006). In an *in vitro* assay, *C. roseus* extract effectively inhibited acetylcholinesterase (AchE). Also, the alkaloid, serpentine exhibited an inhibitory activity against displayed a strong activity against AchE with a IC50 of 0.775μM (Pham et al. 2020). All these researches strongly support the neuroprotective role of *C. roseus* for the management of Alzheimer's and other neurodegenerative diseases. In the *in silico* study of Kalaria and Patel (2020), the alkaloid vindolininol isolated from *C. roseus* demonstrated a greater binding affinity to the spike and papain-like protease-protein of SARS COV-2.

13.2.4 Cinnamomum zeylanicum Blume

Botanical Description

Cinnamons belong to the *Cinnamomum* genus, in the laurel family (Lauraceae). Approximately 250 species are known among this genus, with plants being dispersed worldwide (Vangalapati et al. 2012). Cinnamon, appreciated for centuries for its peculiar flavor and aroma, is the dried inner bark of an evergreen tree native of Sri Lanka, India and China. The spice is obtained from the bark of a tropical tree with two major varieties viz. *C. cassia* and *C. zeylanicum* (Ranasinghe et al. 2013). *C. verum* is also commonly called 'true' cinnamon or Ceylon cinnamon. The plants grow upto 15 meters and have long lance-shaped leaves with small flowers with axial inflorescences, and globose drupe fruits (Ramazani et al. 2020).

Major Bioactive Constituents

The plant has a rich array of phytoconstituents and the phytoprofiles differ in the different plant parts. It has been observed that cinnamaldehyde (Fig. 13.2) and linalool are the main components from the bark of the bark, while the extracts of flower /fruit and leaves, are rich in cinnamyl acetate, and eugenol respectively (Behbahani et al. 2020). Also present in good amount in the bark is benzenoid lactone, in *C. cassia* having a higher content to the tune of 24 times than for *C. verum* (Ananthakrishnan et al. 2018). Cinnamon bark also contains catechins and procyanidins which are potent antioxidants (Peng et al. 2008). The phytoconstituents are depicted in Fig. 13.1.

Medicinal Use

Cinnamaldehyde is one of the chief constituents obtained from *C. zeylanicum* having anti-tyrosinase activity (Marongiu et al. 2007). Mancini-Filho et al. (1998) documented that aqueous, methanolic and ether extracts exhibit considerable antioxidant activities. This was further proved by Shanmugam and Naidu (2000) who reported cinnamaldehyde to inhibit oxidation of fatty acid and peroxidation of lipid. This activity is also exhibited by other flavonoids in cinnamon (Okawa et al. 2001). Cinnamaldehyde and eugenol have been demonstrated by Mateen et al. (2019), to significantly reduce tumor necrosis factor-α and interleukin-6, the proinflammatory cytokine levels in rheumatoid arthritis patients. There is high oxidative damage in such patients and on cinnamon supplementation the antioxidant status was elevated due to increase in levels of catalase, glutathione peroxidase, and superoxide dismutase.

Extract of cinnamon breaks the biofilm thus inhibiting *Solobacterium moorei* growth, thereby reducing hydrogen sulfide production. This activity of cinnamon is used in oral products to regulate bad breath. Being antimicrobial, cinnamon oil can be used in preparation of mouthwash (Mutans 2021). It has been observed that aqueous extract and extracted oils are a strong antimicrobial agent against extensive range of gram negative and gram-positive organisms along with anti-fungal activity (Goñi et al. 2009). In a study, *C. verum* aromatic oil significantly inhibits fungal growth and was effective against of *C. albicans*, *C. tropicalis*, and *C. krusei*, with MIC of 0.064 mg/mL, 0.129 mg/mL and 0.129 mg/mL, respectively (Wang et al. 2020).

Extracts have shown to be effective for treatment of stomach-ache, diarrhea, chest tightness and gastritis (Wang et al. 2020). Obesity leads to headache and migraine and extracts of cinnamon have shown positive effect in treatment of migraine (Zareie et al. 2020).

A topical study by Lee et al. (2005) also reported that 2 -hydroxy cinnamaldehyde extracted from *C. cassia* showed nitric oxide inhibition by obstructing the activation of NF-κB suggesting anti-inflammatory effect. Cinnamophilin, a thromboxane A2 receptor antagonist protects rat brain cells against ischemic damage. These effects were shown to have significant impacts in mitigating brain infarction which may improve neurobehavioral actions (Lee et al. 2009). Cinnamon was observed to reduce the formation of β- amyloid (Aβ) oligomers thereby preventing the toxic effect on pheochromocytoma (PC12) neurons leading to noticeable reduction in plaques and enhancement in the cognitive functions in Alzheimer's patients (Frydman-Marom et al. 2011). Studies have established that the extract can reverse the DJ-1 gene inhibition in neurons exerting neuroprotective effects against Parkinson's disease, attributed to compounds responsible for the modulation of mitochondria and apoptotic functions (Angelopoulou et al. 2021).

Polyphenol type-A polymers, hydroxycinnamic acid and its derivatives such as naphthalene-methyl ester extracted from cinnamon were found to have blood glucose lowering action indicating antidiabetic potential of cinnamon (Kim et al. 2006). Song et al. (2013) documented cinnamic acid and cinnamic aldehyde as potential protective agents against ischemia observed in myocardium, demonstrating anti-CVD effects of cinnamon. Further, cinnamophilin was found to be an inhibitor of vascular smooth muscle cell proliferation mediated by thromboxane receptors that can potentially be used in preventing atherosclerosis (Nie et al. 2004). Cinnamon was also observed to significantly reduce total cholesterol, LDL, triglycerides in human beings (Khan et al. 2003).

Procyanidins and aqueous extracts of cinnamon were observed to be inhibitor of vascular endothelial growth factor subtype 2 kinase activity, leading to inhibition of angiogenesis proving anticancer activity (Lu et al. 2010). It also exhibits significant antiviral efficacy (Aanouz et al. 2020) and is used extensively for treatment and management of lung diseases, including infections, pneumonia as well as malignancy (Lai et al. 2018).

Current Advances

Cinnamon has been recently used for relief of polycystic ovary syndrome. In women with the syndrome, cinnamon capsule administration for a period of 12 weeks demonstrated an improvement of the lipid profile with a decrease in the lipid peroxidation indicated by fall of serum malondialdehyde levels, along with a decrease in total cholesterol and triacylglycerol with an increment in the antioxidant potential (Borzoei et al. 2018).

Hadi et al. (2020) conducted meta-analysis involving 9 randomized clinical trials (641 participants). They observed that both systolic as well as diastolic pressure was significantly reduced post-cinnamon supplementation. The values were significant in participants with a baseline BMI of $\geq$30 kg/m^2 and the dosage being $\leq$2 g/day, over 8 weeks consumption. Cinnamon supplementation affected ALT levels significantly (dosage <1500 mg/day) during clinical trials conducted of both genders (Shekarchizadeh-Esfahani et al. 2021).

Phytoconstituents of cinnamon have also demonstrated a high binding affinity with the SARS CoV-2 spike and main protein. In the study by Prasanth et al. (2021), tenuifolin and pavetannin C1 exhibited a high docking score with a low binding energy of −8.8Kcal/mol.

Anandari et al. (2021), conducted *in-silico* studies on procyanidin compounds found in cinnamon as they are known to possess similarity in activity with insulin, thus facilitating blood sugar regulation. Interaction of Procyanidin A was studied with important enzymes in carbohydrate metabolism, viz. glucosidase and amylase molecular docking between Î ± -glucosidase and Î ± -amylase with procyanidin A was performed using HEX 8.0.0 and was visualized by Discovery Studio. With glucosidase, Procyanidin A interaction was mediated via hydrogen, hydrostatic and hydrophobic bonds, while with amylase hydrogen and hydrophobic bonds were involved. Thus, procyanidin A has excellent ability to inhibit breaking down of glucose in the intestines.

13.2.5 Terminalia arjuna *(Roxb.) Wight and Arn*

Botanical Description

T. arjuna has immense medicinal value and is commonly referred to as Arjuna, belonging to the family Combretaceae with 200 species distributed across the globe. India houses 24 species of *Terminalia*, amongst which most widely distributed as well as used are *T. bialata*, *T. bellirica*, *T. elliptica*, *T. catappa*, *T. porphyrocarpa*, and *T. mantaly*, as they hold immense medicinal properties *T. arjuna* is present from sub-Indo-Himalayan tracts of Uttar Pradesh, Punjab to Deccan plateau, Madhya Pradesh, South Bihar, and on the eastern part in West Bengal and Orissa.

Major Bioactive Constituents

An aqueous bark extract was found to contain calcium salts (23%) along with tannins (16%), while the alcoholic extract contained coloring matter and tannins

(Ghosh 1926). Later alkaloids and glycosides were isolated which also resulted in formation of organic acids and phytosterol (Chopra and Ghosh, 1929). The bark has significant amount of phyoconstituents (Tripathi et al. 1992) (Fig. 13.1). The roots, fruits and the leaves contain largely terpenoids, flavonoids and glycosides (Amalraj and Gopi 2017) . Figure 13.2 represent the major phytoconstituents- Arjunin.

Medicinal Use

T. arjuna is called 'Guardian of the heart' as heart is protected from necrosis mainly due to arjunolic acid (Sumitra et al. 2001). There is protection of oxidative stress in ischemic perfused rats, treated with plant extract (Gauthaman et al. 2001).

Its most popular use is as a cardiotonic agent, though it possesses potential as an antiallergic, antibacterial, anti-HIV, antimicrobial, hypolipidemic, hypocholesterolemic, antioxidant as well as antitumoral (Gupta et al. 2001; Bachaya et al. 2009; Kumar et al. 2013; Kapoor et al. 2014). Plant parts such as stem bark, roots, leaves as well as flowers are used and they possess an array of phytoconstituents providing for the pharmacological properties.

Kokkiripati et al. (2013) investigated the cardioprotective potential of the plant in human cell cultures. Amongst the parameters studied, lipid peroxidation and 3-hydroxy3-methylglutaryl coenzyme A reductase were inhibited along with promotion of catalase and glutathione peroxidase (GPx) activities resulting in reduced ROS generation. On administering varying doses of the aqueous extract to anesthetized dogs, a drop in blood pressure was observed over a 90-minute period (Bhatia et al. 2000). Nammi et al. (2003), found that 70% alcoholic plant extract demonstrated hypotensive effect on dogs on being administered doses of 5–10 mg/kg. Oxidative stress was diffused in presence of the phytoconstituents, and it could treat ischemia in rats (Gauthaman et al. 2001) as the magnitude of myocardial fibrosis and oxidative stress induced by catecholamine showed reduction due to the bark extract in model animals (Kumar et al. 2009).

Bark extracts using alcohol and water are found to contain major arjunic acid, arjunetin and arjungenin and these extracts were assessed for their inhibitory potential against CYP2D6, CYP3A4, and CYP2C9 enzymes by Varghese et al. (2015). They observed that in human liver microsomes the inhibitory activity was appreciable, as IC50 values were lower than 35 mg/mL.

The use of the plant as an antimutagenic and anticarcinogenic was studied by Ahmad et al. (2014), on human lymphocyte culture and albino mice bone marrow cells. The plant extract showed significant reduction in aflatoxin B1 induced metaphase aberrations (59.65% to 36.88% at 72 h), reduction in number of sister chromatid exchanges (15.00 $\pm$ 1.40 to 7.70 $\pm$ 0.50 per cell) and enhancement in replication index (1.33 to 1.55). The antioxidant activity as well as antimutagenic potential of an alcoholic extract was also investigated by Viswanatha et al. (2010) using superoxide radical scavenging activity, DPPH assay, and lipid peroxidation assay. A significant reduction in micronucleus evaluation was noticed for both polychromatic as well as normochromatic erythrocytes with a substantial decline in P/N ratio. Doxorubicin induced cardiotoxicity in rats reduced in presence of fraction of bark obtained with butanol, depicting its cardioprotective effects (Singh et al. 2008).

Study by Mandal et al. (2013) explained the antimicrobial and antioxidative nature of methanolic extract of the bark, wherein the activity was more pronounced against gram negative compared to gram positive bacteria, when tested with *S. aureus*, *E. coli, P. mirabilis*, *Acinetobacter* sp., and *P. aeruginosa*.

A double-blind – placebo controlled study which was randomized, involving 116 patients with coronary artery disease, was conducted to evaluate the inflammatory markers for the remedial potential of *T arjuna* (Kapoor et al. 2014). These patients were on standard medications for at least 3 months, and they were administered either placebo or 500 mg *T. arjuna*, twice a day, along with their regular drug regime. There was a substantial decrease in cytokines that aggravate inflammation, such as c-reactive protein, TNF-α, IL-6, and IL-18 along with levels of serum triglycerides after 3 months and maintained till 6 months, in patients given *T. arjuna* compared to the placebo.

The commercial ayurvedic formulation of *T. arjuna* 'Arjuna Kwatha' was evaluated by Rao et al. (2001) on Stage III hypertensive patients with increased left-ventricular hypertrophy (LV mass). Two groups of patients were considered, one was administered 100 mg/day dose of atenolol while the other group was administered 50 ml/day atenonol, for a period of 6 months along with 'Arjuna Kwatha'. The systolic as well as diastolic blood pressure significantly decreased in both groups investigated, though reduction in LV mass was observed in test group only. Powdered bark was used in another randomized clinical trial with 100 CAD patients divided into 2 groups, one given the powder and other without any treatment. The decrease in systolic and diastolic pressures were 10.28% and 4.8% respectively, with lowering of pulse rate by 4.85% in the study group. Dip in total cholesterol level, serum triglyceride, LDL, VLDL and HDL were 10.2%, 17.9%, 9.59% and 16% and 10.48% respectively (Priya et al. 2019).

Current Advances

A double-blind, randomized and placebo-controlled clinical trial by Maulik et al. (2018) was conducted to assess an aqueous extract of Arjuna bark inpatients having had a cardiac arrest. The functional capacity of the heart and the antioxidant potential was improved though modulation of LVEF was not improvised. Dipeptidyl peptidase-IV inhibitors are potent antidiabetics and this activity of *T. arjuna* was comparable with the expensive synthetic analog-Vildagliptin. Amongst the various phytoconstituents, arjunetin, arjungenin, ellagic acid and arjunic acid showed outstanding inhibitory activity, on basis of *in silico* studies, suggestive of substantial cardioprotective effects found in diabetics (Mohanty et al. 2019). Neuroprotective potential of two Terminalia species, *T. arjuna* and *T. chebula* was evaluated through their potential to possess antioxidant, antiamyloidogenic and anticholinesterase activities. Methanolic extract of bark demonstrated good antioxidant property (597.25 ± 0.5μg of gallic acid equivalent/mg of extract) and acetyl cholinesterase inhibitory activity (IC50 = 73.99 ± 2.11μg/mL, respectively). Molecular docking studies suggested the role of 7-methyl gallic acid as lead compound with desired ADMET properties, suggestive of use in Alzheimer's (Pugazhendhi et al. 2018).

Pingali et al. (2020) conducted a randomized prospective -double-blinded, placebo-controlled clinical study to assess 250 and 500 mg of *T. chebula* water extracts and its effects on oxidative stress and the endothelial dysfunction in IDDM Type 2. It was demonstrated that 500 mg dose improvised endothelial function. Further, the oxidative stress was reduced as glutathione, nitric oxide, and malondialdehyde levels were ameliorated. Risk markers for cardiovascular disorder such as high sensitivity C-reactive protein, lipid profile as well as glycosylated hemoglobin were also significantly improved.

In a docking study to find prospective candidates against SARS-CoV-2, 28 phytoconstituents of *T. arjuna* were assessed. The best candidate was triterpenoid- arjunin, followed by tannin- punicalin and then terflavin A with binding energies of 9.9, 9.8 and 9.8 kcal/mol respectively, on interaction with MPro (Srinivasan and Meenakshi 2021).

13.3 Conclusions

Till the last century the drug discovery process involving active constituents of medicinal aromatic plants was a tedious exercise leading to conclusive results after years of laboratory work. However, with the advancement in dry laboratory techniques, bioinformatics, *in-silico* drug modelling and drug-docking studies this work has been facilitated. High-throughput screening (HTS) wherein many phytoconstituents can be screened within a short time has hastened the drug delivery process. HTS has lead to the development of libraries which facilitate the quick isolation of lead molecules for drug delivery.

Amongst the array of Indian MAPs used, the five plants reviewed are extensively used as they exhibit many pharmacological properties. All the species have a rich array of phytoconstituents providing for development of prophylactic agents against bacterial infections, cancer, and infections of gastrointestinal tract. They posses antioxidant, antihyperlipidemic, antidiabetic, as well as immunomodulatory activities. *In-silico* studies on these MAPs have indicated their efficiency against viruses including SARS-CoV-2.

The immense diversity of MAPs found in India, clubbed with the rich ancient literature of non-conventional therapeutic modalities as ayurveda, unani, siddha amongst many others creates a massive scope of scientifically researching their applications. Thus, though this branch of medicine has prevailed since ancient times, the potential to develop the same even today is immense. This "Midas touch potential" is a treasure and should be shared with one and all.

References

Aanouz I, Belhassan A, El Khatabi K, Lakhlifi T, El Idrissi M, Bouachrine M (2020) Moroccan medicinal plants as inhibitors of COVID-19: computational investigations. J Biomol Str Dyn:1–12

Abou El-Soud NH (2010) Herbal medicine in ancient. Egyp J Med Plants Res 4(2):082–086

Aditi G, Bhandari BS, Rai N (2011) Antimicrobial activity of medicinal plants *Azadirachta indica* A. Juss, *Allium cepa* L. and *Aloe vera* L. Int J Pharm Tech Res:1059–1065

Ahmed MF, Kazim SM, Ghori SS, Mehjabeen SS, Ahmed SR, Ali SM, Ibrahim M (2010) Antidiabetic activity of *Vinca rosea* extracts in alloxan-induced diabetic rats. Int J Endocrinol 1:2010

Ahmad MS, Ahmad S, Gautam B, Arshad M, Afzal M (2014) Terminalia arjuna, a herbal remedy against environmental carcinogenicity: an in vitro and in vivo study. Egypt J Med Hum Genet 15(1):61–68

Allameh A, Abyaneh MR, Shams MR et al (2002) Effects of neem leaf extract on production of aflatoxins and activities of fatty acid synthetase, isocitrate dehydrogenase and glutathione S-transferase in *aspergillus parasiticus*. Mycopathologia 154:79–84. https://doi.org/10.1023/A:1015550323749

Amalraj A, Gopi S (2017) Medicinal properties of *Terminalia arjuna* (Roxb.) Wight & Arn.: a review. J Trad Comp Med 7(1):65–78

Anandari RN, Minnah SK, Widadni VU, Safira D, Fatchiyah F (2021) *In-silico* analysis of Procyanidin type-a extracted from cinnamon for diabetes mellitus type 2 treatment. JSMARTech J Smart Bioprospecting Technol 2(3):92–95

Ananthakrishnan R, Chandra P, Kumar B, Rameshkumar KB (2018) Quantification of coumarin and related phenolics in cinnamon samples from South India using UHPLC-ESI-QqQLIT-MS/MS. Int J Food Prop 21:50–57

Angelopoulou E, Paudel YN, Piperi C, Mishra A (2021) Neuroprotective potential of cinnamon and its metabolites in Parkinson's disease: mechanistic insights, limitations, and novel therapeutic opportunities. J Biochem Mol Toxicol 35(4):e22720

Arumugam A, Agullo P, Boopalan T, Nandy S, Lopez R, Gutierrez C, Narayan M, Rajkumar L (2014) Neem leaf extract inhibits mammary carcinogenesis by altering cell proliferation, apoptosis, and angiogenesis. Cancer Biol Therapy 15(1):26–34

Babulova A, Machova J, Nosalova V (2003) Protective action of vinpocetine against experimentally induced gastric damage in rats. Arzneim Forsch 43:981–985

Bachaya HA, Iqbal Z, Khan MN, Jabbar A, Gilani AH, Din IU (2009) In *vitro* and *in vivo* anthelmintic activity of *Terminalia arjuna* bark. Int J Agric Biol 11:273–278

Barman P, Yadav MC, Kumar H, Meur SK, Rawat M (2009) Antibacterial efficacy of neem oil fractions on clinical isolates of endometritic cows. Indian J Anim Sci:665–668

Behbahani BA, Falah F, Arab FL, Vasiee M, Yazdi FT (2020) Chemical composition and antioxidant, antimicrobial, and antiproliferative activities of *Cinnamomum zeylanicum* bark essential oil. Evid Based Complement Altern Med:1–8

Bhatia J, Bhattacharya SK, Mahajan P, Dwivedi S (2000) Effect of *Terminalia arjuna* on blood pressure of anaesthetised dogs. Indian J Pharmacol 2:159–160

Borzoei A, Rafraf M, Niromanesh S, Farzadi L, Narimani F, Doostan F (2018) Effects of cinnamon supplementation on antioxidant status and serum lipids in women with polycystic ovary syndrome. J Tradit Complement Med 8:128–133

Chattopadhyay RR (2003) Possible mechanism of hepatoprotective activity of *Azadirachta indica* leaf extrac. J Ethnopharmacol 89:217

Chattopadhyay I, Nandi B, Chatterjee R, Biswas K, Bandyopadhyay U, Banerjee RK (2004) Mechanism of antiulcer effect of neem (*Azadirachta indica*) leaf extract: effect on H+-K+-ATPase, oxidative damage and apoptosis. Inflammo Pharmacol 12:153

Chen Q, Li KT, Tian S, Yu TH, Yu LH, Lin HD, Bai DQ (2018) Photodynamic therapy mediated by Aloe-Emodin inhibited angiogenesis and cell metastasis through activating MAPK signaling pathway on HUVECs. Technol Cancer Re Treat 17

Chopra RN, Ghosh S (1929) *Terminalia arjuna*: its chemistry, pharmacology and therapeutic action. Indian Med Gaz 64:70–73

De Oliveira ACL, Tabrez S, Shakil S, Khan MI, Asghar MN, Matias BD, de Carvalho RM (2018) Mutagenic, antioxidant and wound healing properties of *Aloe vera*. J Ethnopharmacol 227: 191–197

Dohroo A, Karnwal A, Ghai M (2016) Recent developments in neem (Azadirachta indica – a. Juss) derived antimicrobial constituents for control of human and plant diseases – a review. Ann Acad Med Siles 70:220–223

Espejel-Nava J, Vega-Avila E, Alarcon-Aguilar F, Contreras-Ramos A, Díaz-Rosas G, Trejo-Aguilar G, Ortega-Camarillo C (2018) A phenolic fraction from *Catharanthus roseus* L. Stems decreases Glycemia and stimulates insulin secretion. Evid Based Complementary Altern Med. https://doi.org/10.1155/2018/7191035

Farha F, Kumar A (2021) A systematic review of *Aloe vera* and its properties. Asian J Pharma Res 11(4)

Frydman-Marom A, Levin A, Farfara D (2011) Orally administrated cinnamon extract reduces *β*-amyloid oligomerization and corrects cognitive impairment in Alzheimer's disease animal models. PLoS One:e16564

Gauthaman K, Maulik M, Kumari R, Manchanda SC, Dinda AK, Maulik SK (2001) Effect of chronic treatment with the bark of *Terminalia arjuna*: a study on the isolated ischemic-reperfused rat heart. J Ethnopharmacol 75:197–201

Ghimeray AK, Jin CW, Ghimire BK, Cho DH (2009) Antioxidant activity and quantitative estimation of azadirachtin and nimbin in *Azadirachta indica* A. Juss grown in foothills of Nepal. African. J Biotechnol 3084–3091

Ghosh S (1926) Annual report of the Calcutta School of Tropical Medicine. Institute of Hygiene and the Carmichel Hospital for Tropical Diseases, Calcutta

Goñi P, López P, Sánchez CR, Gómez-Lus R, Becerril R, Nerín C (2009) Antimicrobial activity in the vapour phase of a combination of cinnamon and clove essential oil. Food Chem:982–989

Gupta A (2019) Association of *Flavonifractor plautii*, a flavonoid-degrading bacterium, with the gut microbiome of colorectal cancer patients in India. Systems https://doi.org/10.1128/msystems.00438-19

Gupta R, Singhal S, Goyle A, Sharma VN (2001) Antioxidant and hypocholesterolemic effects of *Terminalia arjuna* tree-bark powder:a randomized placebo-controlled trial. J Asso Physician India 49:231–235

Hadi A, Campbell MS, Hassani B, Pourmasoumi M, Salehi-sahlabadi A, Hosseini SA (2020) The effect of cinnamon supplementation on blood pressure in adults: a systematic review and meta-analysis of randomized controlled trials. Clin Nutr ESPEN 36:10–16

Hamadama OG, Javeres MNL, Nyemb N, Fabrice MM, Pettang T (2021) Effect of Azadirachta indica and Senna siamea decoction on CD4+ and CD8+ level, toxicological, and antioxidant profile in HIV/AIDS positive persons. J Toxicol 5594505. https://doi.org/10.1155/2021/5594505

Hashmat I, Hussain A, Ajij A (2012) Neem (*Azadirachta indica* A. Juss) – a nature's drugstore: an overview. Int Res J Biol Sci

Hassan KA, Brenda AT, Patrick V, Patrick OE (2011) *In vivo* antidiarrheal activity of the ethanolic leaf extract of *Catharanthus roseus* Linn.(Apocyanaceae) in Wistar rats African. J Pharm Pharmacol 5(15):1797–1180

Hossain MA, Shah MD, Sakari M (2014) Gas chromatography– mass spectrometry analysis of various organic extracts of *Merremia borneensis* from Sabah. Asian Pac J Trop Med 4(8): 637–641

Islas JF, Acosta E, Zuca G, Delgado-Gallegos JL, Moreno-Treviño MG, Escalante B, Moreno-Cuevas JE (2020) An overview of neem (*Azadirachta indica*) and its potential impact on health. J Funct Foods 7:104171. https://doi.org/10.1016/j.jff.2020.104171

ITIS (2022) (Integrated Plant Information System) http://www.itis.gov

Jain S, Rathod N, Nagi R, Sur J, Laheji A, Gupta N, Prasad S (2016) Antibacterial effect of *Aloe vera* gel against oral pathogens: An *in-vitro* study. J Clin Diagn Res 10:41–44

Ji H, Liu Y, He F, An R, Du Z (2018) LC–MS based urinary metabolomics study of the intervention effect of aloe-emodin on hyperlipidemia rats. J Pharm Biomed Anal 156:104–115

Jordan MA (2002) Mechanism of action of antitumor drugs that interact with microtubules and tubulin. Curr Med Chemy-Anti Cancer Agents 2(1):1–17

Kalaria RK, Patel HK (2020) Naturally occurring phytochemical as inhibitors from *Catharanthus roseus*: an *In-silico* approaches for drug development against COVID-19. Research Square

Kapoor D, Vijayvergiya R, Dhawan V (2014) *Terminalia arjuna* in coronary artery disease: ethnopharmacology, pre-clinical, clinical & safety evaluation. J Ethnopharmacol 155: 1029–1045

Khan A, Safdar M, Khan MMA, Khattak KN, Anderson RA (2003) Cinnamon improves glucose and lipids of people with type 2 diabetes. Diabetes:3215–3218

Kim SH, Hyun SH, Choung SY (2006) Anti-diabetic effect of cinnamon extract on blood glucose in db/db mice. J Ethnopharmacol:119–123

Kim K, Chung MH, Park S, Cha J, Baek JH, Lee SY, Choi SY (2018) ER stress attenuation by aloe-derived polysaccharides in the protection of pancreatic β-cells from free fatty acid-induced lipotoxicity. Biochem Biophys Res Commun 500(3):797–803

Kitimu SR, Kirira P, Sokei J, Ochwangi D, Mwitari P, Maina N (2022) Biogenic synthesis of silver nanoparticles using *Azadirachta indica* methanolic bark extract and their anti-proliferative activities against DU-145 human prostate cancer cells African. J Biotech 21(2):64–72. https://doi.org/10.5897/AJB2021.17424

Kokkiripati PK, Kamsala RV, Bashyam L, Manthapuram N, Bitla P, Peddada V, Raghavendra AS, Tetali SD (2013) Stem-bark of *Terminalia arjuna* attenuates human monocytic (THP-1) and aortic endothelial cell activation. J Ethnopharmacol 146(2):456–464

Kumar S, Enjamoori R, Jaiswal A, Ray R, Seth S, Maulik SK (2009) Catecholamine-induced myocardial fibrosis and oxidative stress are attenuated by *Terminalia arjuna* (Roxb.). J Pharm Pharmacol 61:1529–1536

Kumar S, Agrawal D, Patnaik J, Patnaik S (2012) Analgesic effect of neem (*Azadirachta indica*) seed oil on albino rats. Int J Pharma Bio Sc 2:222–225

Kumar PG, Navya K, Ramya EM, Venkataramana M, Anand T, Anilakumar KR (2013) DNA damage protecting and free radical scavenging properties of *Terminalia arjuna* bark in PC-12 cells and plasmid DNA. Free Radic Antioxid 3:35–39

Kumar GR, Devanand G, John BD, Ankit Y, Khursheed O, Sumit M (2014) Preliminary antiplaque efficacy of *Aloe vera* mouthwash on 4 day plaque re-growth model: randomized control trial. Ethiop J Health Sci 24:139–144

Kumar A, Malik R, Giri P, Parveen N (2015) Up-to-date review on therapeutic interior of *Catharanthus roseus* for anticancer and antidiabetic activities. J Phytomed 1(1):1–13

Kumar S, Yadav M, Yadav A, Rohilla P, Yadav JP (2017) Antiplasmodial potential and quantification of aloin and aloe-emodin in *Aloe vera* collected from different climatic regions of India. BMC Complement Altern Med 17:369

Kumar S, Singh B, Singh R (2022) *Catharanthus roseus* (L.) G. Don: a review of its ethnobotany, phytochemistry, ethnopharmacology and toxicities. J Ethnopharmacol 284(10):114647

Lahare RP, Yadav HS, Dashahre AK, Bisen YK (2020) An updated review on phytochemical and pharmacological properties of *Catharanthus rosea* Saudi. J Med Pha

Lai K, Shen H, Zhou X, Qiu Z, Cai S, Huang K, Wang Q, Wang C, Lin J, Hao C, Kong L, Zhang S, Chen Y, Luo W, Jiang M, Xie J, Zhong N (2018) Clinical practice guidelines for diagnosis and management of cough-Chinese thoracic society (CTS) asthma consortium. J Thorac Dis 10(11): 6314–6351

Lee SH, Lee SY, Son DJ (2005) Inhibitory effect of 2′-hydroxycinnamaldehyde on nitric oxide production through inhibition of NF-κB activation in RAW 264.7 cells. Biochem Pharmacol:791–799

Lee EJ, Chen H-Y, Hung YC (2009) Therapeutic window for cinnamophilin following oxygen-glucose deprivation and transient focal cerebral ischemia. Exp Neurol:74–83

Lu J, Zhang K, Nam SR, Anderson A, Jove R, Wen W (2010) Novel angiogenesis inhibitory activity in cinnamon extract blocks VEGFR2 kinase and downstream signaling. Carcinogenesis:481–488

Mahmoud DA, Hassanein NM, Yousuf KA, Abouzeid MA (2011) Antifungal activity of different neem leaf extracts and the nimonol against some important human pathogens Braz. J Microbiol:1007–1016

Mancini-Filho J, Van-Koiij A, Mancini DA, Cozzolino FF, Torres RP (1998) Antioxidant activity of cinnamon (*Cinnamomum zeylanicum*, Breyne) extracts. Boll Chim Farm 137(11):443–447

Mandal S, Patra A, Samanta A (2013) Analysis of phytochemical profile of *Terminalia arjuna* bark extract with antioxidative and antimicrobial properties. Asian Pac J Trop Biomed 3:960–966

Marongiu B, Piras A, Porcedda S et al (2007) Supercritical CO_2 extract of *Cinnamomum zeylanicum*: chemical characterization and antityrosinase activity. J Agric Food Chem: 10022–10027

Mateen S, Rehman T, Shahzad S, Naeem SS, Faizy AF, Khan AQ, Khan MS, Husain FM, Moin S (2019) Anti-oxidant and anti-inflammatory effects of cinnamaldehyde and eugenol on mononuclear cells of rheumatoid arthritis patients. Eur J Pharmacol 852:14–24

Máthé A, Khan I (2021) Medicinal and aromatic plants of India Vol 1: introduction to medicinal and aromatic plants in India. In: Máthé Á, Khan IA (eds) Medicinal and aromatic plants of India Vol. 1. Medicinal and aromatic plants of the world, vol 8. Springer, Cham. https://doi.org/10.1007/978-3-030-98701-5_3

Maulik SK, Seth S, Dua P, Bhargava B, Wilson V (2018) Clinical trail of a standardised water extract of the stem bark of *Terminalia arjuna* in heart failure In Proceedings for annual meeting of the Japanese pharmacological society WCP2018 (The 18th World Congress of Basic and Clinical Pharmacology) (pp. PO1–2) Japanese Pharmacological Society

Mbah AU, Udeinya IJ, Shu EN (2007) Fractionated neem leaf extract is safe and increases CD4+ cell levels in HIV/AIDS patients. Am J Ther 14:369–374

Mehrotra N (2021) Herbs that heal: natures pharmacy. Ann Phytomed 10(1):6–22

Milutinovici RA, Chioran D, Buzatu R, Macasoi I, Razvan S, Chioibas R, Soica C (2021) Vegetal compounds as sources of prophylactic and therapeutic agents in dentistry. Plan Theory 10(10): 2148. https://doi.org/10.3390/plants10102148

Mohanty IR, Borde M, Kumar SC, Maheshwari U (2019) Dipeptidyl peptidase IV inhibitory activity of *Terminalia arjuna* attributes to its cardioprotective effects in experimental diabetes: *in silico*, *in vitro* and *in vivo* analyses. Phytomedicine 57:158–165

Moneim AE (2014) *Azadirachta indica* attenuates cisplatin-induced neurotoxicity in rats. Ind J Pharma 46(3):316–321

Mutans S (2021) Evaluation of the antibacterial effect of cinnamon extract on *Streptococcus mutans* Al-Azhar. Dent J Girls:123–128

Naik MR, Bhattacharya A, Behera R, Agrawal D, Dehury S, Kumar S (2014) Study of anti-inflammatory effect of neem seed oil (*Azadirachta indica*) on infected albino rats. J Health Res Rev 1(3):66–69

Nalimu F, Oloro J, Kahwa I, Ogwang PE (2021) Review on the phytochemistry and toxicological profiles of *Aloe vera* and *Aloe ferox*. Futur J Pharm Sci 7:145. https://doi.org/10.1186/s43094-021-00296-2

Nammi S, Gudavalli R, Babu BS, Lodagala DS, Boini KM (2003) Possible mechanism of hypotension produced by 70% alcoholic extract of *Terminalia arjuna* (L.) in anesthetized dogs. BMC Comp Alt Med 3(1):1–4

Narkhede RR, Pise AV, Cheke RS, Shinde SD (2020) Recognition of natural products as potential inhibitors of COVID-19 main protease (Mpro): in-silico evidences. Nat Prod Bioprospect 10: 297–306

Natarajan V, Pushkala S, Karuppiah VP, Prasad PV (2002) Anti dermatophytic activity of *Azardirachta indica* (neem) by invitro study. Indian J Pathol Microbiol 45:311

Nayak BS, Pereira LMP (2006) *Catharanthus roseus* flower extract has wound healing activity in Sprague Dawley rats. BMC Complement Altern Med 6:41–46

Nayak A, Ranganathan N, Sowmya GB, Kishore B, Kudalkar M (2011) Evaluation of antibacterial and anticandidial efficacy of aqueous and alcoholic effect of neem (*Azadirachta indica*): an I study. Int J Res Ayurveda Pharm 2:230–235

Nesari TM, Bhardwaj A, ShriKrishna R, Ruknuddin G, Ghildiyal S, Das A, Pandey AK, Chaudhary N, Soman G, Barde M (2021) Neem (*Azadirachta indica* a. Juss) capsules for prophylaxis of COVID-19 infection: a pilot, double-blind, randomized controlled trial. Altern Ther Health Med 27(S1):196–203

Nie D, Che M, Zacharek A, Qiao Y, Li L, Li X, Lamberti M, Tang K, Cai Y, Guo Y, Grignon D, Honn KV (2004) Differential expression of thromboxane synthase in prostate carcinoma: role in tumor cell motility. Am J Pathol:429–439

Okawa M, Kinjo J, Nohara T, Ono M (2001) DPPH (1,1-diphenyl-2-Picrylhydrazyl) radical scavenging activity of flavonoids obtained from some medicinal plants. Biol Pharm Bull: 1202–1205

Omóbòwálé TO, Oyagbemi AA, Adejumobi OA, Orherhe EV, Amid AS, Adedapo A, Yakubu MA (2016) Preconditioning with *Azadirachta indica* ameliorates cardiorenal dysfunction through reduction in oxidative stress and extracellular signal regulated protein kinase signalling. J Ayurveda Integr Med 7(4):209–217

Pai MR, Acharya LD, Udupa NJ (2004) Evaluation of antiplaque activity of *Azadirachta indica* leaf extract gel a 6-week clinical study. Ethnopharmacol 90:99

Panahi Y, Khedmat H, Valizadegan G, Mohtashami R, Sahebkar A (2015) Efficacy and safety of *Aloe vera* syrup for the treatment of gastroesophageal reflux disease: a pilot randomized positive-controlled trial. J Tradit Chin Med 35:632–636

Parida MM, Upadhyay C, Pandya G (2002) Possible anti-viral effects of neem (*Azadirachta indica*) on dengue virus. AMJ Ethnopharmacol 79:273

Parida P, Paul F, Chakravorty D (2020) Nature to nurture- identifying phytochemicals from Indian medicinal plants as prophylactic medicine by rational screening to be potent against multiple drug targets of SARS-CoV-2. ChemRxiv, pp 1–17

Park CH, Son HU, Yoo CY, Lee SH (2017) Low molecular-weight gel fraction of *Aloe vera* exhibits gastroprotection by inducing matrix metalloproteinase-9 inhibitory activity in alcohol-induced acute gastric lesion tissues. Pharm Biol 55:2110–2115

Patel Y, Vadgama V, Baxi S, Tripathi CB (2011) Evaluation of hypolipidemic activity of leaf juice of *Catharanthus roseus* (Linn.) G. Donn. In Guinea pigs. Acta Pol Pharm 68(6):927–935

Patil PJ, Ghosh JS (2010) Antimicrobial activity of *Catharanthus roseus* –a detailed study British. J Pharmacol Toxicol 1(1):40–44

Peng X, Cheng KW, Ma J (2008) Cinnamon bark proanthocyanidins as reactive carbonyl scavengers to prevent the formation of advanced glycation endproducts. J Agri Food Chem:1907–1911

Petrovska B (2012) Historical review of medicinal plants' usage. Pharmacognosy Rev 6(11):1–5

Pham Hong NT, Vuong QV, Bowyer MC, Scarlett CJ (2020) Phytochemicals derived from *Catharanthus roseus* and their health benefits. Technologies 8(4):80. https://doi.org/10.3390/technologies8040080

Pingali U, Sukumaran D, Nutalapati C (2020) Effect of an aqueous extract of *Terminalia chebula* on endothelial dysfunction, systemic inflammation, and lipid profile in type 2 diabetes mellitus: a randomized double-blind, placebo-controlled clinical study. Phytother Res 34(12):3226–3235

Polaquini SR, Svidzinski TI, Kemmelmeir C, Gasparetto A (2006) Effect of aqueous extract from neem on hydrophobicity, biofilm formation and adhesion in composite resin by *Candida albicans*. Arch Oral Biol 51(6):482–490. https://doi.org/10.1016/j.archoralbio.2005.11.007

Prakoso YA (2018) The effects of *Aloe vera* cream on the expression of $CD4^{+}$ and $CD8^{+}$ lymphocytes in skin wound healing. J Trop Med. https://doi.org/10.1155/2018/6218303
Pramanik KK, Singh AK, Alam M, Kashyap T, Mishra P, Panda AK, Mishra R (2016) Reversion-inducing cysteine-rich protein with Kazal motifs and its regulation by glycogen synthase kinase 3 signaling in oral cancer. Tumor Biol:15253–15264
Prasanth DS, Murahari M, Chandramohan V, Panda SP, Atmakuri LR, Guntupalli C (2021) *In silico* identification of potential inhibitors from cinnamon against main protease and spike glycoprotein of SARS CoV-2. J Biomol Struct Dyn 39(13):4618–4632
Prashant GM, Chandu GN, Murulikrishna KS, Shafiulla MD (2007) The effect of mango and neem extract on four organisms causing dental caries: *Streptococcus mutans*, *streptococcus slivavius*, *Streptococcus mitis*, and *streptococcus Sanguis*: an in vitro study. Indian J Dent Res 18:148–151
Priya N, Mathur KC, Sharma A, Agrawal RP, Agarwal V, Acharya J (2019) Effect of *Terminalia arjuna* on total platelet count and lipid profile in patients of coronary artery disease. Adv Hum Biol 9:98–101
Pugazhendhi A, Rajamohamed BS, Devi KP, Natarajan S (2018) Assessment of antioxidant, anticholinesterase and antiamyloidogenic effect of *Terminalia chebula*, *Terminalia arjuna* and its bioactive constituent 7-methyl gallic acid – an in vitro and in silico studies. J Mol Liquids 257:69–81
Rahmani AH, Alzohairy MA, Khan MA, Aly SM (2014) Therapeutic implications of black seed and its constituent thymoquinone in the prevention of cancer through inactivation and activation of molecular pathways. Evid Based Complement Alternat Med 13:18
Ramazani E, Yazdfazeli M, Emami SA, Mohtashami L, Javadi B, Asili J, Tayarani-Najaran Z (2020) Protective effects of *Cinnamomum verum*, *Cinnamomum cassia* and cinnamaldehyde against 6-OHDA-induced apoptosis in PC12 cells. Mol Biol Rep 47:2437–2445
Ranasinghe P, Pigera S, Sirimal Premakumara GA, Galappaththy P, Constantine GR, Katulanda P (2013) Medicinal properties of 'true' cinnamon (*Cinnamomum zeylanicum*): a systematic review. BMC Complement Altern Med 13:275
Rao BCS, Singh RH, Tripathi K (2001) Effect of *Terminalia arjuna* (W&A) on regression of LVH in hypertensives: a clinical study. J Res Ayurveda Siddha 22(3–4):216–227
Saito M, Tanaka M, Misawa E, Yao R, Nabeshima K, Yamauchi K, Furukawa F (2016) Oral administration of *Aloe vera* gel powder prevents UVB-induced decrease in skin elasticity via suppression of overexpression of MMPs in hairless mice. Biosci Biotech Biochem 80: 1416–1424
Sánchez M, Elena G-B, Iglesias I, Gómez-Serranillos MP (2020) Pharmacological update properties of *Aloe vera* and its major active constituents. Molecules 25:1324. https://doi.org/10.3390/molecules25061324
Sarkar L, Oko LB, Gupta SC, Bubak A, Das B, Gupta P, Safiriyu AA, Singhal C, Neogi U, Bloom D, Banerjee A, Mahalingam R, Cohrs RJ, Koval M, Shindler KS, Pal D, Nagel M, Sarma JD (2022) *Azadirachta indica A.* Juss bark extract and its Nimbin isomers restrict β-coronaviral infection and replication. Virology 569:13–28
Shanmugam S, Naidu A (2000) Antioxidant activity of selected India spices. Prostaglandins Leukot Essent Fatty Acids 62:107–110
Sharma P, Tomar L, Bachwani M, Bansal V (2011) Review on neem (*Azadirachta indica*): thousand problem one solution. Int Res J Pharm 2:97–102
Shekarchizadeh-Esfahani P, Heydarpour F, Izadi F, Jalili C (2021) The effect of cinnamon supplementation on liver enzymes in adults: a systematic review and meta-analysis of randomized controlled trials. Complement Ther Med 58. https://doi.org/10.1016/j.ctim.2021.102699
Shilpa G, Renjitha J, Saranga R, Sajin FK, Nair MS, Joy B, Priya S (2017) Epoxyazadiradione purified from the *Azadirachta indica* seed induced mitochondrial apoptosis and inhibition of NFκB nuclear translocation in human cervical cancer cells. Phytother Res:1892–1902
Singh G, Singh AT, Abrahama A (2008) Protective effects of *Terminalia arjuna* against doxorubicin-induced cardiotoxicity. J Ethnopharmacol 117:123–129

Sithisarn P, Supabphol R, Gritsanapan W (2005) Antioxidant activity of Siamese neem tree (VP 1209). J Ethnopharmacol:109–112

Song F, Li H, Sun J, Wang S (2013) Protective effects of cinnamic acid and cinnamic aldehyde on isoproterenol-induced acute myocardial ischemia in rats. J Ethnopharmacol:125–130

Srinivasan R, Meenakshi R (2021) *In silico* screening and molecular docking of phytochemicals of *Terminalia arjuna* and *Terminalia chebula* against the Covid-19 main protease. J Adv Sci Res 12(3):1–10

Størsrud S, Pontén I, Simrén M (2015) A pilot study of the effect of *Aloe barbadensis* mill. extract (AVH200®) in patients with irritable bowel syndrome: a randomized, double-blind, placebo-controlled study. J Gastrointestin Liver Dis 24:275–280

Sumitra M, Manikandan P, Kumar DA, Arutselvan N, Balakrishna K, Manohar BM, Puvanakrishnan R (2001) Experimental myocardial necrosis in rats: role of arjunolic acid on platelet aggregation, coagulation, and antioxidant status. Mol Cell Biochem 224:135–142

Taheri JB, Azimi S, Rafieian N, Akhavan Zanjani H (2011) Herbs in dentistry. Int Dent J 61: 287–296

Tamizhazhagan V, Pugazhendy K, Sakthidasan V, Jayanthi C (2017) Preliminary screening of phytochemical evaluation selected plant of *Pisonia alba*. J Biol Res 2(4):63–66

Teplicki E, Ma Q, Castillo DE, Zarei M, Hustad AP, Chen J, Li J (2018) The effects of *Aloe vera* on wound healing in cell proliferation, migration, and viability. Wounds 30:263–268

Tripathi VK, Pandey VB, Udupa KN, Rucker G (1992) Arjunolitin, a triterpene glycoside from *Terminalia arjuna*. Phytochemistry 31:349–351

Trybus W, Krol T, Trybus E, Stachurska A, Kopacz-Bednarska A, Krol G (2018) Induction of mitotic catastrophe in human cervical cancer cells after administration of aloe-emodin. Anticancer Res 38:2037–2044

Uppuluri LLPB, Garge VN, Kadam VJ (2015) Evaluation of anti-angiogenesis activity of neem root using zebra fish model. Int J Pharma Sci Res:2437–2440

Upreti J, Ali S, Basir SF (2013) Effect of lower doses of vanadate in combination with *Azadirachta indica* leaf extract on hepatic and renal antioxidant enzymes in streptozotocin-induced diabetic rats. Biol Trace Elem Res:202–209

Vangalapati M, Sree Satya N, Surya Prakash D, Avanigadda S (2012) A review on pharmacological activities and clinical effects of cinnamon species. Res J Pharm, Biol Chem Sci:653–663

Varghese A, Savai J, Pandita N, Gaud R (2015) *In vitro* modulatory effects of *Terminalia arjuna*, arjunic acid, arjunetin and arjungenin on CYP3A4, CYP2D6 and CYP2C9 enzyme activity in human liver microsomes. Toxicol Rep 2:806–816

Viswanatha SGL, Vaidya SK, Ramesh C, Krishnadas N, Rangappa S (2010) Antioxidant and antimutagenic activities of bark extract of Terminalia arjuna. Asian Pac J Trop Med 3(12): 965–970

Wang J, Su B, Jiang H, Cui N, Yu Z, Yang Y, Sun Y (2020) Traditional uses, phytochemistry and pharmacological activities of the genus *Cinnamomum* (Lauraceae): a review. J Pre Proof Fitoter:104675

Wei W, Jia Y, Hui C (2017) Radiotherapy plus procarbazine, lomustine, and vincristine versus radiotherapy alone for glioma: a meta-analysis of randomized controlled trials. Int J Clin Exp Med 10:6810–6818

Werawatganon D, Linlawan S, Thanapirom K, Somanawat K, Klaikeaw N, Rerknimitr R, Siriviriyakul P (2014) *Aloe vera* attenuated liver injury in mice with acetaminophen-induced hepatitis. BMC Complement Altern Med 14:229

Xiang H, Cao F, Ming D, Zheng Y, Dong X, Zhong X, Wang L (2017) Aloe-emodin inhibits *Staphylococcus aureus* biofilms and extracellular protein production at the initial adhesion stage of biofilm development. Appl Microbiol Biotechnol 101:6671–6681

Yerima MB, Jodi SM, Oyinbo K, Maishanu HM, Farouq AA, Junaidu AU, Al-Mustapha MN, Shinkafi AL (2012) Effect of neem extracts (*Azadirachta indica*) on bacteria isolated from adult mouth. Nigerian J Basic Appl Sci 20(1):64–67

Zareie A, Bagherniya M, Sharma M, Khorvash F, Hasanzadeh A, Askari G (2020) Effects of cinnamon on anthropometry status and headache disability of migraine patients: a randomized double-blind placebo-controlled trial. J Headache Pain 34(11):2945–2952

Zheng W, Wang SY (2001) Antioxidant activity and phenolic compounds in selected herbs. J Agri Food Chem 49(11):5165–5170

Chapter 14
Conservation Strategies for Indian Medicinal Plants

Saidaiah Pidigam, Amarapalli Geetha, Someshwar Rao Pandravada, Suhail Khan, Vishnukiran Thuraga, Natarajan Sivaraj, and Kattula Nagaraju

Abstract With her ca. 7% biodiversity contribution to the world's diversity, India is ranked among the mega diverse countries. Around 8000 Indian medicinal plant species have utilization as core raw drugs in the Indian Systems of Medicines. Around 65% health care is met from traditional medicines available in India as per the World Health Organization report. Remarkably, in India, natural forests provide the base for the medicinal plants. More than 90 per cent of Indian medicinal plant species are threatened due to their over exploitation, unscientific collection and harvesting methods. Around 1000 species of medicinal plants are also in threat across India, in different eco-systems. Conservation practices and sustainable use of these medicinal plants is vital in India for livelihood security, their history, culture, ecological balance maintenance and ensuring health. *In-situ* conservation practices

S. Pidigam (✉)
Department of Genetics and Plant Breeding, College of Horticulture, Sri Konda Laxman Telangana State Horticultural University, Mojerla, Telangana, India

A. Geetha
Department of Crop Physiology, College of Agriculture, Professor Jayashankar Telangana State Agricultural University, Palem, Telangana, India

S. R. Pandravada · N. Sivaraj
Division of Plant Genetic Resources and Economic Botany, ICAR-National Bureau of Plant Genetic Resources, Hyderabad, Telangana, India

S. Khan
Department of Biotechnology, University School of Biotechnology, Guru Gobind Singh Indraprastha University, Dwaraka, New Delhi, India

V. Thuraga
Department of Genetics and Plant Breeding, College of Horticulture, Sri Konda Laxman Telangana State Horticultural University, Mojerla, Telangana, India

Department of Plant Breeding, Swedish University of Agricultural Sciences, Uppsala, Sweden

K. Nagaraju
Department of Vegetable Science, College of Horticulture, Sri Konda Laxman Telangana State Horticultural University, Mojerla, Telangana, India

Á. Máthé, I. A. Khan (eds.), *Medicinal and Aromatic Plants of India Vol. 2*,
Medicinal and Aromatic Plants of the World 9,
https://doi.org/10.1007/978-3-031-44914-7_14

and *ex-situ* conservation strategies are well established to conserve the medicinal plant species in India. Biosphere reserves, national parks, wild life sanctuaries, sacred groves and other protected areas forms are the examples of '*in-situ*' methods of conservation in place in India. Medicinal plant conservation must go hand in hand by government and collectors, cultivators. However, for effective conservation of medicinal plants, Indian policy framework and laws needs to be further strengthened. In the current review, we focused on biodiversity hotspots, mega-centres, micro-centres of endemic medicinal plants and the IUCN red listed medicinal and aromatic plants (MAPs) of India. *In-situ* conservation like gene banks, biosphere reserves, national parks, sacred grooves and *ex-situ* conservation strategies such as botanic gardens, parks, gene banks (seed bank, field gene bank, pollen bank, DNA based libraries, *etc.*) etc., are also reviewed.

Keywords *Conservation strategies* · *in-situ* · *ex-situ* · gene banks · India · medicinal plants

14.1 Introduction

More than 30% of India's land, including areas like the Himalayas, the Western Ghats, and the Andaman and Nicobar Islands, has not yet been thoroughly explored or documented in terms of its flora. The biodiversity of India is facing significant threats, including both natural causes such as competition between species and imbalances within species, as well as human-driven factors such as deforestation for agriculture, mining, and urbanization, overuse of natural resources, and introduction of non-native species. These threats have led to the decline of several species of plants, including *Eremostachys superba*, *Frerea indica*, *Aitchisoniella himalayensis*, *Monoselenium tenerum*, *Sewardiella tuberifera*, and *Stephensoniella brevipedunculata*. According to estimates, over 26,000 plant species are endangered globally (Sharma et al. 2014).

Biodiversity hotspots around the world, including India, have a significant amount of genetic resources. India is particularly rich in biodiversity, with around 17,000–18,000 species of flowering plants, accounting for 7% of global biodiversity (Pushpangadan and Nair 2001). Within India, there are around 8000 species of medicinal plants that are distributed across 386 families and 2200 genera of flowering plants. These plants are the primary source of raw materials for Indian Systems of Medicine (Ijinu et al. 2022), which have a long history of using plant-based medicine, described in ancient Vedic literature. Over the past few years, there has been a significant increase in the recognition of 10–18% of the total medicinal plant biodiversity (50,000 plants) in the pharmaceutical industry. There is also a growing demand for herbal products in India and abroad, resulting in a rapid increase in the trade of plants, plant parts, and value-added products. India is the second-largest exporter of medicinal plants in the world, with over 6000 species of medicinal plants found in the country. The World Health Organization estimates that almost 65% of India's population depends on traditional medicine for their healthcare needs. The majority of medicinal plants are found in natural forests, which are particularly important given the increasing global demand for herbal products. It is predicted that the global market for herbal products will reach USD 5 trillion by 2050 (Anonymous 2019).

India possesses a wide range of agro-climatic, ecological, and edaphic conditions, which have influenced its vegetation and floristic composition. It is recognized as one of the world's 12 leading centers of biodiversity, with 16 different agro-climatic zones, 10 vegetation zones, 25 biotic provinces, and about 426 habitats of specific species. The Indian subcontinent is home to approximately 45,000 plant species (Trivedi 2006), which is almost 20% of the global species. However, over 90% of medicinal plants in India are facing the threat of over-exploitation, unsustainable collection, or unskilled harvesting (Kumari et al. 2011). It is estimated that around 1000 medicinal plant species may be under threat in different ecosystems across India (Frlhtenvis 2016). There are approximately 3500 types of medicinal and aromatic plants (MAP) belonging to both higher and lower plant groups are harvested from around 17 million hectares of forest land in India. India is rich in all three levels of biodiversity - species, genetic, and habitat diversity (Mukerji 2006) and a significant portion of it is traded, generating an annual turnover of US$ 2.5 billion. The rising demand and destructive harvesting practices are not only endangering the survival of many species but also the livelihoods of the people who depend on them (UNU-IAS 2012). The conservation and sustainable use of medicinal plants are crucial for the 4635 ethnic communities in India, their history and culture, as well as for ensuring health, livelihood security, and ecological balance (Máthé and Khan 2022).

14.2 Biodiversity Hotspots in India

India is a country with diverse geographical regions, climatic conditions, and ecological habitats that are home to rich floral diversity. It has nearly 8% of the world's flora, of which 28% are endemic to the country (Myers et al. 2000; Bapat et al. 2008; Lakshminarasimhan et al. 2020). According to Conservation International, there are 36 biodiversity hotspots in the world that support more than 50% of the global endemic plant species and nearly 43% of the bird, mammal, reptile, and amphibian species. India is one of the 17 mega-diversity countries in the world (Williams 2001)and is estimated to have a total of 18,800 species of angiosperms, 82 species of gymnosperms, 1307 species of pteridophytes, and 2786 species of bryophytes, as well as many species of fungi, algae, lichens, viruses, and bacteria. India has four biodiversity hotspots: Himalaya, Indo-Burma, Sunderland, and Western Ghats (Fig. 14.1).

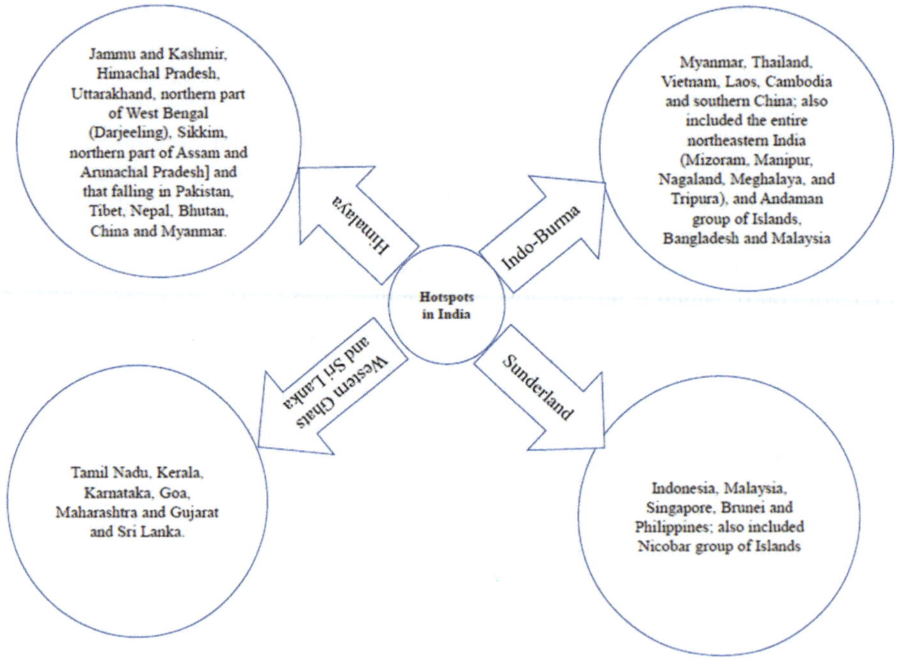

Fig. 14.1 The ecosystem profiles of four hotspot regions India

14.3 Megacentres and Microcentres of Endemic Plants in Indian Subcontinent

The plant life of India, specifically flowering plants, has a notable feature of having a high number of unique species found only in India, second only to Australia. A botanist named Nayar (1996) has identified three major areas and 25 smaller areas within India where these endemic plant species are particularly diverse and distributed (Fig. 14.2).

14.4 Distribution of Medicinal Plants in India

An overall analysis of the distribution of medicinal plants reveals that MAPs are spread out in a wide variety of habitats and elements within the landscape. A significant proportion, approximately 70%, of India's medicinal plants can be found in tropical regions, particularly in different types of forests located in areas such as the Western and Eastern ghats, Vindhyas, Chotta Nagpur plateau, Aravalis, and Himalayas. Although less than 30% of medicinal plants are present in the temperate and alpine regions at higher altitudes, these areas do contain species that

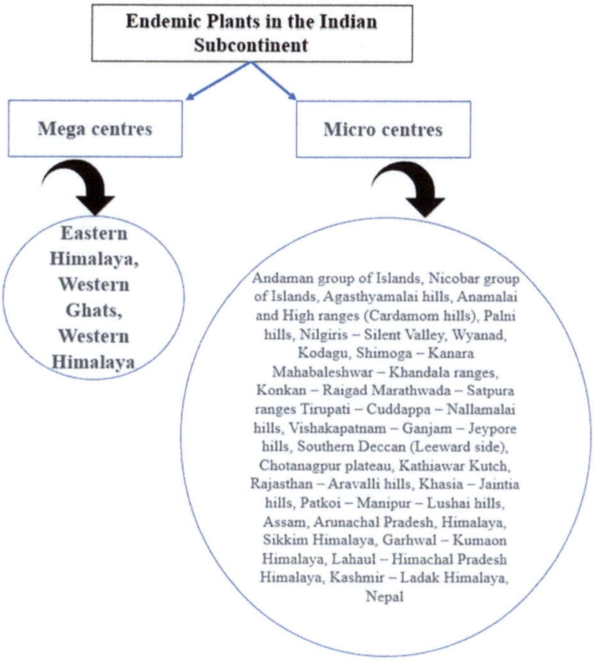

Fig. 14.2 Mega- and Microcentres of Endemic Plants in the Indian Subcontinent. (Source: Nayar (1996))

are of significant medicinal value. Studies suggest that a greater percentage of the known medicinal plants are located in dry and deciduous vegetation as compared to evergreen or temperate habitats.

Medicinal plants are present in a variety of habitats and have different growth patterns, with one third being trees, another third being shrubs, herbs, grasses and climbers making up the rest, with only a small proportion being lower plants such as lichens, ferns, and algae. Most medicinal plants are higher flowering plants, belonging to 386 families and 2200 genera, with Asteraceae, Euphorbiacae, Lamiaceae, Fabaceae, Rubiaceae, Poaceae, Acanthaceae, Rosaceae and Apiaceae being the families with the largest number of medicinal plant species. Of these, Asteraceae has the highest number of species, with 419 in total. Although the industry uses over 800 medicinal plant species, less than 20 of them are commercially cultivated, with the majority being collected from the wild, which poses a threat to the genetic stock and diversity of medicinal plants due to destructive harvesting practices. The Deccan Peninsula is estimated to have the highest number of medicinal plants, followed by the Himalayan region, with desert and coastal areas having the least number of medicinal plants (Fig. 14.3).

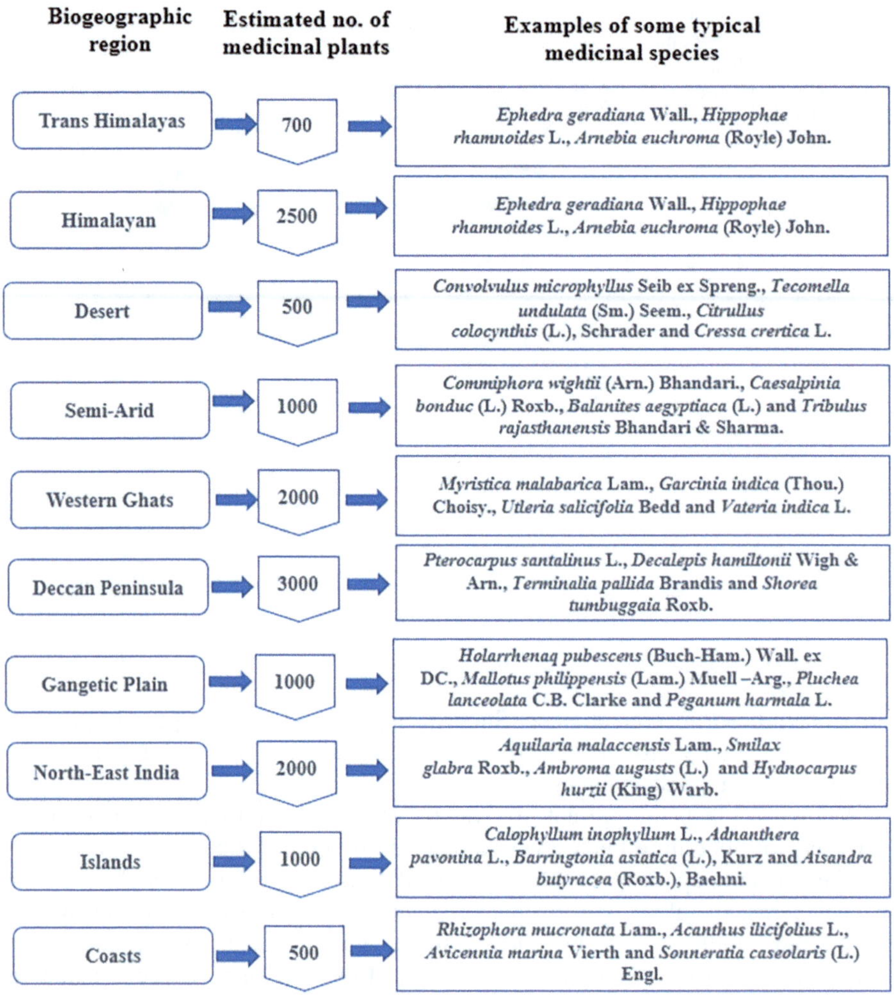

Fig. 14.3 Medicinal plants: species diversity and representative species of different biogeographic zones of India. (Ved et al. 2001)

14.5 IUCN Red Listed Medicinal Plants of India

In 1998, the IUCN Red List of Threatened Plants was created and includes over 8000 endangered species (Walter and Gillett 1998). This is a global effort that assesses the conservation status of plant, animal, and fungi species. The 2015 update (Fig. 14.4) added 44 medicinal plant species from India, and these were classified as vulnerable, endangered, and critically endangered. Critically endangered indicates that the species' numbers have already decreased or will decrease by 80% within three generations. Endangered species have or will decrease by 50% and vulnerable

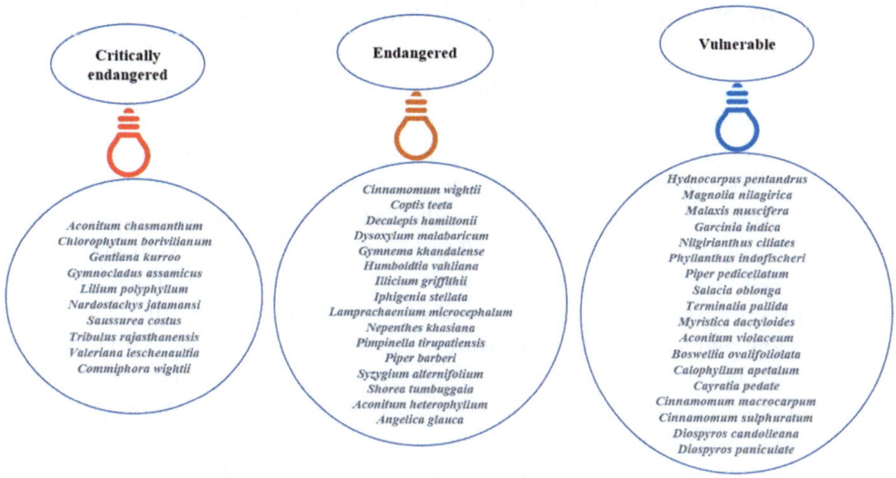

Fig. 14.4 IUCN Red List medicinal plants category, 2015

species by 30%. In India's recent update, ten medicinal plant species were listed as critically endangered, 16 as endangered, and 18 as vulnerable. For instance, *Aconitum chasmanthum*, a poisonous herb found in the Himalayan region of India and Pakistan, is listed as critically endangered due to habitat loss, road construction, and overexploitation. *Lilium polypyllum* is also critically endangered due to unregulated collection, habitat loss, agricultural invasions, and increased human habitation.

In the last 20 years, the Conservation Breeding Specialist Group (CBSG), India in collaboration with the Foundation for Revitalization of Local Health Traditions (FRLHT) have organized 14 CAMP workshops, which covered 17 states in India. The workshops aimed to conduct quick assessments of medicinal plants species that are of high conservation value in various regions and states of India.

The promotion and awareness of plant products is crucial, and various methods such as press reports, advertising, education, and scientific reports should be used. To conserve and maintain the gene pool of medicinal plants, their germplasms should be collected and preserved in gene banks using biotechnological methods, as well as vegetative propagation in conservation centers. Funding for botanical gardens is available from various institutes and universities. Tribal communities should be educated on the significance of medicinal plants. Despite existing regulations and policies, effective action has been limited for many endangered species. Effective policies regarding conservation, cultivation, education, capacity building, research, regulations, and trade need to be drafted, as suggested by Gopi et al. (2018), who reviewed the IUCN red-listed medicinal plants of India, including examples of critically endangered species, such as *Heterophyllum* sp., *Aquilaria malaccensis*, *Adhatoda beddomei*, and *Nardostachys jatamansi*, which are facing continuous exploitation from the wild.

14.6 Conservation of Medicinal Plants in India

Medicinal plant conservation is difficult due to the vast range of habitats and geographical distribution of taxa. The long-term usage of medicinal PGRs is linked to their conservation. Some initiatives in this regard include *in situ* and *ex situ* kinds of conservation programs including gene banks, sustainable harvesting, and BioTrade. The 'BioTrade is launched with UNCTAD in tune with the activities with biodiversity preservation in trade of goods and services derived from biodiversity. Regulation of use of germplasm, their survey and use for commercialization, getting IPRs, transfer of results, and accessed biological resources are all examples of protection and conservation measures. At international level, CBD and the Nagoya protocol on access, sharing of benefits and nationally, through domestic laws regulating access to genetic resources, ABS principles are developed. As a result, their conservation threats and eventual uses are diversified, and users include both local rural people and far-flung city dwellers. However, it is commonly acknowledged that medicinal plant conservation can be accomplished through a combined approach that balances *in situ* and *ex situ* conservation techniques. To succeed, medicinal plant conservation must work across various domains, bringing together dissimilar parties and mutually honoring distinct stakeholder interests.

The conservation plan (IUCW, UNEP, and WWF, 1980) defines conservation as "the management of human use of biodiversity to provide the maximum sustainable benefit to current generations while maintaining its potential to meet the needs and aspirations of future generations" (Baricevic et al. 2015). Scientific techniques as well as social measures can be used to combat the conservation of wild medicinal plants or any other vulnerable species. The basic two techniques are *in-situ* conservation and *ex-situ* conservation, in addition to the scientific conservation strategies such as legislation for medicinal plants in India.

Rajasekharan and Rao (2019) investigated the present condition of conservation and utilization of horticulture PGRs including medicinal category, focusing on modern conservation techniques in relation to biotechnological approaches, as used in India and elsewhere. Investigation also presented on legal and biotechnological elements of horticulture genetic resources, as well as crucial components of sustained maintenance and re-use. In addition, Baruah (2015) and Gupta (2018) reported on R&D trends in high-demand species, efficient usage, and legislation -actions for PGR related protection in Indian sub-continent. The Scoping Paper on "Protection of Traditional Knowledge in India (2018)" by FITM maps the protection of TK and associated PGRs. Medicinal PGRs, on the other hand, are primarily included in the larger umbrella of PGRs and are not investigated separately.

14.7 Legislation on Medicinal Plant Trade in India

The WHO, IUCN, and WWF issued Guidelines on the Conservation of Medicinal Plants, which stated that legislation should prohibit not just the collecting, but also the possession and trading of medicinal plants acquired in the wild. This is crucial for enforcement since catching an offender in the act of gathering a protected plant is difficult but catching him or her afterwards in possession of the plant is considerably easy. Landowners should be prohibited from collecting plant material on their property as part of the legislation. When traditional practitioners' access to a plant is restricted, steps should be taken to provide them with alternative species that contain equivalent components. Practitioners should be involved in the conservation of vulnerable plants (WHO, IUCN and WWF 1993).

In India, as such no exclusive policies or regulations in place to protect medicinal plants that thrive in forests. Existing forestry rules provide protection for the environment. The following are regulations enacted by the Indian government for forest conservation, which preserve the flora of wild herbs directly or indirectly. To prevent over-use of medical plants of wild nature by collectors and to regulate medicinal plant exports, the country established a number of special nature reserves to safeguard wild medicinal plants and assign employees to monitor them (WHO, IUCN and WWF 1993). The Forest Department of Tamil Nadu is executing for India's endemic and vulnerable flora. Several Indian species have become extinct in the last 10 years. More than 1200 plants that are only found in India are thought to be under threat in varying degrees. According to the BD Act of 2002, the Indian Union Govt. has the authority to declare any species threatened and impose regulations on it. Due to the fact that forests are on the concurrent list, the federal government can only provide guidance to the states. Only six medicinal plants are protected under the Indian Wildlife Protection Act of 1972, and they are still protected after 44 years. The Wildlife Protection Act is comprehensive enough to safeguard animal species, yet it only applies to six plant species. The Forest Conservation Acts of 1980 and 1988 are solely concerned with habitat preservation. Instead of putting these medical plants on the endangered list and prohibiting their usage, they should be grown in greater numbers. Banning discourages the business and reduces production, putting it in jeopardy as a result of increased illegal commerce that cannot be controlled.'

Habitat destruction is another danger that medicinal plants confront. Medicinal plants are protected under the Forest (Conservation) Act of 1980 and the Wildlife (Protection) Act of 1972. However, many medicinal plants grow outside of protected areas and because coordinated strategy for medicinal plants does not exist, many of them simply vanished. The lack of a concerted conservation plan within protected areas could also result in the depletion of this unique resource. Today, the Indian pharmaceutical sector collects over 95% wild plants (FRLHT 1997). Two thirds of collections are roots, barks or its traces, and stems, as well as the entire plant in some cases, resulting in harmful harvesting. This method, if not adequately regulated, could result in the reduction of valuable plants.

The want of legislative and policy support for wild harvesting is the main problem in medicinal plants conservation. It data is rarely avialable or shocased at the national level because production and as well as consumption is at a subsistence level. That leads to lower emphasis on govt. planning, analysis and budget allocations (Vantomme in Anon. 2002). Listing of medicinal plant species to the Convention on International Trade in Endangered Species would be helpful to governments to frame working legislation on their harvest and trade even in protected areas. The policy should be written with the diverse user groups of medicinal plants in mind. It must take into account the fact that non-commercial users outnumber business users by a substantial margin. The policy framework should promote in-*situ* and *ex-situ* conservation, as well as horticulture. Given the long-term availability of medicinal plants and the current demands of user groups, this is necessary. The strategy would have to consider ways to increase financial resources as well as incentives for conservation. To make implementation easier, the policy should examine existing field institutions, encourage their improvement, and, if required, consider the potential of establishing new organizations. Most essential, the policy must consider the legal and regulatory framework surrounding medicinal plants. Several initiatives are underway to develop such a policy. In January 1997, the FRLHT made a groundbreaking endeavor. In Madras, the FRLHT hosted national level consultation on medicinal plants to detailing national policy recommendations.

The Central Government signals the states about the danger of extinction under Section 38 of the BD Act, 2002. To promote and expand the medicinal plants industry, the NMPB devised many policies, strategies, and programs. On July 6, 2020, ICAR-NBPGR and NMPB signed a Memorandum of Understanding (MoU) for the safe conservation of MAPs at ICAR-National NBPGR's Gene Bank (ICAR-NBPGR 2020). Government of India has signed a few international accords in addition to its national policy. One of such is CITES which protects approximately 5000 animals species and 28,000 plantspecies by over-use by international traders. The MOEF, Government of India listed 113 species (CITES 2017) mandating 'Certificate of Cultivation' or 'Legal Procurement Certificate' from the Forest Department's authorization for exports. Currently, 12 medicinal plants from India are listed in the CITES. Similarly, as a part of EXIM Policy, around 29 plants are prohibited for exports. Further, parts of the plants, derivatives, and extracts sourced from wild plants are out of legal.

14.7.1 **In-situ** *Conservation of Medicinal and Aromatic Plants in India*

In-situ conservation refers to the preservation of a species in its natural habitat or the place where it grows naturally. It encompasses gene banks and gene sanctuaries, as well as biosphere reserves, national parks, sacred locations, and sacred grooves. This type of conservation covers around 10 million hectares, or 4.5 percent of India's land area (Singh and Chowdhery 2002), and had around 2706 establishments of this type. In addition, India has between 100,000 and 150,000 sacred groves (Kandari et al. 2014). There are 870 protected places in India, including 104 national parks, 551 animal sanctuaries, 127 community reserves, and 88 conservation reserves (WIIENVIS 2019). *In situ* conservation of PGRs through community biodiversity management was compiled in detail by De Boef et al. (ed.) (2013).

14.7.1.1 Biosphere Reserves

In India, there are 12 biosphere reserves. The Nanda Devi Biosphere Reserve, a World Heritage site in the Indian Himalaya separated by a buffer zone of 1612 kilo meters, is located in the state of Uttar Pradesh. In the buffer zone, 17 Bhotiya towns are interspersed as patches in the matrix of natural ecosystems ranging from temperate forests to snow-capped peaks, covering 21% of the total area of the buffer zone. The reserve is home to a collection of medicinal plants. e.g., *Aconitum heterophyllum*, *Podophyllum hexandrum*, *Dactylorhiza hatagirea*, *Nardostachys grandiflora*, *Taxus baccata* and *Picrorhiza kurrooa* (Maikhuri et al. 2001). Tourists who enjoy adventure and environment are drawn to the area. In 1992, the reserve was designated as a UNESCO World Heritage Site. Eight medicinal plant species were grown on 4% of private farmland, developed as an indigenous activity in reaction to limits on customary rights to collect in the wild and attempts to meet the market's growing demand for medical items. 70% of the total land area in medicinal plant cultivation was devoted to *Allium humile* and *Allium stracheyi*. Except for *Allium stracheyi*, all farmed species were naturally regenerating in forests and pasture lands. Carum carvi produced the highest economic returns among the farmed species, followed by *Allium humile* (Nautiyal et al. 2001). This biosphere has been designated as a model reserve for long-term medicinal plant use. The majority of plant species used in traditional healthcare and herbal industries (90 percent) is brought from their natural habitats (i.e., subalpine and alpine meadows) in the Himalayan region, which has an impact on the population density of economically valuable MAPs and plant diversity in the alpine ecosystem. The growing demand for MAPs on the local, regional, and international markets has drawn scientists and pharmaceutical companies' attention to their protection through cultivation. Maikhuri et al. 2017 emphasized the importance of selecting potential MAPs species with high economic value for large-scale cultivation and domestication in order to link MAP-based livelihoods to conservation. Through the Herbal Research Development Institute (HRDI), Uttarakhand, the Uttarakhand government has been establishing a program to promote growth of some selected medicinal plant species with high conservation value and commercial potential. One of the policy's main goals is to encourage the cultivation of threatened species, which will alleviate wildlife pressure. The G.B. Pant National Institute of Himalayan Environment and Sustainable Development supporting medicinal plant cultivation/conservation and value addition for the period 1996–2014. Unemployment in Uttarakhand's mountain area, which is one of the state's most serious concerns and could be reduced to some extent by involving unemployed young people in small-scale entrepreneurship based on medicinal plants. However, the plants' ecological state must be determined. *In situ* and *ex situ* conservation and cultivation strategies that decrease anthropogenic pressure on wild habitat and preserve MAPs. *In-situ* medicinal plant conservation in India could be achieved with participation of people who lives in or around the protected forest areas.

14.7.1.2 Sacred Groves

One of the key strategies for conserving natural resources is the concept and beliefs of sacred trees and forest groves (Kandari et al. 2014). SGs, sacred groves are small woodland areas left undisturbed by locals in order for the local village folk deities to maintain them. Natural museums of giant trees, treasure houses of endangered species, dispensaries of medicinal plants, gene banks of economic species, are some of the other names given to them (Bhagwat et al. 2005; Manikandan et al. 2011). In India, there are said to be between 100,000 and 150,000 sacred groves (Malhotra et al. 2007). The state of Himachal Pradesh has been reported to have the largest number of SGs (5000), followed by Kerala and Chhattisgarh. According to several research reports, these SGs can support and conserve various wildlife and vegetation (Singh et al. 2010). The SGs are thought to contain a wealth of medicinal and aromatic plants. All Indians are aware that Lord Hanuman (a Hindu god) used Sanjeevani found in the Himalayan region to resuscitate Lakshman, Lord Rama's brother. In various parts of India, a vast number of medicinal plants that are an essential component of SGs are to be rotected, as evidenced by Pandit and Bhakat 2007. The preservation of medicinal plants is an important aspect of these people's sustainable relationship with nature (King-Oliver et al. 1997). Medicinal plants like as mint, coriander, and fenugreek are claimed as vital component of SGs and the preservation of the Himalayan ecology up to now. Documentation of all sacred groves should be prioritized, as should promoting benefit sharing for the conservation and maintenance of the groves. Stringent legislation, with appropriate punishment, is also required for protection of sacred groves, as well as the conservation of medicinal plants.

14.7.2 Ex-situ *Conservation of Medicinal and Aromatic Plants in India*

Conservation of medicinal plants outside of their natural habitat can be achieved through cultivating in botanic gardens, parks, and etc., beside the plant propagule conservation in various gene banks *viz.,* seed, pollen, genomic libraries and cryopreservation. In order to promote the medicinal plants sector and to develop coordination among various departments towards implementation of support policies/

programs in improvement of medicinal plants sector, the Government of India established the National Medicinal Plants Board (NMPB) on 24th November 2000, which is working under the guidance of ministry of AYUSH.

Ex-situ conservation deals with maintenance of economically important germplasm through collection, preservation multiplication and distribution of materials (Bhattacharyya et al. 2006). The germplasm material is stored and maintained by ICAR through National Gene bank. Currently, 5756 accessions of medicinal plants representing 578 species of medicinal plants were conserved at the NGB. The cryopreservation of the germplasm was established through cryo-bank at NBPGR for preservation through desiccation sensitive seeds, vegetative tissues, pollen and selected orthodox seed species. The CSIR-CIMAP is actively working on medicinal and aromatic plants cultivation and varietal improvement & agro-technologies, gene bank development and bio-village mission for cultivation and increasing productivity. Twenty-five and seven improved medicinal plants and aromatic plants corresponding to 14 and six species of MAPs were identified and released. The target of the conservation, inventorization, and quantification of medicinal plants for commercial use was shared by the National Medicinal Plants Board (NMPB), inferring that, India framed sufficient agencies for protection, research and development of medicinal PGRs (Anonymous 2019).

14.7.2.1 Botanical Gardens

Botanical gardens are vital in *ex-situ*conservation, especially for those which have imminent threat of extinction. The preamble of Article 9 of the Convention on Biological Diversity describes the importance of establishing Botanic Gardens, as a complementary approach to *in-situ* conservation. The aim is to conserve threatened plant species and taxa of the country of their origin (https://bsi.gov.in/page/en/ex-situ-conservation). India constitutes big network of 140 botanical gardens, with guidance of 33 universities only for 24% of the botanical gardens. Around 4900 flowering plant species which are endemic are conserved in North-East India, the Western Ghats, north-west Himalayas and the Andaman and Nicobar Islands. It represents two of the 18 hot spots identified in the world. Botanical survey of India also listed 1500 rare and threatened species which requires conservation.

Tropical Botanical Gardens & Research Institute (TGBRI) is also working in these directions. *Rauvolfia serpentina* (Indian snakeroot, Sarpaghanda) is one of the medicinally important plant sources for the alkaloid resperine with crucial role in treating hypertension, is found in India, Pakistan and South East Asia, has revolutionized the treatment dated back to 1950s and is still in use today. Resperine is chemically similar to serotonin and is used in the treatment of severe mental illnesses, due to its role in sedation. Distribution of the snakeroot was confined to <5000 km^2 in most of tropical India during 1998, with wild population occupancy of <500 km^2 (Mamgain et al. 1998). Due to the limited availability of the resources, the export was banned by the Indian government (Indian Ministry of Commerce) in 1994, and in Nepal in 2001, leading to reduced trade compared to past decades

(CITES 2005). The quick reduction and isolated nature of snakeroot populations indicates that wild genetic stock is severely reduced. Despite of the reduced populations, India noticed successful snakeroot cultivation for many decades, and continues to invest in this area. According to plant search database, *Rauvolfia serpentina* is in cultivation at 22 botanic gardens. With the aim of developing a strategy for restoring the wild populations, BGCI is encouraging the policy framework (Hawkins 2008).

- To improve stock of the medicinal and aromatic plants, Krishna Mahavidyalaya Botanic Garden is growing ~130 species of medicinal plants in pot cultures (Hawkins 2008).
- Malabar Botanic Garden is promoting the cultivation of the state's native medicinal plants. In the last 2 years >200 farmers have been trained to promote the increased cultivation and trade of medicinal plants (Hawkins 2008).
- The Indian Council of Forest Research has commenced a program for developing 'Vanaspati Van' and cultivation of medicinal plants.
- Biodiversity in the North-Western Himalayan region of Himalayas was undertaken by G. B. Pant Institute of Himalayan Environment and Development.
- In Kerala, Tropical Botanical Garden and Research Institute is working on conservation, sustainable use of the medicinal plants of peninsular India through display garden, field gene bank, *in vitro* gene bank and seed gene bank.

14.7.2.2 Gene Banks

The concept of establishing field gene banks (FGBs) of plants facilitates for long term preservation of the genetic variability) of species. FGBs provide an easy and ready access to the PGRs, for characterization, evaluation and utilization, with conservation as seeds, *in vitro* or cryopreservation. FGB is highly efficient in conservation of vegetatively propagated genotypes than *in vitro.* TBGRI initiated the field gene banks MAPs under the G-15-GBMAP. The DBT, Government of India, and TBGRI have enough research for successful establishment MAPs FGBs. The Red listed species and endemic species are to be conserved in gene banks first. The world's second largest refurbished state-of-the-art National Gene Bank is established at ICAR-NBPGR, New Delhi. The National Gene Bank established in the year 1996 to preserve the seeds of Plant Genetic Resources (PGR) for future generations, has the capacity to preserve about 1 million germplasm in the form of seeds. Around 4.52 lakh accessions, of which 2.7 lakh are Indian germplasm are conserved by it (https://pib.gov.in/PressReleaseIframePage.aspx?PRID=1746485). The DBT had supported establishment of 3 gene banks viz., at NBPGR Campus, CIMAPs, Lucknow and TBFRI in Thiruvananthapuram. TBGRI field gene bank covers 3x10^4 accessions equivalent to 250 MAPs species (includes 100 endemic, rare and endangered) and the one at IIHR, Bangalore has targeted RET medicinal plants.

The present FGB at IIHR established the following species accessions of RET medicinal plants: *Kaempferia galanga* (11), *Kaempferia galanga* (8), *Holostemma adakodien* (4), *Oroxylum indicum* (9), *Embelia ribes* (2), *Celastrus paniculatus* (23), *Decalepis hamiltonii* (6), *Aristolochia tagala* (3), *Saraca asoca* (6), *Alpinia galanga* (19), *Kaempferia galang* (26), *Rauvolfia serpentiana* (10). The FGB is completely organic in management and chemical sprays and fertilizers were avoided (https://www.iihr.res.in/field-gene-bank-ret-medicinal-plants). MSSBG conserving 300 plants in a year span through nursery for RET, economical and ethno-botanical plants since 2006. The garden has fostered matured fruit bearing RET and medicinal plants in 41-acre campus as a result of the conservation activities since 2006 (https://mssbg.mssrf.org/programs-services/287-2/). *Gentiana kurroo* Royle, a critically endangered medicinal plant species of India is conserved in the *In Vitro* Gene bank at ICAR- NBPGR, New Delhi, India (Sharma et al. 2021). State Forest Departments of Andhra Pradesh, Karnataka, Kerala, Tamil Nadu and Maharashtra, in consultation with FRLHT established 54 forest genebank sites called Medicinal Plant Conservation Areas. The area is about 200 ha to 500 ha each. The gene banks harbour 45% of populations of flowering and medicinal plants including 70% of the red-listed related to Peninsular India. MPCAs are identified as 'study sites' for threatened species recovery research (https://www.fao.org/3/ad871e/ad871e09.htm).

14.7.2.3 Nursery Network

The medicinal plant nurseries ensure immediate availability of plants and planting materials to user groups. Planting material for 40 odd species of MAPs is made available within the ICAR and CSIR (CIMAP) network. In South India, FRLHT made a network of 55 supply nurseries. Local management committees are formed within each conservation area to manage nurseries in addition to prevent fires, grazing, or other destructive activities (UNDP, 2012). Till 2021, 89 medicinal plant nurseries aimimg to produce quality planting material for cultivation are established (http://164.100.24.220/loksabhaquestions/annex/176/AS277.pdf).

14.7.2.4 Home Gardens

Indian kitchen and herbal home garden are the seats for health care system through herbal home gardens (kitchen gardens). Home gardens are focuses on medicinal plant propagation and introduction programs intended to encourage the use of traditional remedies by making the plant sources more accessible (Agelet et al. 2000). The important medicinal plans are *Andrographis paniculata*, *Withania somnifera*, *Aloe vera*, *Ocimum sanctum*, *Ocimum basilicum*, *Piper longum*, *Bacopa monnieri*, *Centella asiatica*, *Tinospora cordifolia*, *Adathoda vasica*, *Eclipta prostrata*, *Chlorophytum borivilianum*, *Cassia senna*, *Gymnema sylvestris*, *Coleus aromaticus*, *Coleus forskolii*, *Catharanthus roseus*, *Phyllanthus niruri*, *Bryophyllum pinnatum, Mentha arvensis*, Lemon-scented basil, Lemongrass, Ginger, Turmeric, Citronella, Rosemary, Thyme, *Costus* spp. and Stevia etc. can be planted in the home gardens. UNDP and GoI initiative started a program named as 'Conservation of Medicinal Plants for Health and Livelihood Security' to revitalize local health traditions. It is implemented in eight states - Karnataka, Tamil Nadu, Andhra Pradesh, Kerala, Maharashtra, West Bengal, Rajasthan and Orissa. The project operates through home gardens that are strengthening the traditional medicinal system within families; also strengthening the communities' self-reliance in health care (https://www.in.undp.org/content/india/en/home/climate-and-disaster-reslience/successstories/conservation_of_medicinalplants.html). Herbal Garden Joginder Nagar situated at Mandi near Baijnath block of Kangra for the ex-situ conservation of medicinal plants. The herbal garden developed agro-techniques of medicinal plants for making cultivation of medicinal plants. Around 28 tree species are maintained in the garden among which *Cinnamomum tamala* has 312 maintained plants and *Emblica officinalis* Gaertn has 144 plants for conservation. The National Medicinal Plants Board, Ministry of AYUSH, supporting a scheme on "Conservation, Development and Sustainable Management of Medicinal Plants" to establish Herbal Gardens such as Home Herbal Gardens, School Herbal Gardens, Institutional/ Public Herbal Gardens and Herbal Gardens of State and National Importance. Govt. of India established approximately 292 herbal gardens across 30 states (http://164.100.24.220/loksabhaquestions/annex/176/AS277.pdf).

14.7.2.5 Cultivation of Medicinal and Aromatic Plants

Formal texts on the propagation of medicinal plants is >10% and agro-technology >1% of the identified plants in the globe (Khan and Khanum 1998). Out of 400 medicinal plants species in use by the Indian herbal industry, >20 species are in cultivation (Uniyal et al. 2000). This demands the commercial cultivation of medicinal plants. The agro-technology of 40 odd species is developed by ICAR-Agricultural University System. Dabur, Zandu, Indian Herbs, Arya Vaidya Shala and Arya Vaidya Pharmacy etc. took initiatives for cultivation. NABARD is also providing financial support for cultivation and processing of medicinal plants. With respect to economic viability, many highly endangered MAPs are unfit for cultivation. The National Medicinal Plants Board (NMPB) and CSIR-CIMAP, Lucknow are supporting the production of quality planting material of medicinal plants. Due to continued collection and increasing market demand, numerous plant species are threatened with extinction. High risks, transaction costs and lack of trust are the constraints for the smallholder producers to be in cultivation of medicinal plants (Petra et al. 2006).

From marketing angle, domestication and cultivation had advantages over wild harvest for plant-based medicines: cultivation provides reliable botanical identification, a steady source of raw material with a scope for selection and development of commercially desirable genotypes; post-harvest handling with quality controls thereby certified organic or biodynamic certification are easy (Máthé 2015).

14.7.2.6 Community Based Enterprises

Community-based approaches to conservation of medicinal plants are implemented by NGOs and international funding agencies. Three community-based enterprises are known in south India are: Gandhigram Trust, Dindigul; Preemade development Society, Kerala, and VGKK in B.R. Hills, Mysore (http://www.ikisan.com/medicinal-plants-introduction.html). NGOs such as RCMPCC showcased the community-based and local approaches at various levels of MAPs marketing. Sustainable harvesting combined with cultivation can improve yields of medicinal plants (Kala 2005). In this direction, the MPCA ensures autonomous development of a rural community through the people's income. It's also leads to decision making to use their assets and resources with the long-term supply of resources (Kala 2009).

14.7.2.7 Re-Introduction of Threatened Species

Reintroduction of plant species that are threatened is the way to re-entry of threatened plants into the area amenable for its growth support or the area of its actual threat. It is possible with efficient propagation techniques such plants grown in large quantities to reintroduce them in their natural habitat. Several threatened medicinal plant are conserved using this approach. The species conserved are: *Ceropegia fantastica* Sedgtvick (Ravikanth et al. 2018), *Syzygium travancorium* Gamble and *Vanda coerulea* Griff ex. Lindl (Seeni and Latha 2000) and *the vast list is provided by* Krishnan et al. 2011. In 2019, the DBT started several programs which are species-specific targeting highly threatened Indian species (during the past three decades). The species belongs to 101 genera and 64 families comprising 50 herbs, 42 trees, 24 orchids 24 and 14 each of shrubs and climbers. The results are fruitful (DBT 2019).

14.8 Medicinal Plant Conservation Centers in India

Several programs are in implementation in India for medicinal plants conservation by major institutions. They are:

- Council of Scientific and Industrial Research, CSIR, through its institutes like Central Institute of Medicinal and Aromatic Plants, Lucknow, National Botanical Research Institute, Lucknow, Central Drug Research Institute, Lucknow.
- The Indian Council of Agricultural Research, ICAR, along with its constituent programs or institutes such as All India Coordinated Research Project on Medicinal and Aromatic Plants in association with the National Bureau of Plant Genetic Research, National Research Center for Medicinal and Aromatic Plants and Indian Institute of Horticultural Research is actively working on medicinal plan conservation.

- Botanical Survey of India, BSI, of the Department of Environment and Forests, Government of India, established experimental gardens at different geographic regions such as Dehradun, Allahabad, Shillong, Pune, Coimbatore, Port Blair, Jodhpur, Gangtok and Itanagar.

Sanjappa (2004) reported that 10 experimental botanical gardens are employed in rehabilitating medicinal plants which are under serious threat. The biological diversity of medicinal plants of North-Western Himalayan region and peninsular India are conserved by G. B. Pant Institute of Himalayan Environment and Development, Ministry of Environment and Forest, Government of India and Tropical Botanical Garden and Research Institute (TBGRI), Kerala respectively. Major activities such as in *vitro* gene bank, seed gene bank, and field gene bank and display garden are developed by TBGRI. Similarly, the Tropical Forest Research Institute at Jabalpur encouraging cultivation of medicinal trees. DBT, India has established three cryo--preservation gene banks for 2707 medicinal plants. Many Government and non-Government Organizations (NGOs), private organizations and industries are also promoting commercial cultivation of medicinal plants (Nair 2002; Purohit and Vyas 2004) in addition to farmers, including ayurvedic practitioners (Bhattacharyya et al. 2006), in India.

14.8.1 ICAR-NBPGR

ICAR-NBPGR is the nodal agency for all the activities such as introduction, collection, conservation, documentation, evaluation and distribution of plant genetic resources (PGR), including medicinal plants, in India. It has 10 regional stations, 59 National Active Germplasm Sites (Singh et al. 2016). The National Gene Bank of ICAR-NBPGR has four types of conservation facilities i.e., seed gene bank, cryo-gene bank, in vitro gene bank and field gene bank. At ICAR-NBPGR, 8071 accessions of MAPs in seed gene bank, 178 accessions in *in vitro* gene bank and 1041 accessions in cryo-gene bank are being conserved (Singh and Pandey 2019; Sharma et al. 2020). NBPGR assembled in-house generated *de novo* genome-wide transcripts of *Tinospora cordifolia* and *Andrographis paniculata*. The transcripts of both the medicinal plants will serve as 'spring-board' for expediting research in the area of gene discovery, candidate gene determination, marker development etc. This will results in biosynthetic genes pathways responsible for synthesis of bio-active compounds and also provide useful information regarding markers linked to inherent chemo-diversity (http://www.nbpgr.ernet.in/med_plant/index.html). The MAPs collections conserved at National Gene bank (NGB), ICAR-NBPGR, New Delhi as on 31 October 2021 in the Seed Gene bank (−18 °C) are 8534, *in vitro* Gene bank (25 °C) and cryo-gene bank (−196 °C) conserved are 185, and 2194 number of genomic resources in the DNA (http://www.nbpgr.ernet.in/Research_Projects/Base_Collection_in_NGB.aspx).

14.8.2 National Medicinal Plants Board

The National Medicinal Plants Board was set up in 2000 coordinates all matters relating to medicinal plants, including drawing up policies and strategies for conservation, proper harvesting, cost-effective cultivation, research and development. It provide guidance in the formulation of proposals, schemes and programs etc. to be taken by agencies having access to land for cultivation and infrastructure for collection, storage transportation of medicinal plants; promotion of ex-situ and in-situ cultivation and conservation of medicinal plants; setting up of database on medicinal plants, dissemination of information and facilitating prevention of patents on plants used in traditional systems; development of protocols for cultivation and quality control and encouraging the protection of Patent Rights and IPR.

As a part of community-based enterprise, a large number of projects (n = 4254) have been sanctioned on the different aspects of the medicinal plants under both promotional and contractual farming schemes of NMPB (Kala and Sajwan 2007). Around 83% projects under contractual farming and the rest under promotional scheme were sanctioned. NMPB funding schemes are covering in 32 States and Union Territories of the country. Under the latter scheme, herbal gardens (220) are developed followed by dissemination of information on medicinal plants sector (192) and R&D (111). Developing agro-techniques to tissue culture, intercropping, molecular studies and pathological studies of medicinal plants are involved in R& D projects. The Forest Department are given major projects. About 120 medicinal plant species were in promotional schemes, while majority of projects are sanctioned for mixed-cropping of medicinal plants (Kala and Sajwan 2007). NMPB has supported 89,626 acres of land across the country (79,444 acres for cultivation of prioritized species) and > 25 species of medicinal plants covereing10, 182 acres of land. *Plantago ovata* in the largest area (23,348 acres), followed by *Emblica officinalis* (13,300 acres) and *Cassia angustifolia* (12,635 acres) are cultivated. *Aloe vera* and *Stevia rebaudiana* are two important medicinal plant species other than prioritized supported for cultivation in relatively large area (Kala and Sajwan 2007). NMPB supported schools for developing herbal gardens to encourage the young brains on the sustainable development of medicinal plant species (Kala and Sajwan 2006) in addition to make networking mode with the multi-institutional involvement to avoid duplication in research and development on medicinal plants.

Under the Environment and Energy Budget a total of US$11.4 million was allocated for the conservation of medicinal plants, from 2008 to 2015. Collaborating Government Counterparts were MOEF; NMBP; State Forest Departments and State Medicinal Plants Boards of Arunachal Pradesh, Chhattisgarh and Uttarakhand and the NBA along with Indira Gandhi National Forest Academy and FRLHT for conservation of medicinal plants. Arunachal Pradesh in north-eastern India, Chhattisgarh in central India and Uttarakhand in North-Western India represent India's medicinal and aromatic plant diversity and are home for over 30 globally significant medicinal plants (UNDP 2015) are supported by UNDP for long-term conservation and sustainable use of India's medicinal plants into forest management policy and practice.

14.9 Digitalization of MAPs Information

Easily retrievable resources are crucial for planned conservation of medicinal plants on the basis of their state of exploitation. Digitization of information in the form of electronic databases is highly useful to study the medicinal plants for better understanding. This is an emerging field in germplasm conservation; hence, several groups are working in this area. Rajasri (2006) reviewed the sources of MAPs digitalization in India. The Bioinformatics Centre of TBGRI has developed centralized online databases called 'Plant Info', which provides data on endemic medicinal plants and trees of Kerala. Another database 'Garden Info' contains data on plants conserved in TBGRI, and 'Seed Pack' is a database on the seed bank of TBGRI. Indian Medicinal Plants National Network of Distributed Databases (INMEDPLAN) is network of several Indian organizations with multidisciplinary aspects like botanical, horticultural, pharmacological and other information pool of medicinal plants by sharing their resources. The information is available on request to the network secretariat at FRLHT, Bangalore. FRLHT's online databases are 'Encyclopedia of Indian Medicinal Plants' (7361 plant species) and 'Medicinal Plant Conservation Concern' (880 species of traded medicinal plants of India). National Institute of Science Communication & Information Resources (NISCAIR) has developed a computerized version of the bimonthly abstracting journal with 30,000 records data in electronic form in CDROMs. CDRI, Lucknow developed the Natural Products Database, NAATS; Central Institute of Medicinal and Aromatic Plants has developed databases two data bases *viz.*, REFMAP (References on Medicinal and Aromatic Plants) and MAPI (Major Aromatic Plants of India) with the information on medicinal and aromatic plants. The Environmental Information System Centre on Floral Diversity of BSI, Kolkata, developed three data bases i.e., 'COBOMAN' on rare and threatened plants of India, 'Medicinal Plants' on important medicinal plants, and 'CITES Plants' restricted plants for export (Rajasri 2006).

14.10 Conclusions

India is one of the world's most plant-diverse countries. The vital source of pharmaceuticals used in traditional systems of medicine in India is constituted by almost 8000 medicinal plant species. The usage of these MAPs is a continuous tradition in India for long term industrial survival. Ayurvedic, Unani, Siddha, and Tibetan medicinal systems are noticed in India. The exploitation of medicinal plants outside of protected zones is unregulated. The majority of species that are prohibited from export are under endangered status. The elaboration of national policy is the precondition of any concerted activities to sustainable use of Indian medicinal plants. Both *in-situ* and *ex-situ* conservation policy environment coupled with the production needs to be promoted further in this direction. The strategy would also have to consider ways to increase financial resources as well as conservation incentives. The stringent legislation aiming sustainable harvesting of wild MAPs should be in place, in view of 65% of India's population depends on medicinal plant based traditional medicines. The introduction and implementation of permit systems at local, regional and national seems to be a necessity for uncompromised conservation of Indian medicinal plants.

References

Agelet A, Bonet MÀ, Vallés J (2000) Homegardens and their role as a main source of medicinal plants in mountain regions of Catalonia (Iberian Peninsula). Econ Bot 54(3):295–309

Anonymus (2019) Protection of medicinal plant genetic resources in India. Scoping Paper No. 4, November 2019, P.16

Bapat VA, Yadav SR, Dixit GB (2008) Rescue of endangered plants through biotechnological applications. Nat Acad Sci Lett 31(7–8):201–210

Baricevic D, Máthé Á, Bartol T (2015) Conservation of wild crafted medicinal and aromatic plants and their habitats. In: Máthé Á (ed) Medicinal and aromatic plants of the world. Scientific, production, commercial and utilization aspects. Springer Science+Business Media, Dordrecht, pp 131–144

Baruah A (2015) Medicinal and aromatic plants. EBH Publishers

Bhagwat SA, Kushalappa CG, Williams PH, Brown ND (2005) A landscape approach to biodiversity conservation of sacred groves in the Western Ghats of India. Conserv Biol 19(6): 1853–1862

Bhattacharyya R, Bhattacharya S, Chaudhuri S (2006) Conservation and documentation of the medicinal plant resources of India. In: Human exploitation and biodiversity conservation. Springer, Dordrecht, pp 365–377. https://doi.org/10.1007/978-1-4020-5283-5

Convention on International Trade in Endangered Species of wild flora and fauna (C.I.T.E.S.) (2005)

Convention on International Trade in Endangered Species of wild flora and fauna (C.I.T.E.S.) (2017)

DBT (2019) Department of Biotechnology. Conservation of Threatened Plants of India. http://www.dbtindia.gov.in/conservation-threatened-plants-india. Accessed on 4 May 2020

De Boef WS, Subedi A, Peroni N, Thijssen M, O'Keeffe E (2013) Community biodiversity management. Earthscan Routledge, London

Foundation for the Revitalization of Local Health Traditions (1997) Conserving a National Resource. Need for a National Policy and National Program on Medicinal Plants Conservation Draft of Madras Consultation (unpublished)

FRLHTENVIS (2016) Centre on Medicinal Plants. Medicinal plants under threat. http://envis.frlht.org/overview.html. Accessed on 6 Apr 2020

Gopi DK, Mattummal R, Narayana SKK, Parameswaran S (2018) IUCN red listed medicinal plants of siddha. J Res Siddha Med 1(1):15–22

Gupta V (2018) Indian plant genetic resources of medicinal value (Peter KV, ed), pp 267–287

Hawkins B (2008) Plants for life: medicinal plant conservation and botanic gardens. Botanic Gardens Conservation International, Richmond

https://dmapr.icar.gov.in/Research/CropImprovement.html

https://www.in.undp.org/content/india/en/home/operations/projects/closed/mainstreaming_conservationandsustainableuseofmedicinalplantdiver.html

https://www.nmpb.nic.in/content/functions-board

ICAR-NBPGR (2020) Memorandum of Undertaking signed between National Medicinal Plants Board and ICAR-National Bureau of Plant Genetic Resources on 06–07–2020. http://www.nbpgr.ernet.in/News_Details/aid/249.aspx. Accessed on 23 July 2020

Ijinu TP, George V, Pushpangadan P (2022) History of research on medicinal plants in India. In: Máthé Á, Khan IA (eds) Medicinal and aromatic plants of India Vol. 1. Medicinal and aromatic plants of the world, vol 8. Springer, Cham. https://doi.org/10.1007/978-3-030-98701-5_2

Kala CP (2005) Indigenous uses, population density, and conservation of threatened medicinal plants in protected areas of the Indian Himalayas. Conserv Biol 19(2):368–378

Kala CP (2009) Medicinal plants conservation and enterprise development. Med Plants Int J Phytomed Relat Ind 1(2):79–95

Kala CP, Sajwan BS (2006) Herbal gardens in schools. Curr Sci 91(11):1442–1443

Kala CP, Sajwan BS (2007) Revitalizing Indian systems of herbal medicine by the National Medicinal Plants Board through institutional networking and capacity building. Curr Sci:797–806

Kandari LS, Bisht VK, Bhardwaj M, Thakur AK (2014) Conservation and management of sacred groves, myths and beliefs of tribal communities: a case study from North-India. Environ Syst Res 3(1):1–10

Khan IA, Khanum A (1998) Role of biotechnology in medicinal and aromatic plants

King EIO, Viji C, Narasimhan D (1997) Sacred groves: traditional ecological heritage. Int J Ecol Environ Sci 23(4):463–470

Krishnan PN, Decruse SW, Radha RK (2011) Conservation of medicinal plants of Western Ghats, India and its sustainable utilization through in vitro technology. Vitro Cell Dev Biol-Plant 47(1): 110–122

Kumari P, Joshi GC, Tewari LM (2011) Diversity and status of ethno-medicinal plants of Almora district in Uttarakhand, India. Int J Biodivers Conserv 3(7):298–326

Lakshminarasimhan P, Gnanasekaran G, Murthy GVS, Arisdason W, Karthigeyan K, Roy DK, Krishna G, Bhattacharya J, Albertson WD, Venu P, Ghosh T, Debnath HS, Panja Kundu D (2020) In: Dash SS, Mao AA (eds) Flora of India, an annotated checklist, vol 2, Kolkata, pp 1034–1081

Maikhuri RK, Nautiyal S, Rao KS, Saxena KG (2001) Conservation policy–people conflicts: a case study from Nanda Devi biosphere reserve (a world heritage site), India. Forest Policy Econ 2(3–4):355–365

Maikhuri RK, Phondani PC, Rawat LS, Jha NK, Maletha A, Bahuguna YM, Kandari LS (2017) Conservation and management strategies of medicinal plant resources through action research approaches in Indian Himalaya. Iran J Sci Technol Trans A Sci 41(3):771–777

Malhotra KC, Gokhale Y, Chatterjee S, Srivastava S (2007) Sacred groves in India. Aryan Books International, New Delhi

Mamgain SK, Goel AK, Sharma SC (1998) Conservation of assessment of some important threatened medicinal plants of India. J Non-Timber Forest Prod 5:1–9

Manikandan P, Venkatash DR, Muthuchelian K (2011) Conservation and management of sacred groves in Theni district, Tamil Nadu, India. J Biosci Res 2(2):76–80

Máthé Á (2015) Botanical aspects of medicinal and aromatic plants. In: Máthé Á (ed) Medicinal and aromatic plants of the world. Scientific, production, commercial and utilization aspects. Springer Science+Business Media, Dordrecht, pp 13–34

Máthé Á, Khan IA (2022) Introduction to medicinal and aromatic plants in India. In: Máthé Á, Khan IA (eds) Medicinal and aromatic plants of India Vol. 1. Medicinal and aromatic plants of the world, vol 8. Springer, Cham. https://doi.org/10.1007/978-3-030-98701-5_1

Mukerji AK (2006) Evolution of good governance through forest policy reform in India. In: JFM at crossroads: future strategy and action program for institutionalizing community forestry, workshop pre-prints. International Centre for Community Forestry and Indian Institute of Forest Management, pp 17–24

Myers N, Mittermeier RA, Mittermeier CG, Fonseca GA, Kent J (2000) Biodiversity hotspots for conservation priorities. Nature 403:853–858. https://doi.org/10.1038/35002501

Nair GK (2002) KAU plans to develop medicinal plants. The Hindu Business Line – Financial Daily from the Hindu Group of Publication May 17, Friday

Nautiyal S, Maikhuri RK, Rao KS, Saxena KG (2001) Medicinal plant resources in Nanda Devi biosphere Reserve in the Central Himalaya. J Herbs Species Med Plants 8:47–64
Nayar MP (1996) Hotspots of endemic plants of India, Nepal and Bhutan. Tropical Botanic Garden and Research Institute, Thiruvananthapuram
Pandit PK, Bhakat RK (2007) Conservation of bio-diversity and ethnic culture through sacred groves in Midnapur District, West Bengal, India. Indian Forester 133(3):323–344
Petra VD, Alam G, de Steenhuijsen Piters B (2006) Developing a sustainable medicinal plant chain in India. Linking people, markets and values. In: Ruben R, Slingerland M, Nijhoff H (eds) Agro food chains and networks for development. Springer, Netherlands, pp 191–237
Purohit SS, Vyas SP (2004) Medicinal plant cultivation: a scientific approach: including processing and financial guidelines. Agrobios (India)
Pushpangadan P, Nair KN (2001) Future of systematics and biodiversity research in India: need for a national consortium and national agenda for systematic biology research. Curr Sci 80(5): 631–638
Rajasekharan PE, Ramanatha Rao V (eds) (2019) Conservation and utilization of horticultural genetic resources. Springer, Singapore
Rajasri B (2006) Conservation and documentation of the medicinal plant resources of India. Biodivers Conserv 15:2705–2717. https://doi.org/10.1007/s10531-005-6974-4
Ravikanth G, Jagadish MR, Vasudeva R, Shaanker RU, Aravind NA (2018) Recovery of critically endangered plant species in India: need for a comprehensive approach. Curr Sci 114(3):504–511
Sanjappa M (2004) Medicinal plants and their conservation with reference to peninsular India. Seminar proceedings of national seminar on medicinal plants, plant products & patents, 30th September–3rd October, Kolkata
Seeni S, Latha PG (2000) *In vitro* multiplication and eco rehabilitation of the endangered blue Vanda. Plant Cell Tissue Organ Cult 61(1):1–8
Sharma P, Rana JC, Devi U, Randhawa SS, Kumar R (2014) Floristic diversity and distribution pattern of plant communities along altitudinal gradient in Sangla Valley, northwest Himalaya. Sci World J 2014
Sharma N, Pandey R, Gowthami R (2020) *In vitro* conservation and cryopreservation of threatened medicinal plants of India. In: Rajasekharan PE, Wani SH (eds) Conservation and utilization of threatened medicinal plants. Springer, New York, pp 181–228. https://doi.org/10.1007/978-3-030-39793-7
Sharma N, Gowthami R, Devi SV, Malhotra EV, Pandey R, Agrawal A (2021) Cryopreservation of shoot tips of *Gentiana kurroo* Royle–a critically endangered medicinal plant of India. Plant Cell, Tissue Organ Cult (PCTOC) 144(1):67–72
Singh NP, Chowdhery HJ (2002) Biodiversity conservation in India. In: Das AP (ed) Perspective of plant biodiversity. Department of Botany, North Bengal University, West Bengal, pp 501–527. 9–11 November 2000, Bishen Singh Mahendra Pal Singh, Dehra Dun
Singh N, Pandey S (2019) Conservation of plant genetic resources. In: Pandey CD, Koul AK, Vimala Devi S, Singh N, Radhamani J, Pandey S, Jacob SR, Aravind J, Gore PG, Gupta V (eds) International training programme on management of plant genetic resources for officers from directorate of seed testing and certification ministry of agriculture, Baghdad. Republic of Iraq, ICAR-NBPGR, New Delhi, p 27
Singh H, Husain T, Agnihotri P (2010) Haat Kali sacred grove, central Himalaya, Uttarakhand. Curr Sci 98(3):290
Singh JP, Kumar S, Venkatesan K, Kulloli RN (2016) Conservation status and utilization of *Caralluma edulis*: an important threatened medicinal plant species of the Thar Desert, India. Genet Resour Crop Evol 63(4):721–732. https://doi.org/10.1007/s10722-016-0366-3
Trivedi PC (2006) Medicinal plants: traditional knowledge. IK International Pvt Ltd.
United Nations Development Programme (2012) Medicinal plants conservation Centre, India. Equator Initiative Case Study Series, New York
United Nations Development Programme (2015) Available at mainstreaming_conservation_sustainable_use_medicinal_plant_diversity_factsheet_project%20(2).pdf

Uniyal RC, Uniyal MR, Jain P (2000) Cultivation of medicinal plants in India: a reference book. TRAFFIC-India

UNU-IAS (2012) Biodiversity, traditional knowledge and community health: strengthening Linkages

Ved DK, Prathima CL, Morton N, Shnakar D (2001) Conservation of India's medicinal plant diversity through a novel approach of establishing a network of in situ gene banks. In: Uma Shaankar R, Ganeshaiah KN, Bawa KS (eds) Forest genetic resources: status, threats and conservation strategies. Oxford and IBH Publishing Co., New Delhi, pp 183–195

Walter KS, Gillett HJ (eds) (1998) 1997 IUCN red list of threatened plants. IUCN, Gland

WHO, IUCN & WWF (1993) Guidelines on the conservation of medicinal plants. – Gland & Geneva

WII-ENVIS (2019). Database available at http://wiienvis.nic.in/Database/PA_Updates_7814.aspx

Williams J (2001) Biodiversity theme report Australia state of the environment report (Theme Report). Published by CSIRO on Behalf of the Department of the Environment and Heritage. Available at http://www.environment.gov.au/node/21579

MIX
Papier aus verantwortungsvollen Quellen
Paper from responsible sources
FSC® C105338

If you have any concerns about our products,
you can contact us on
ProductSafety@springernature.com

In case Publisher is established outside the EU,
the EU authorized representative is:
Springer Nature Customer Service Center GmbH
Europaplatz 3, 69115 Heidelberg, Germany

Printed by Libri Plureos GmbH
in Hamburg, Germany